INTRODUCTION TO CELL BIOLOGY

Edited by
Kristy June

ArclerPress

Introduction to Cell Biology

Edited by Kristy June

ArclerPress

Published by

Arcler Press LLC

708 3rd Avenue, 6th Floor

New York

NY 10017

United States of America

www.arclerpress.com

ISBN: 978-1-68094-528-7 (Hardback)

Library of Congress Control Number: 0000000000

About the Editor

Kristy June

Kristy June is a Doctor of Medicine who also masters copyediting and proofreading scientific and academic research papers. She finished Bachelor of Science in Biology Major in Ecology and Entomology in the University of the Philippines Mindanao as her undergraduate degree where she started engaging herself into writing, editing and proofreading articles and books for publications. She is currently working as both a Medical Doctor and an Editor

Table of Contents

2 CELL THEORY

3 CYTOSKELETON

4 ENDOMEMBRANE SYSTEM

6 RNA

7 CYTOPLASM

8 PROTEASOME

List of Abbreviations

DNA	Deoxyribonucleic Acid
IBD	Identical-By-Descent
ECM	Cell-Cell And Cell-Extracellular Matrix
CNS	Central Nervous System
ECM	*'Extra-Cellular Matrix'*
LCAM	L-Cell Adhesion Molecule
LIF	Leukemia Inhibitory Factor
LAD-I	Leukocyte Adhesion Deficiency-I
TSA	Trichostatin A
GTP	Guanosine Triphosphate
MAPS	Microtubule-Associated Proteins
MTOCS	Microtubule-Organizing Centres
TTL	Tubulin—Tyrosine Ligase
ATP	Adenosine Triphosphate
ER	Endoplasmic Reticulum
MAM	Mitochondria-Associated Er Membrane
MTDNA	Mitochondrial Dna
MTERS	Mitochondrial Estrogen Receptors
MERRF	Myoclonic Epilepsy With Ragged Red Fibers
ROS	Reactive Oxygen Species
RER	Rough Endoplasmic Reticulum
SER	Smooth Endoplasmic Reticulum
TIM	The Inner Membrane
TEM	Transmission Electron Microscopy
UCPS	Uncoupling Proteins
ADP	Adenosine Diphosphate
ATP	Adenosine Triphosphate
LSC	Long Single Copy Section
SSC	Short Single Copy Section
MRNA	Messenger Rna
RNA	Ribonucleic Acid
RRNA	Ribosomal Rna

TRNA	Transfer Rna
APC	Anaphase Promoting Complex
ARF	Auxin-Response Factor
CDKS	Cyclin-Dependent Kinases
MHC	Major Histocompatibility Complex
PQC	Protein Quality Control
RB	Retinoblastoma
RA	Rheumatoid Arthritis
UBA	Ubiquitin-Associated
UPS	Ubiquitin-Proteasome System

Preface

Cell biology is a branch of biology that studies the different structures and functions of the cell and focuses mainly on the idea of the cell as the basic unit of life. Cell biology explains the structure, organization of the organelles they contain, their physiological properties, metabolic processes, signaling pathways , life cycle, and interactions with their environment. Focusing on the cell permits a detailed understanding of the tissues and organisms that cells compose. Some organisms have only one cell, while others are organized into cooperative groups with huge numbers of cells. On the whole, cell biology focuses on the structure and function of a cell, from the most general properties shared by all cells, to the unique, highly intricate functions particular to specialized cells. There are several main subfields within cell biology. One is the study of cell energy and the biochemical mechanisms that support cell metabolism. As cells are machines unto themselves, the focus on cell energy overlaps with the pursuit of questions of how energy first arose in original primordial cells, billions of years ago. Another subfield of cell biology concerns the genetics of the cell and its tight interconnection with the proteins controlling the release of genetic information from the nucleus to the cell cytoplasm. Yet another subfield focuses on the structure of cell components, known as subcellular compartments. Cutting across many biological disciplines is the additional subfield of cell biology, concerned with cell communication and signaling, concentrating on the messages that cells give to and receive from other cells and themselves. And finally, there is the subfield primarily concerned with the cell cycle, the rotation of phases beginning and ending with cell division and focused on different periods of growth and DNA replication. Many cell biologists dwell at the intersection of two or more of these subfields as our ability to analyze cells in more complex ways expands.

Content Coverage

Chapter One introduces the basic concepts of cell biology. Knowing the components of cells and how cells work is fundamental to all biological sciences.

Chapters Two provide the feature of cell theory; the basic structural, functional, and biological unit of all known living organisms. A cell is the smallest unit of life that can replicate independently, and cells are often called the "building blocks of life".

Chapter Three discusses about the cytoskeleton of a cell is made up of microtubules, actin filaments, and intermediate filaments. These structures give the cell its shape and help organize the cell's parts. In addition, they provide a basis for movement and cell division.

Chapter Four and *Five* present the endomembrane system and endosymbiont. In a cell, the endomembrane system is made up of both the endoplasmic reticulum and the Golgi apparatus. Endosymbionts are smaller symbiotic partners living inside a host organism, establishing endosymbiosis.

Chapters Six and *Seven* discuss about RNA structure and key discoveries in RNA biology, and the history and nature of cytoplasm, respectively.

Chapter Eight provides the knowledge of proteasome; a multisubunit enzyme complex that plays a central role in the regulation of proteins that control cell-cycle progression and apoptosis, and has therefore become an important target for anticancer therapy.

CHAPTER 1

Cell Biology

Chapter Contents

- **Introduction**
- **Anatomy**
- **Cellular Processes**
- **Origins**
- **Evolutionary History of Life**

INTRODUCTION

The cell (from Latin *cella*, meaning "small room") is the basic structural, functional, and biological unit of all known living organisms. A cell is the smallest unit of life that can replicate independently, and cells are often called the "building blocks of life". The study of cells is called cell biology.

Cell

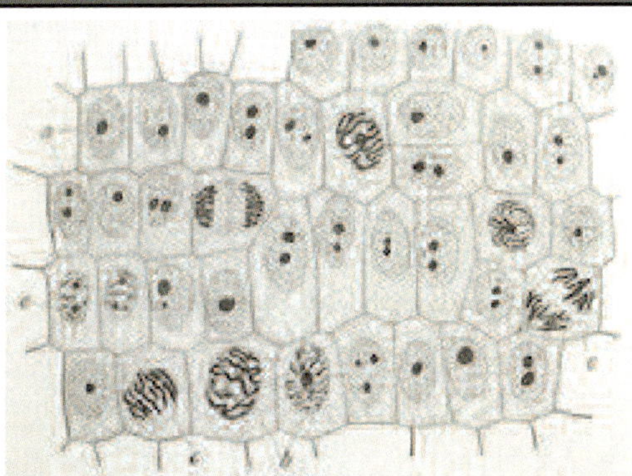

Onion (Allium) cells in different phases of the cell cycle

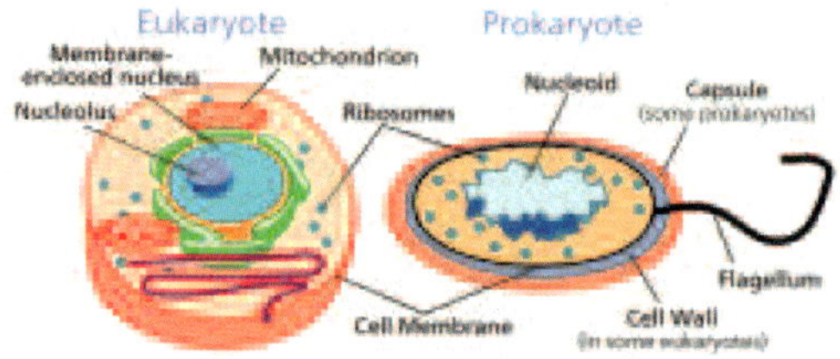

A eukaryotic cell (left) and a prokaryotic cell (right).

Identifiers	
TH	H1.00.01.0.00001
FMA	68646
Anatomical terminology	

Structure of an Animal Cell

Cells consist of cytoplasm enclosed within a membrane, which contains many biomolecules such as proteins and nucleic acids. Organisms can be classified as unicellular (consisting of a single cell; including bacteria) or multicellular (including plants and animals). While the number of cells in plants and animals varies from species to species, humans contain more than 10 trillion (10^{12}) cells. Most plant and animal cells are visible only under a microscope, with dimensions between 1 and 100 micrometres.

The cell was discovered by Robert Hooke in 1665, who named the biological unit for its resemblance to cells inhabited by Christian monks in a monastery. Cell theory, first developed in 1839 by Matthias Jakob Schleiden and Theodor Schwann, states that all organisms are composed of one or more cells, that cells are the fundamental unit of structure and function in all living organisms, that all cells come from preexisting cells, and that all cells contain the hereditary information necessary for regulating cell functions and for transmitting information to the next generation of cells. Cells emerged on Earth at least 3.5 billion years ago.

ANATOMY

Comparison of features of prokaryotic and eukaryotic cells		
	Prokaryotes	**Eukaryotes**
Typical organisms	bacteria, archaea	protists, fungi, plants, animals
Typical size	~ 1–5 μm	~ 10–100 μm
Type of nucleus	nucleoid region; no true nucleus	true nucleus with double membrane
DNA	circular (usually)	linear molecules (chromosomes) with histone proteins
RNA/protein synthesis	coupled in the cytoplasm	RNA synthesis in the nucleus protein synthesis in the cytoplasm
Ribosomes	50S and 30S	60S and 40S
Cytoplasmic structure	very few structures	highly structured by endomembranes and a cytoskeleton
Cell movement	flagella made of flagellin	flagella and cilia containing microtubules; lamellipodia and filopodia containing actin
Mitochondria	none	one to several thousand
Chloroplasts	none	in algae and plants
Organization	usually single cells	single cells, colonies, higher multicellular organisms with specialized cells
Cell division	binary fission (simple division)	mitosis (fission or budding) meiosis
Chromosomes	single chromosome	more than one chromosome
Membranes	cell membrane	Cell membrane and membrane-bound organelles

Cells are of two types, eukaryotic, which contain a nucleus, and prokaryotic, which do not. Prokaryotes are single-celled organisms, while eukaryotes can be either single-celled or multicellular.

Prokaryotic Cells

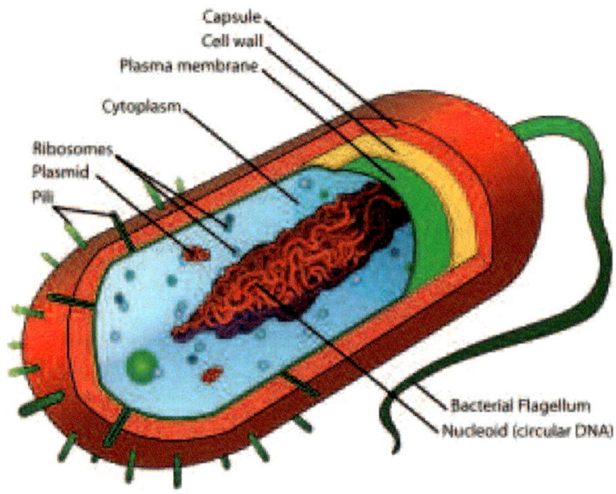

Structure of a typical prokaryotic cell

Prokaryotic cells were the first form of life on Earth, characterised by having vital biological processes including cell signaling and being self-sustaining. They are simpler and smaller than eukaryotic cells, and lack membrane-bound organelles such as the nucleus. Prokaryotes include two of the domains of life, bacteria and archaea. The DNA of a prokaryotic cell consists of a single chromosome that is in direct contact with the cytoplasm. The nuclear region in the cytoplasm is called the nucleoid. Most prokaryotes are the smallest of all organisms ranging from 0.5 to 2.0 μm in diameter.

A prokaryotic cell has three architectural regions:

- Enclosing the cell is the cell envelope – generally consisting of a plasma membrane covered by a cell wall which, for some bacteria, may be further covered by a third layer called a capsule. Though most prokaryotes have both a cell membrane and a cell wall, there are exceptions such as *Mycoplasma* (bacteria) and *Thermoplasma* (archaea) which only possess the cell membrane layer. The envelope gives rigidity to the cell and separates the interior of the cell from its environment, serving as a protective filter. The cell wall consists of peptidoglycan in bacteria, and acts as an additional barrier against exterior forces. It also prevents the cell from expanding and bursting (cytolysis) from osmotic pressure due to a hypotonic environment. Some eukaryotic cells (plant cells and fungal cells) also have a cell wall.
- Inside the cell is the cytoplasmic region that contains the genome (DNA), ribosomes and various sorts of inclusions. The genetic material is freely found in the cytoplasm. Prokaryotes can carry extrachromosomal DNA elements called plasmids, which are usually circular. Linear bacterial plasmids have been identified in several species of spirochete bacteria, including members of the genus Borrelia notably *Borrelia burgdorferi*, which causes Lyme disease. Though not forming a *nucleus*, the DNA is condensed in a *nucleoid*. Plasmids encode additional genes, such as antibiotic resistance genes.

- On the outside, flagella and pili project from the cell's surface. These are structures (not present in all prokaryotes) made of proteins that facilitate movement and communication between cells.

Eukaryotic Cells

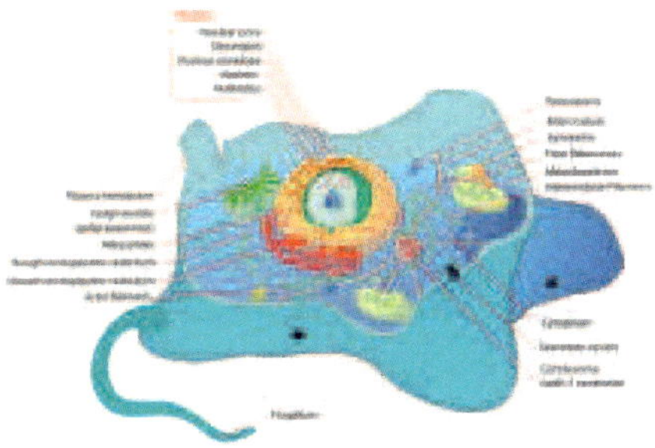

Structure of a typical animal cell

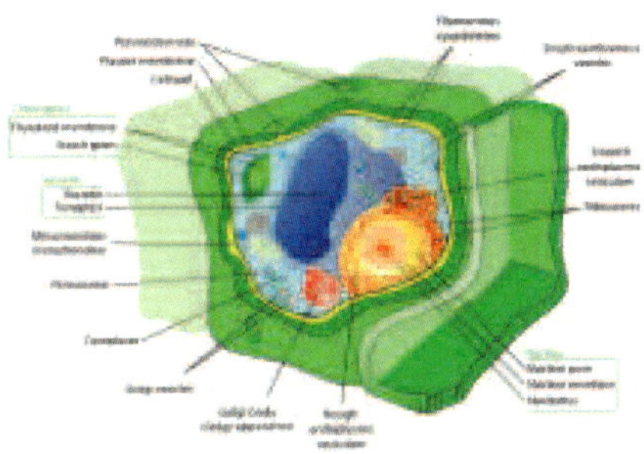

Structure of a typical plant cell

Plants, animals, fungi, slime moulds, protozoa, and algae are all eukaryotic. These cells are about fifteen times wider than a typical prokaryote and can be as much as a thousand times greater in volume. The main distinguishing feature of eukaryotes as compared to prokaryotes is compartmentalization: the presence of membrane-bound organelles (compartments) in which specific metabolic activities take place. Most important among these is a cell nucleus, an organelle that houses the cell's DNA. This nucleus gives the eukaryote its name, which means "true kernel (nucleus)". Other differences include:

- The plasma membrane resembles that of prokaryotes in function, with minor differences in the setup. Cell walls may or may not be present.
- The eukaryotic DNA is organized in one or more linear molecules, called chromosomes, which are associated with histone proteins. All chromosomal DNA is stored in the *cell nucleus*, separated from the cytoplasm by a membrane. Some eukaryotic organelles such as mitochondria also contain some DNA.

- Many eukaryotic cells are ciliated with *primary cilia*. Primary cilia play important roles in chemosensation, mechanosensation, and thermosensation. Cilia may thus be "viewed as a sensory cellular antennae that coordinates a large number of cellular signaling pathways, sometimes coupling the signaling to ciliary motility or alternatively to cell division and differentiation."
- Motile cells of eukaryotes can move using *motile cilia* or *flagella*. Motile cells are absent in conifers and flowering plants. Eukaryotic flagella are less complex than those of prokaryotes.

Subcellular components

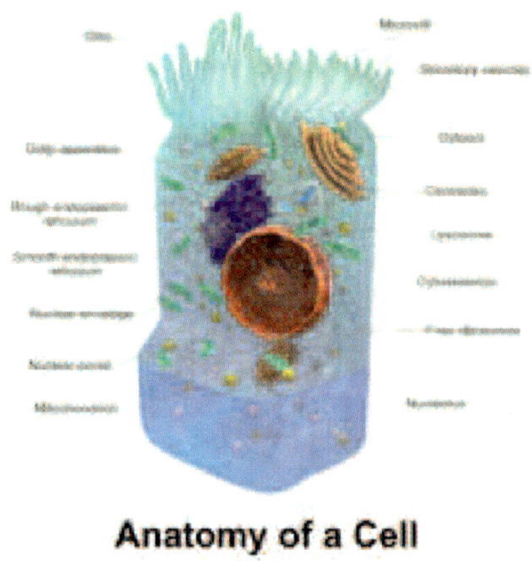

Illustration depicting major structures inside a eukaryotic animal cell

All cells, whether prokaryotic or eukaryotic, have a membrane that envelops the cell, regulates what moves in and out (selectively permeable), and maintains the electric potential of the cell. Inside the membrane, the cytoplasm takes up most of the cell's volume. All cells (except red blood cells which lack a cell nucleus and most organelles to accommodate maximum space for hemoglobin) possess DNA, the hereditary material of genes, and RNA, containing the information necessary to build various proteins such as enzymes, the cell's primary machinery. There are also other kinds of biomolecules in cells. This article lists these primary components of the cell, then briefly describes their function.

Membrane

The cell membrane, or plasma membrane, is a biological membrane that surrounds the cytoplasm of a cell. In animals, the plasma membrane is the outer boundary of the cell, while in plants and prokaryotes it is usually covered by a cell wall. This membrane serves to separate and protect a cell from its surrounding environment and is made mostly from a double layer of phospholipids, which are amphiphilic (partly hydrophobic and partly hydrophilic). Hence, the layer is called a phospholipid bilayer, or sometimes a fluid mosaic membrane. Embedded within this membrane is a variety of protein molecules that act as channels and pumps that move different molecules into and out of the cell. The membrane is said to be 'semi-permeable', in that it can either let a substance (molecule or ion) pass through freely, pass through to a limited extent or not pass through at all. Cell surface membranes also contain receptor proteins that allow cells to detect external signaling molecules such as hormones.

Cytoskeleton

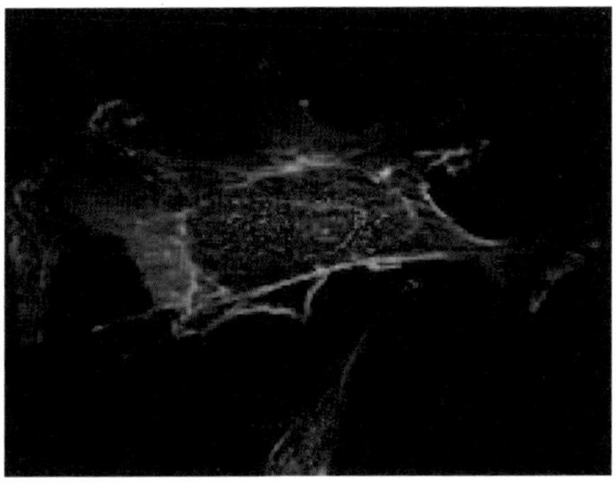

A fluorescent image of an endothelial cell. Nuclei are stained blue, mitochondria are stained red, and microfilaments are stained green.

The cytoskeleton acts to organize and maintain the cell's shape; anchors organelles in place; helps during endocytosis, the uptake of external materials by a cell, and cytokinesis, the separation of daughter cells after cell division; and moves parts of the cell in processes of growth and mobility. The eukaryotic cytoskeleton is composed of microfilaments, intermediate filaments and microtubules. There are a great number of proteins associated with them, each controlling a cell's structure by directing, bundling, and aligning filaments. The prokaryotic cytoskeleton is less well-studied but is involved in the maintenance of cell shape, polarity and cytokinesis. The subunit protein of microfilaments is a small, monomeric protein called actin. The subunit of microtubules is a dimeric molecule called tubulin. Intermediate filaments are heteropolymers whose subunits vary among the cell types in different tissues. But some of the subunit protein of intermediate filaments include vimentin, desmin, lamin (lamins A, B and C), keratin (multiple acidic and basic keratins), neurofilament proteins (NF - L, NF - M)

Genetic material

Two different kinds of genetic material exist: deoxyribonucleic acid (DNA) and ribonucleic acid (RNA). Cells use DNA for their long-term information storage. The biological information contained in an organism is encoded in its DNA sequence. RNA is used for information transport (e.g., mRNA) and enzymatic functions (e.g., ribosomal RNA). Transfer RNA (tRNA) molecules are used to add amino acids during protein translation.

Prokaryotic genetic material is organized in a simple circular DNA molecule (the bacterial chromosome) in the nucleoid region of the cytoplasm. Eukaryotic genetic material is divided into different, linear molecules called chromosomes inside a discrete nucleus, usually with additional genetic material in some organelles like mitochondria and chloroplasts (see endosymbiotic theory).

A human cell has genetic material contained in the cell nucleus (the nuclear genome) and in the mitochondria (the mitochondrial genome). In humans the nuclear genome is divided into 46 linear DNA molecules called chromosomes, including 22 homologous chromosome pairs and a pair of sex chromosomes. The mitochondrial genome is a circular DNA molecule distinct from the nuclear DNA. Although the mitochondrial DNA is very small compared to nuclear chromosomes, it codes for 13 proteins involved in mitochondrial energy production and specific tRNAs.

Foreign genetic material (most commonly DNA) can also be artificially introduced into the cell by a process called transfection. This can be transient, if the DNA is not inserted into the cell's genome, or stable, if it is. Certain viruses also insert their genetic material into the genome.

Organelles

Organelles are parts of the cell which are adapted and/or specialized for carrying out one or more vital functions, analogous to the organs of the human body (such as the heart, lung, and kidney, with each organ performing a different function). Both eukaryotic and prokaryotic cells have organelles, but prokaryotic organelles are generally simpler and are not membrane-bound.

There are several types of organelles in a cell. Some (such as the nucleus and golgi apparatus) are typically solitary, while others (such as mitochondria, chloroplasts, peroxisomes and lysosomes) can be numerous (hundreds to thousands). The cytosol is the gelatinous fluid that fills the cell and surrounds the organelles.

Eukaryotic

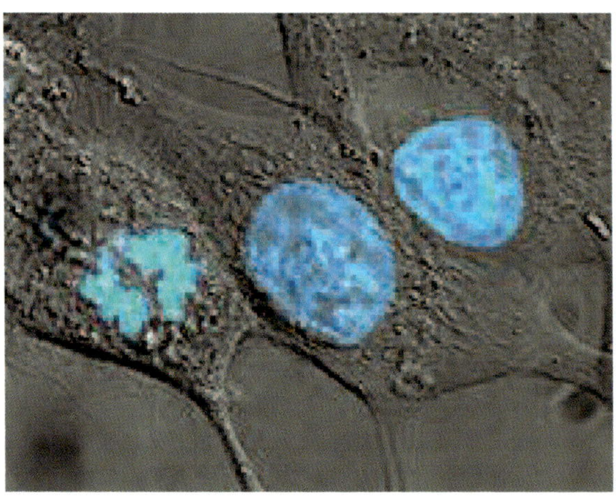

Human cancer cells with nuclei (specifically the DNA) stained blue. The central and rightmost cell are in interphase, so the entire nuclei are labeled. The cell on the left is going through mitosis and its DNA has condensed.

- **Cell nucleus**: A cell's information center, the cell nucleus is the most conspicuous organelle found in a eukaryotic cell. It houses the cell's chromosomes, and is the place where almost all DNA replication and RNA synthesis (transcription) occur. The nucleus is spherical and separated from the cytoplasm by a double membrane called the nuclear envelope. The nuclear envelope isolates and protects a cell's DNA from various molecules that could accidentally damage its structure or interfere with its processing. During processing, DNA is transcribed, or copied into a special RNA, called messenger RNA (mRNA). This mRNA is then transported out of the nucleus, where it is translated into a specific protein molecule. The nucleolus is a specialized region within the nucleus where ribosome subunits are assembled. In prokaryotes, DNA processing takes place in the cytoplasm.
- **Mitochondria and Chloroplasts**: generate energy for the cell. Mitochondria are self-replicating organelles that occur in various numbers, shapes, and sizes in the cytoplasm of all eukaryotic cells. Respiration occurs in the cell mitochondria, which generate the cell's energy by oxidative phosphorylation, using oxygen to release energy stored in cellular nutrients (typically pertaining to glucose) to generate ATP. Mitochondria multiply by binary fission, like prokaryotes. Chloroplasts can only be found in plants and algae, and they capture the sun's energy to make carbohydrates through photosynthesis.

Diagram of an endomembrane system

- **Endoplasmic reticulum**: The endoplasmic reticulum (ER) is a transport network for molecules targeted for certain modifications and specific destinations, as compared to molecules that float freely in the cytoplasm. The ER has two forms: the rough ER, which has ribosomes on its surface that secrete proteins into the ER, and the smooth ER, which lacks ribosomes. The smooth ER plays a role in calcium sequestration and release.
- **Golgi apparatus**: The primary function of the Golgi apparatus is to process and package the macromolecules such as proteins and lipids that are synthesized by the cell.
- **Lysosomes and Peroxisomes**: Lysosomes contain digestive enzymes (acid hydrolases). They digest excess or worn-out organelles, food particles, and engulfed viruses or bacteria. Peroxisomes have enzymes that rid the cell of toxic peroxides. The cell could not house these destructive enzymes if they were not contained in a membrane-bound system.
- **Centrosome**: the cytoskeleton organiser: The centrosome produces the microtubules of a cell – a key component of the cytoskeleton. It directs the transport through the ER and the Golgi apparatus. Centrosomes are composed of two centrioles, which separate during cell division and help in the formation of the mitotic spindle. A single centrosome is present in the animal cells. They are also found in some fungi and algae cells.
- **Vacuoles**: Vacuoles sequester waste products and in plant cells store water. They are often described as liquid filled space and are surrounded by a membrane. Some cells, most notably *Amoeba*, have contractile vacuoles, which can pump water out of the cell if there is too much water. The vacuoles of plant cells and fungal cells are usually larger than those of animal cells.

Eukaryotic and Prokaryotic

- **Ribosomes**: The ribosome is a large complex of RNA and protein molecules. They each consist of two subunits, and act as an assembly line where RNA from the nucleus is used to synthesise proteins from amino acids. Ribosomes can be found either floating freely or bound to a membrane (the rough endoplasmatic reticulum in eukaryotes, or the cell membrane in prokaryotes).

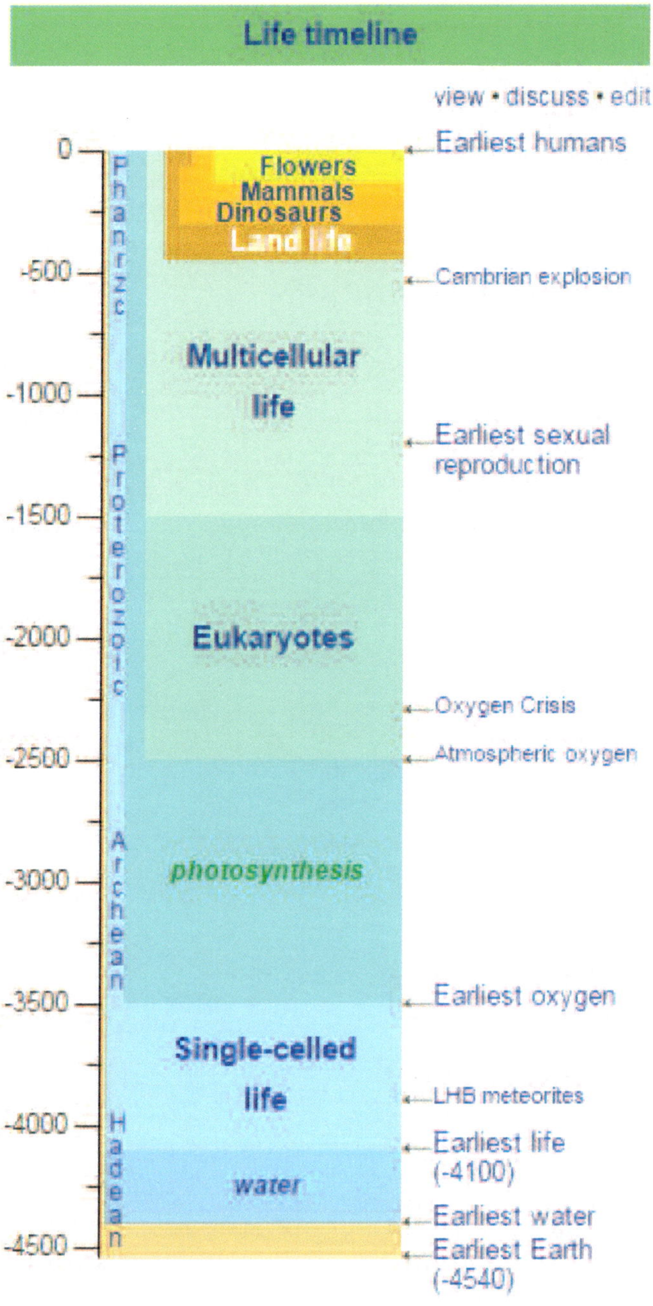

Structures outside the cell membrane

Many cells also have structures which exist wholly or partially outside the cell membrane. These structures are notable because they are not protected from the external environment by the semipermeable cell membrane. In order to assemble these structures, their components must be carried across the cell membrane by export processes.

Cell Wall

Many types of prokaryotic and eukaryotic cells have a cell wall. The cell wall acts to protect the cell mechanically and chemically from its environment, and is an additional layer of protection to the cell membrane. Different types of cell have cell walls made up of different materials; plant cell walls are primarily made up of cellulose, fungi cell walls are made up of chitin and bacteria cell walls are made up of peptidoglycan.

Prokaryotic

Capsule

A gelatinous capsule is present in some bacteria outside the cell membrane and cell wall. The capsule may be polysaccharide as in pneumococci, meningococci or polypeptide as *Bacillus anthracis* or hyaluronic acid as in streptococci. Capsules are not marked by normal staining protocols and can be detected by India ink or methyl blue; which allows for higher contrast between the cells for observation.[:87]

Flagella

Flagella are organelles for cellular mobility. The bacterial flagellum stretches from cytoplasm through the cell membrane(s) and extrudes through the cell wall. They are long and thick thread-like appendages, protein in nature. A different type of flagellum is found in archaea and a different type is found in eukaryotes.

Fimbria

A fimbria also known as a pilus is a short, thin, hair-like filament found on the surface of bacteria. Fimbriae, or pili are formed of a protein called pilin (antigenic) and are responsible for attachment of bacteria to specific receptors of human cell (cell adhesion). There are special types of specific pili involved in bacterial conjugation.

CELLULAR PROCESSES

Growth and Metabolism

Main articles: Cell growth and Metabolism

Between successive cell divisions, cells grow through the functioning of cellular metabolism. Cell metabolism is the process by which individual cells process nutrient molecules. Metabolism has two distinct divisions: catabolism, in which the cell breaks down complex molecules to produce energy and reducing power, and anabolism, in which the cell uses energy and reducing power to construct complex molecules and perform other biological functions. Complex sugars consumed by the organism can be broken down into simpler sugar molecules called monosaccharides such as glucose. Once inside the cell, glucose is broken down to make adenosine triphosphate (ATP), a molecule that possesses readily available energy, through two different pathways.

Replication

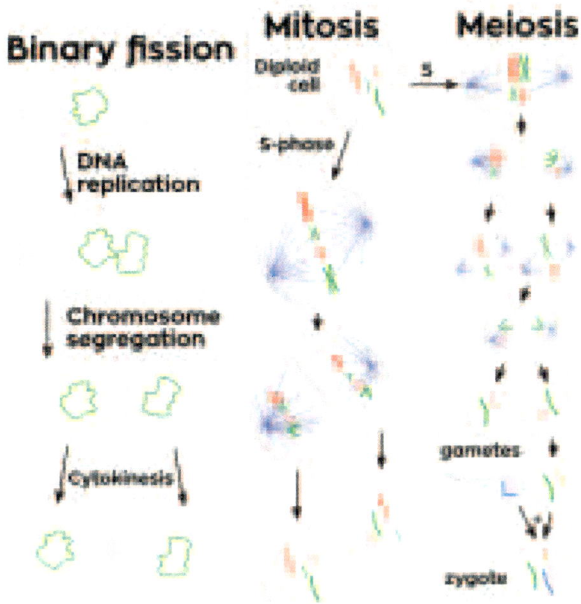

Bacteria divide by binary fission, while eukaryotes divide by mitosis or meiosis.

Cell division involves a single cell (called a *mother cell*) dividing into two daughter cells. This leads to growth in multicellular organisms (the growth of tissue) and to procreation (vegetative reproduction) in unicellular organisms. Prokaryotic cells divide by binary fission, while eukaryotic cells usually undergo a process of nuclear division, called mitosis, followed by division of the cell, called cytokinesis. A diploid cell may also undergo meiosis to produce haploid cells, usually four. Haploid cells serve as gametes in multicellular organisms, fusing to form new diploid cells.

DNA replication, or the process of duplicating a cell's genome, always happens when a cell divides through mitosis or binary fission. This occurs during the S phase of the cell cycle.

In meiosis, the DNA is replicated only once, while the cell divides twice. DNA replication only occurs before meiosis I. DNA replication does not occur when the cells divide the second time, in meiosis II.Replication, like all cellular activities, requires specialized proteins for carrying out the job.

Protein Synthesis

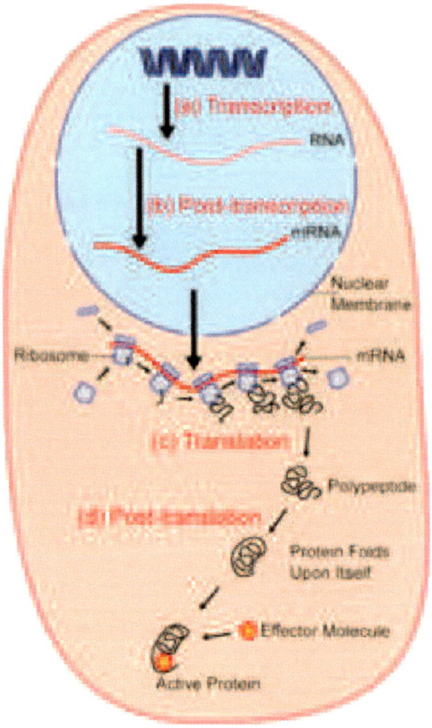

An overview of protein synthesis.

Within the nucleus of the cell (*light blue*), genes (DNA, *dark blue*) are transcribed into RNA. This RNA is then subject to post-transcriptional modification and control, resulting in a mature mRNA (*red*) that is then transported out of the nucleus and into the cytoplasm (*peach*), where it undergoes translation into a protein. mRNA is translated by ribosomes (*purple*) that match the three-base codons of the mRNA to the three-base anti-codons of the appropriate tRNA. Newly synthesized proteins (*black*) are often further modified, such as by binding to an effector molecule (*orange*), to become fully active.

Cells are capable of synthesizing new proteins, which are essential for the modulation and maintenance of cellular activities. This process involves the formation of new protein molecules from amino acid building blocks based on information encoded in DNA/RNA. Protein synthesis generally consists of two major steps: transcription and translation.

Transcription is the process where genetic information in DNA is used to produce a complementary RNA strand. This RNA strand is then processed to give messenger RNA (mRNA), which is free to migrate through the cell. mRNA molecules bind to protein-RNA complexes called ribosomes located in the cytosol, where they are translated into polypeptide sequences. The ribosome mediates the formation of a polypeptide sequence based on the mRNA sequence. The mRNA sequence directly relates to the polypeptide sequence by binding to transfer RNA (tRNA) adapter molecules in binding pockets within the ribosome. The new polypeptide then folds into a functional three-dimensional protein molecule.

Movement or Motility

Unicellular organisms can move in order to find food or escape predators. Common mechanisms of motion include flagella and cilia.

In multicellular organisms, cells can move during processes such as wound healing, the immune response and cancer metastasis. For example, in wound healing in animals, white blood cells move to the wound site to kill the microorganisms that cause infection. Cell motility involves many receptors, crosslinking, bundling, binding, adhesion, motor and other proteins. The process is divided into three steps – protrusion of the leading edge of the cell, adhesion of the leading edge and de-adhesion at the cell body and rear, and cytoskeletal contraction to pull the cell forward. Each step is driven by physical forces generated by unique segments of the cytoskeleton.

Multicellularity

Cell Specialization

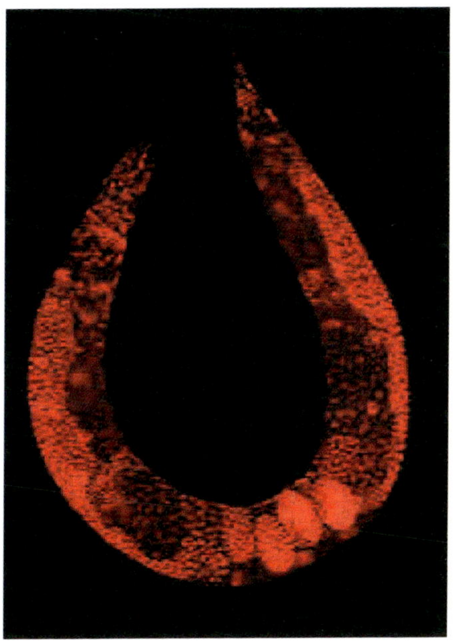

Staining of a *Caenorhabditis elegans* which highlights the nuclei of its cells.

Multicellular organisms are organisms that consist of more than one cell, in contrast to single-celled organisms.

In complex multicellular organisms, cells specialize into different cell types that are adapted to particular functions. In mammals, major cell types include skin cells, muscle cells, neurons, blood cells, fibroblasts, stem cells, and others. Cell types differ both in appearance and function, yet are genetically identical. Cells are able to be of the same genotype but of different cell type due to the differential expression of the genes they contain.

Most distinct cell types arise from a single totipotent cell, called a zygote, that differentiates into hundreds of different cell types during the course of development. Differentiation of cells is driven by different environmental cues (such as cell–cell interaction) and intrinsic differences (such as those caused by the uneven distribution of molecules during division).

Origin of Multicellularity

Multicellularity has evolved independently at least 25 times, including in some prokaryotes, like cyanobacteria, myxobacteria, actinomycetes, *Magnetoglobus multicellularis* or *Methanosarcina*. However, complex multicellular organisms evolved only in six eukaryotic groups: animals, fungi, brown algae, red algae, green algae, and plants. It evolved repeatedly for plants (Chloroplastida), once or twice for animals, once for brown algae, and perhaps several times for fungi, slime molds, and red algae.Multicellularity may have evolved from colonies of interdependent organisms, from cellularization, or from organisms in symbiotic relationships.

The first evidence of multicellularity is from cyanobacteria-like organisms that lived between 3 and 3.5 billion years ago. Other early fossils of multicellular organisms include the contested Grypania spiralis and the fossils of the black shales of the Palaeoproterozoic Francevillian Group Fossil B Formation in Gabon.

The evolution of multicellularity from unicellular ancestors has been replicated in the laboratory, in evolution experiments using predation as the selective pressure.

ORIGINS

The origin of cells has to do with the origin of life, which began the history of life on Earth.

Origin of the first cell

Stromatolites are left behind by cyanobacteria, also called blue-green algae. They are the oldest known fossils of life on Earth. This one-billion-year-old fossil is from Glacier National Park in the United States.

There are several theories about the origin of small molecules that led to life on the early Earth. They may have been carried to Earth on meteorites (see Murchison meteorite), created at deep-sea vents, or synthesized by lightning in a reducing atmosphere (see Miller–Urey experiment). There is little experimental data defining what the first self-replicating forms were. RNA is thought to be the earliest self-replicating molecule, as it is capable of both storing genetic information and catalyzing chemical reactions (see RNA world hypothesis), but some other entity with the potential to self-replicate could have preceded RNA, such as clay or peptide nucleic acid.

Cells emerged at least 3.5 billion years ago. The current belief is that these cells were heterotrophs. The early cell membranes were probably more simple and permeable than modern ones, with only a single fatty acid chain per lipid. Lipids are known to spontaneously form bilayered vesicles in water, and could have preceded RNA, but the first cell membranes could also have been produced by catalytic RNA, or even have required structural proteins before they could form.

Origin of Eukaryotic Cells

The eukaryotic cell seems to have evolved from a symbiotic community of prokaryotic cells. DNA-bearing organelles like the mitochondria and the chloroplasts are descended from ancient symbiotic oxygen-breathing proteobacteria and cyanobacteria, respectively, which were endosymbiosed by an ancestral archaean prokaryote.

There is still considerable debate about whether organelles like the hydrogenosome predated the origin of mitochondria, or vice versa: see the hydrogen hypothesis for the origin of eukaryotic cells.

Multicellular Organism

Multicellular organisms are organisms that consist of more than one cell, in contrast to unicellular organisms.

All species of animals, land plants and most fungi are multicellular, as are many algae, whereas a few organisms are partially uni- and partially multicellular, like slime molds and social amoebae such as the genus *Dictyostelium*.

Multicellular organisms arise in various different ways, for example by cell division or by aggregation of many single cells. Colonial organisms are the result of many identical individuals joining together to form a colony. However, it can often be hard to separate colonial protists from true multicellular organisms, because the two concepts are not distinct; colonial protists have been dubbed pluricellular' rather than multicellular'.

Evolutionary History

Occurrence

Multicellularity has evolved independently at least 46 times, including in some prokaryotes, like cyanobacteria, myxobacteria, actinomycetes, *Magnetoglobus multicellularis* or *Methanosarcina*. However, complex multicellular organisms evolved only in six eukaryotic groups: animals, fungi, brown algae, red algae, green algae, and land plants. It evolved repeatedly for Chloroplastida (green algae and land plants), once or twice for animals, once for brown algae, three times in the fungi (chytrids, ascomycetes and basidiomycetes) and perhaps several times for slime molds and red algae. The first evidence of multicellularity is from cyanobacteria-like organisms that lived 3–3.5 billion years ago. To reproduce, true multicellular organisms must solve the problem of regenerating a whole organism from germ cells (i.e. sperm and egg cells), an issue that is studied in evolutionary developmental biology. Animals have evolved a considerable diversity of cell types in a multicellular body (100–150 different cell types), compared with 10–20 in plants and fungi.

Loss of Multicellularity

Loss of multicellularity occurred in some groups. Fungi are predominantly multicellular, though early diverging lineages are largely unicellular (e.g. Microsporidia) and there have been numerous reversions to unicellularity across fungi (e.g. *Saccharomycotina*, *Cryptococcus*, and other yeasts). It may also have occurred in some red algae (e.g. *Porphyridium*), but it is possible that they are primitively unicellular). Loss of multicellularity is also considered probable in some green algae (e.g. *Chlorella vulgaris* and some Ulvophyceae). In other groups, generally parasites, a reduction of multicellularity occurred, in number or types of cells (e.g. the myxozoans, multicellular organisms, earlier thought to be unicellular, are probably extremely reduced cnidarians).

Cancer

Multicellular organisms, especially long-living animals, face the challenge of cancer, which occurs when cells fail to regulate their growth within the normal program of development. Changes in tissue morphology can be observed during this process. Cancer in animals (metazoans) has often been described as a loss of multicellularity. There is a discussion about the possibility of existence of cancer in other multicellular organisms or even in protozoa. For example, plant galls have been characterized as tumors but some authors argue that plants do not develop cancer.

Separation of Somatic and Germ Cells

In some multicellular groups, which are called Weismannists, a separation between a sterile somatic cell line and a germ cell line evolved. However, Weismannist development is relatively rare (e.g. vertebrates, arthropods, *Volvox*), as great part of species have the capacity for somatic embryogenesis (e.g. land plants, most algae, many invertebrates).

Hypotheses for Origin

One hypothesis for the origin of multicellularity is that a group of function-specific cells aggregated into a slug-like mass called a grex, which moved as a multicellular unit. This is essentially what slime molds do. Another hypothesis is that a primitive cell underwent nucleus division, thereby becoming a syncytium. A membrane would then form around each nucleus (and the cellular space and organelles occupied in the space), thereby resulting in a group of connected cells in one organism (this mechanism is observable in Drosophila). A third hypothesis is that as a unicellular organism divided, the daughter cells failed to separate, resulting in a conglomeration of identical cells in one organism, which could later develop specialized tissues. This is what plant and animal embryos do as well as colonial choanoflagellates.

Because the first multicellular organisms were simple, soft organisms lacking bone, shell or other hard body parts, they are not well preserved in the fossil record. One exception may be the demosponge, which may have left a chemical signature in ancient rocks. The earliest fossils of multicellular organisms include the contested Grypania spiralis and the fossils of the black shales of the Palaeoproterozoic Francevillian Group Fossil B Formation in Gabon (Gabonionta). The Doushantuo Formation has yielded 600 million year old microfossils with evidence of multicellular traits.

Until recently, phylogenetic reconstruction has been through anatomical (particularly embryological) similarities. This is inexact, as living multicellular organisms such as animals and plants are more than 500 million years removed from their single-cell ancestors. Such a passage of time allows both divergent and convergent evolution time to mimic similarities and accumulate differences between groups of modern and extinct ancestral species. Modern phylogenetics uses sophisticated techniques such as alloenzymes, satellite DNA and other molecular markers to describe traits that are shared between distantly related lineages.

The evolution of multicellularity could have occurred in a number of different ways, some of which are described below:

The Symbiotic Theory

This theory suggests that the first multicellular organisms occurred from symbiosis (cooperation) of different species of single-cell organisms, each with different roles. Over time these organisms would become so dependent on each other they would not be able to survive independently, eventually leading to the incorporation of their genomes into one multicellular organism. Each respective organism would become a separate lineage of differentiated cells within the newly created species.

This kind of severely co-dependent symbiosis can be seen frequently, such as in the relationship between clown fish and Riterri sea anemones. In these cases, it is extremely doubtful whether either species would survive very long if the other became extinct. However, the problem with this theory is that it is still not known how each organism›s DNA could be incorporated into one single genome to constitute them as a single species. Although such symbiosis is theorized to have occurred (e.g. mitochondria and chloroplasts in animal and plant cells—endosymbiosis), it has happened only extremely rarely and, even then, the genomes of the endosymbionts have retained an element of distinction, separately replicating their DNA during mitosis of the host species. For instance, the two or three symbiotic organisms forming the composite lichen, although dependent on each other for survival, have to separately reproduce and then re-form to create one individual organism once more.

The Cellularization (syncytial) Theory

This theory states that a single unicellular organism, with multiple nuclei, could have developed internal membrane partitions around each of its nuclei. Many protists such as the ciliates or slime molds can have several nuclei, lending support to this hypothesis. However, the simple presence of multiple nuclei is not enough to support the theory. Multiple nuclei of ciliates are dissimilar and have clear differentiated functions. The macronucleus serves the organism's needs, whereas the micronucleus is used for sexual reproduction with exchange of genetic material. Slime molds syncitia form from individual amoeboid cells, like syncitial tissues of some multicellular organisms, not the other way round. To be deemed valid, this theory needs a demonstrable example and mechanism of generation of a multicellular organism from a pre-existing syncytium.

The Colonial Theory

The Colonial Theory of Haeckel, 1874, proposes that the symbiosis of many organisms of the same species (unlike the symbiotic theory, which suggests the symbiosis of different species) led to a multicellular organism. At least some, it is presumed land-evolved, multicellularity occurs by cells separating and then rejoining (e.g. cellular slime molds) whereas for the majority of multicellular types (those that evolved within aquatic environments), multicellularity occurs as a consequence of cells failing to separate following division. The mechanism of this latter colony formation can be as simple as incomplete cytokinesis, though multicellularity is also typically considered to involve cellular differentiation.

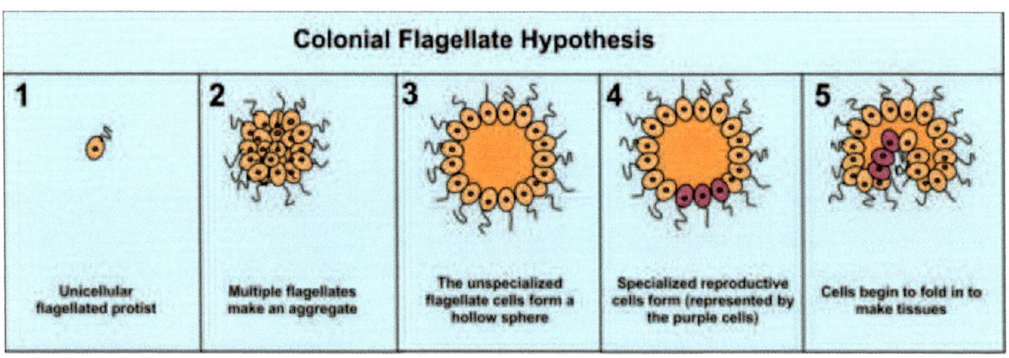

The advantage of the Colonial Theory hypothesis is that it has been seen to occur independently in 16 different protoctistan phyla. For instance, during food shortages the amoeba Dictyostelium groups together in a colony that moves as one to a new location. Some of these amoeba then slightly differentiate from each other. Other examples of colonial organisation in protista are Volvocaceae, such as Eudorina and Volvox, the latter of which consists of up to 500–50,000 cells (depending on the species), only a fraction of which reproduce. For example, in one species 25–35 cells reproduce, 8 asexually and around 15–25 sexually. However, it can often be hard to separate colonial protists from true multicellular organisms, as the two concepts are not distinct; colonial protists have been dubbed pluricellular' rather than multicellular'.

The Synzoospore theory

Some authors suggest that the origin of multicellularity, at least in Metazoa, occurred due to a transition from temporal to spatial cell differentiation, rather than through a gradual evolution of cell differentiation, as affirmed in Haeckel's Gastraea theory.

GK-PID

About 800 million years ago, a minor genetic change in a single molecule called guanylate kinase protein-interaction domain (GK-PID) may have allowed organisms to go from a single cell organism to one of many cells.

The Role of Viruses

Genes borrowed from viruses have recently been identified as playing a crucial role in the differentiation of multicellular tissues and organs and even in sexual reproduction, in the fusion of egg cell and sperm. Such fused cells are also involved in metazoan membranes such as those that prevent chemicals crossing the placenta and the brain body separation. Two viral components have been identified. The first is syncytin, which came from a virus. The second identified in 2007 is called EFF1, which helps form the skin of *Caenorhabditis elegans*, part of a whole family of FF proteins. Felix Rey, of the Pasteur Institute in Paris has constructed the 3D structure of the EFF1 protein and shown it does the work of linking one cell to another, in viral infections. The fact that all known cell fusion molecules are viral in origin suggests that they have been vitally important to the inter-cellular communication systems that enabled multicellularity. Without the ability of cellular fusion, colonies could have formed, but anything even as complex as a sponge would not have been possible.

Advantages

Multicellularity allows an organism to exceed the size limits normally imposed by diffusion: single cells with increased size have a decreased surface-to-volume ratio and have difficulty in absorbing sufficient nutrients and transporting them throughout the cell. Multicellular organisms thus have the competitive advantages of an increase in size without its limitations. It also permits increasing complexity by allowing the differentiation of numerous cellular lineages within an organism. Being multicellular also benefits with having a longer life span, thus still living even when a single-cell dies.

EVOLUTIONARY HISTORY OF LIFE

The **evolutionary history of life** on Earth traces the processes by which living and fossil organisms have evolved since life appeared on the planet, until the present day. Earth formed about 4.5 Ga (billion years) ago and there is evidence that life appeared within

0.5 billion years. The similarities between all present-day organisms indicate the presence of a common ancestor from which all known species have diverged through the process of evolution. More than 99 percent of all species, amounting to over five billion species, that ever lived on Earth are estimated to be extinct. Estimates on the number of Earth›s current species range from 10 million to 14 million,of which about 1.9 million are estimated to have been named and 1.6 million documented in a central database to date.

The earliest evidence for life on Earth is graphite found to be a biogenic substance in 3.7 billion-year-old metasedimentary rocks discovered in western Greenland and microbial mat fossils found in 3.48 billion-year-old sandstone discovered in Western Australia. More recently, in 2015, remains of biotic life' were found in 4.1 billion-year-old rocks in Western Australia.According to one of the researchers, If life arose relatively quickly on Earth ... then it could be common in the universe.'

Microbial mats of coexisting bacteria and archaea were the dominant form of life in the early Archean and many of the major steps in early evolution are thought to have taken place within them. The evolution of photosynthesis, around 3.5 Ga, eventually led to a buildup of its waste product, oxygen, in the atmosphere, leading to the great oxygenation event, beginning around 2.4 Ga. The earliest evidence of eukaryotes (complex cells with organelles) dates from 1.85 Ga, and while they may have been present earlier, their diversification accelerated when they started using oxygen in their metabolism. Later, around 1.7 Ga, multicellular organisms began to appear, with differentiated cells performing specialised functions. Sexual reproduction, which involves the fusion of male and female reproductive cells (gametes) to create a zygote in a process called fertilization is, in contrast to asexual reproduction, the primary method of reproduction for the vast majority of macroscopic organisms, including almost all eukaryotes (which includes animals and plants). However the origin and evolution of sexual reproduction remain a puzzle for biologists though it did evolve from a common ancestor that was a single celled eukaryotic species. Bilateria, animals with a front and a back, appeared by 555 Ma (million years ago).

The earliest land plants date back to around 450 Ma, although evidence suggests that microorganisms formed the earliest terrestrial ecosystems, at least 2.9 Ga. Microorganisms are thought to have paved the way for the inception of land plants in the Phanerozoic. Land plants were so successful that they are thought to have contributed to the Late Devonian extinction event. Ediacara biota appear during the Ediacaran period, while vertebrates, along with most other modern phyla originated about 525 Ma during the Cambrian explosion.During the Permian period, synapsids, including the ancestors of mammals, dominated the land, but most of this group became extinct in the Permian–Triassic extinction event 252.17 Ma. During the recovery from this catastrophe, archosaurs became the most abundant land vertebrates; one archosaur group, the dinosaurs, dominated the Jurassic and Cretaceous periods. After the Cretaceous–Paleogene extinction event 66 Ma killed off the non-avian dinosaurs, mammals increased rapidly in size and diversity. Such mass extinctions may have accelerated evolution by providing opportunities for new groups of organisms to diversify.

Earliest History of Earth

Main article: History of the Earth

The oldest meteorite fragments found on Earth are about 4.54 billion years old; this, coupled primarily with the dating of ancient lead deposits, has put the estimated age of Earth at around that time.The Moon has the same composition as Earth's crust but does not contain an iron-rich core like the Earth's. Many scientists think that about 40 million years later a body the size of Mars struck the Earth, throwing into orbit crust material that formed the Moon. Another hypothesis is that the Earth and Moon started to coalesce at the same time but the Earth, having much stronger gravity than the early Moon, attracted almost all the iron particles in the area.

Until 2001, the oldest rocks found on Earth were about 3.8 billion years old, leading scientists to estimate that the Earth›s surface had been molten until then. Accordingly, they named this part of Earth›s history the Hadean, whose name means hellish.'However, analysis of zircons formed 4.4 Ga indicates that Earth's crust solidified about 100 million years after the planet's formation and that the planet quickly acquired oceans and an atmosphere, which may have been capable of supporting life.

Evidence from the Moon indicates that from 4 to 3.8 Ga it suffered a Late Heavy Bombardment by debris that was left over from the formation of the Solar System, and the Earth should have experienced an even heavier bombardment due to its stronger gravity. While there is no direct evidence of conditions on Earth 4 to 3.8 Ga, there is no reason to think that the Earth was not also affected by this late heavy bombardment. This event may well have stripped away any previous atmosphere and oceans; in this case gases and water from comet impacts may have contributed to their replacement, although volcanic outgassing on Earth would have supplied at least half. However, if subsurface microbial life had evolved by this point, it would have survived the bombardment.

Earliest Evidence for Life on Earth

The earliest identified organisms were minute and relatively featureless, and their fossils look like small rods, which are very difficult to tell apart from structures that arise through abiotic physical processes. The oldest undisputed evidence of life on Earth, interpreted as fossilized bacteria, dates to 3 Ga. Other finds in rocks dated to about 3.5 Ga have been interpreted as bacteria, with geochemical evidence also seeming to show the presence of life 3.8 Ga. However, these analyses were closely scrutinized, and non-biological processes were found which could produce all of the "signatures of life" that had been reported. While this does not prove that the structures found had a non-biological origin, they cannot be taken as clear evidence for the presence of life. Geochemical signatures from rocks deposited 3.4 Ga have been interpreted as evidence for life, although these statements have not been thoroughly examined by critics.

Origins of life on Earth

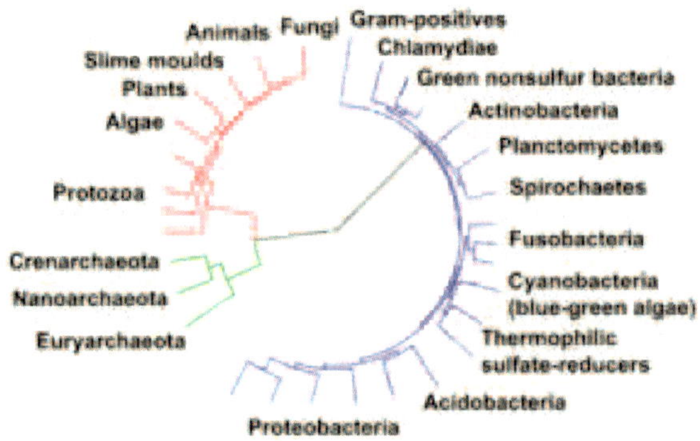

Evolutionary tree showing the divergence of modern species from their common ancestor in the center. The three domains are colored, with bacteria blue, archaea green, and eukaryotes red.

Further information: Evidence of common descent, Common descent, and Homology (biology)

Biologists reason that all living organisms on Earth must share a single last universal ancestor, because it would be virtually impossible that two or more separate lineages could have independently developed the many complex biochemical mechanisms common to all living organisms. As previously mentioned the earliest organisms for which fossil evidence is available are bacteria. The lack of fossil or geochemical evidence for earlier organisms has left plenty of scope for hypotheses, which fall into two main groups: 1) that life arose spontaneously on Earth or 2) that it was seeded' from elsewhere in the Universe.

Life "seeded" from Elsewhere

Panspermia does not explain how life arose in the first place, but simply examines the possibility of it coming from somewhere other than the Earth. The idea that life on Earth was "seeded" from elsewhere in the Universe dates back at least to the Greek philosopher Anaximander in the sixth century BCE. In the twentieth century it was proposed by the physical chemist Svante Arrhenius, by the astronomers Fred Hoyle and Chandra Wickramasinghe, and by molecular biologist Francis Crick and chemist Leslie Orgel.

There are three main versions of the seeded from elsewhere' hypothesis: from elsewhere in our Solar System via fragments knocked into space by a large meteor impact, in which case the most credible sources are Mars and Venus; by alien visitors, possibly as a result of accidental contamination by microorganisms that they brought with them; and from outside the Solar System but by natural means.

Experiments in low Earth orbit, such as EXOSTACK, demonstrated that some microorganism spores can survive the shock of being catapulted into space and some can survive exposure to outer space radiation for at least 5.7 years. Scientists are divided over the likelihood of life arising independently on Mars, or on other planets in our galaxy.

Independent Emergence on Earth

Life on Earth is based on carbon and water. Carbon provides stable frameworks for complex chemicals and can be easily extracted from the environment, especially from carbon dioxide. There is no other chemical element whose properties are similar enough to carbon›s to be called an analogue; silicon, the element directly below carbon on the periodic table, does not form very many complex stable molecules, and because most of its compounds are water-insoluble, it would be more difficult for organisms to extract. The elements boron and phosphorus have more complex chemistries, but suffer from other limitations relative to carbon. Water is an excellent solvent and has two other useful properties: the fact that ice floats enables aquatic organisms to survive beneath it in winter; and its molecules have electrically negative and positive ends, which enables it to form a wider range of compounds than other solvents can. Other good solvents, such as ammonia, are liquid only at such low temperatures that chemical reactions may be too slow to sustain life, and lack water›s other advantages. Organisms based on alternative biochemistry may, however, be possible on other planets.

Research on how life might have emerged from non-living chemicals focuses on three possible starting points: self-replication, an organism›s ability to produce offspring that are very similar to itself; metabolism, its ability to feed and repair itself; and external cell membranes, which allow food to enter and waste products to leave, but exclude unwanted substances. Research on abiogenesis still has a long way to go, since theoretical and empirical approaches are only beginning to make contact with each other.

Replication first: RNA world

Even the simplest members of the three modern domains of life use DNA to record their "recipes' and a complex array of RNA and protein molecules to "read" these instructions and use them for growth, maintenance and self-replication. The discovery that some RNA molecules can catalyze both their own replication and the construction of proteins led to the hypothesis of earlier life-forms based entirely on RNA. These ribozymes could have formed an RNA world in which there were individuals but no species, as mutations and horizontal gene transfers would have meant that the offspring in each generation were quite likely to have different genomes from those that their parents started with. RNA would later have been replaced by DNA, which is more stable and therefore can build longer genomes, expanding the range of capabilities a single organism can have. Ribozymes remain as the main components of ribosomes, modern cells› protein factories.'

Although short self-replicating RNA molecules have been artificially produced in laboratories, doubts have been raised about where natural non-biological synthesis of RNA is possible. The earliest ribozymes' may have been formed of simpler nucleic acids such as PNA, TNA or GNA, which would have been replaced later by RNA.

In 2003, it was proposed that porous metal sulfide precipitates would assist RNA synthesis at about 100 °C (212 °F) and ocean-bottom pressures near hydrothermal vents. Under this hypothesis, lipid membranes would be the last major cell components to appear and, until then, the protocells would be confined to the pores.

Metabolism first: Iron–sulfur world

A series of experiments starting in 1997 showed that early stages in the formation of proteins from inorganic materials including carbon monoxide and hydrogen sulfide could be achieved by using iron sulfide and nickel sulfide as catalysts. Most of the steps required temperatures of about 100 °C (212 °F) and moderate pressures, although one stage required 250 °C (482 °F) and a pressure equivalent to that found under 7 kilometres (4.3 mi) of rock. Hence it was suggested that self-sustaining synthesis of proteins could have occurred near hydrothermal vents.

Membranes first: Lipid world

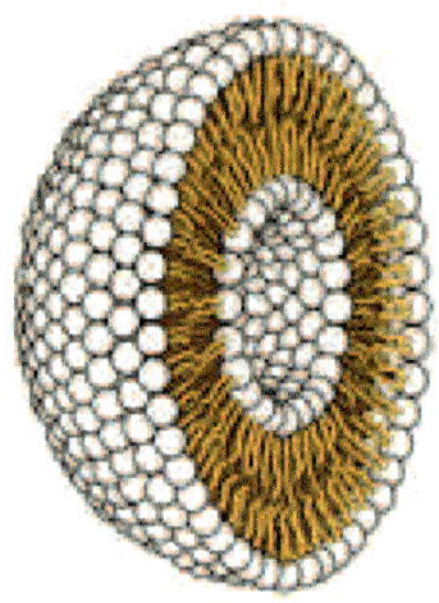

= water-attracting heads of lipid molecules

= water-repellent tails

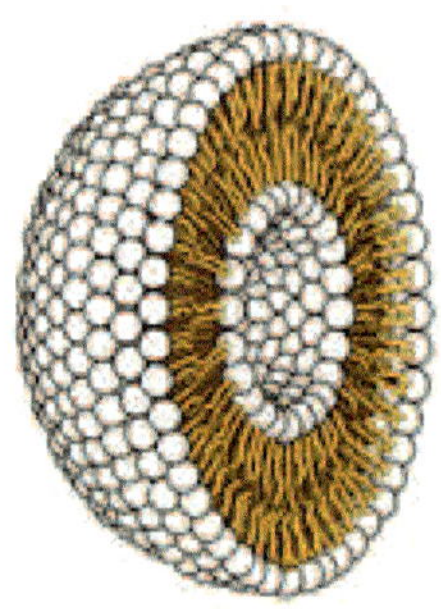

Cross-section through a liposome

It has been suggested that double-walled "bubbles" of lipids like those that form the external membranes of cells may have been an essential first step. Experiments that simulated the conditions of the early Earth have reported the formation of lipids, and these can spontaneously form liposomes, double-walled "bubbles," and then reproduce themselves. Although they are not intrinsically information-carriers as nucleic acids are, they would be subject to natural selection for longevity and reproduction. Nucleic acids such as RNA might then have formed more easily within the liposomes than they would have outside.

The clay Hypothesis

RNA is complex and there are doubts about whether it can be produced non-biologically in the wild. Some clays, notably montmorillonite, have properties that make them plausible accelerators for the emergence of an RNA world: they grow by self-replication of their crystalline pattern; they are subject to an analog of natural selection, as the clay "species" that grows fastest in a particular environment rapidly becomes dominant; and they can catalyze the formation of RNA molecules. Although this idea has not become the scientific consensus, it still has active supporters.

Research in 2003 reported that montmorillonite could also accelerate the conversion of fatty acids into "bubbles," and that the "bubbles" could encapsulate RNA attached to the clay. These "bubbles" can then grow by absorbing additional lipids and then divide. The formation of the earliest cells may have been aided by similar processes.

A similar hypothesis presents self-replicating iron-rich clays as the progenitors of nucleotides, lipids and amino acids.

Environmental and Evolutionary Impact Of Microbial Mats

Modern stromatolites in Shark Bay, Western Australia

Microbial mats are multi-layered, multi-species colonies of bacteria and other organisms that are generally only a few millimeters thick, but still contain a wide range of chemical environments, each of which favors a different set of microorganisms. To some extent each mat forms its own food chain, as the by-products of each group of microorganisms generally serve as food' for adjacent groups.

Stromatolites are stubby pillars built as microorganisms in mats slowly migrate upwards to avoid being smothered by sediment deposited on them by water. There has been vigorous debate about the validity of alleged fossils from before 3 Ga, with critics arguing that so-called stromatolites could have been formed by non-biological processes. In 2006, another find of stromatolites was reported from the same part of Australia as previous ones, in rocks dated to 3.5 Ga.

In modern underwater mats the top layer often consists of photosynthesizing cyanobacteria which create an oxygen-rich environment, while the bottom layer is oxygen-free and often dominated by hydrogen sulfide emitted by the organisms living there. It is estimated that the appearance of oxygenic photosynthesis by bacteria in mats increased biological productivity by a factor of between 100 and 1,000. The reducing agent used by oxygenic photosynthesis is water, which is much more plentiful than the geologically produced reducing agents required by the earlier non-oxygenic photosynthesis. From this point onwards life itself produced significantly more of the resources it needed than did geochemical processes. Oxygen is toxic to organisms that are not adapted to it, but greatly increases the metabolic efficiency of oxygen-adapted organisms. Oxygen became a significant component of Earth's atmosphere about 2.4 Ga. Although eukaryotes may have been present much earlier, the oxygenation of the atmosphere was a prerequisite for the evolution of the most complex eukaryotic cells, from which all multicellular organisms are built. The boundary between oxygen-rich and oxygen-free layers in microbial mats would have moved upwards when photosynthesis shut down overnight, and then downwards as it resumed on the next day. This would have created selection pressure for organisms in this intermediate zone to acquire the ability to tolerate and then to use oxygen, possibly via endosymbiosis, where one organism lives inside another and both of them benefit from their association.

Cyanobacteria have the most complete biochemical toolkits' of all the mat-forming organisms. Hence they are the most self-sufficient of the mat organisms and were well-adapted to strike out on their own both as floating mats and as the first of the phytoplankton, providing the basis of most marine food chains.

Diversification of Eukaryotes

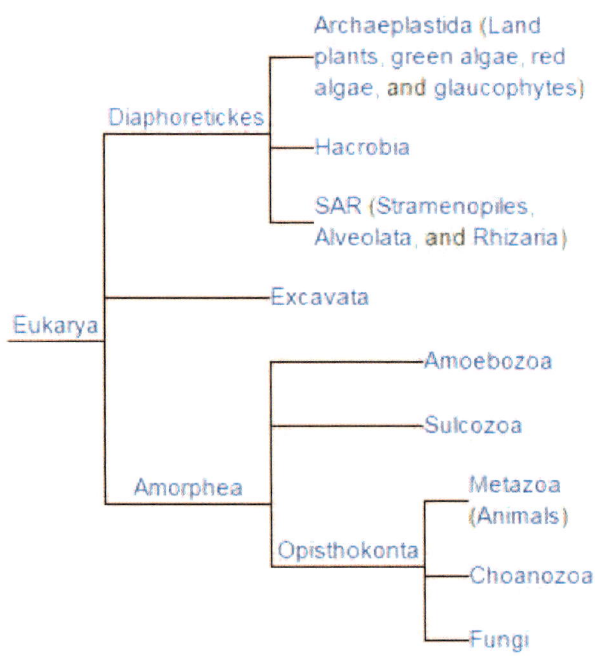

One possible family tree of eukaryotes

Chromatin, Nucleus, Endomembrane System, and Mitochondria

Eukaryotes may have been present long before the oxygenation of the atmosphere, but most modern eukaryotes require oxygen, which their mitochondria use to fuel the production of ATP, the internal energy supply of all known cells. In the 1970s it was proposed and, after much debate, widely accepted that eukaryotes emerged as a result of a sequence of endosymbiosis between prokaryotes.' For example: a predatory microorganism invaded a large prokaryote, probably an archaean, but the attack was neutralized, and the attacker took up residence and evolved into the first of the mitochondria; one of these chimeras later tried to swallow a photosynthesizing cyanobacterium, but the victim survived inside the attacker and the new combination became the ancestor of plants; and so on. After each endosymbiosis began, the partners would have eliminated unproductive duplication of genetic functions by re-arranging their genomes, a process which sometimes involved transfer of genes between them. Another hypothesis proposes that mitochondria were originally sulfur- or hydrogen-metabolising endosymbionts, and became oxygen-consumers later. On the other hand, mitochondria might have been part of eukaryotes› original equipment.

There is a debate about when eukaryotes first appeared: the presence of steranes in Australian shales may indicate that eukaryotes were present 2.7 Ga; however, an analysis in 2008 concluded that these chemicals infiltrated the rocks less than 2.2 Ga and prove nothing about the origins of eukaryotes. Fossils of the algae *Grypania* have been reported in 1.85 billion-year-old rocks (originally dated to 2.1 Ga but later revised), and indicates that eukaryotes with organelles had already evolved. A diverse collection of fossil algae were found in rocks dated between 1.5 and 1.4 Ga. The earliest known fossils of fungi date from 1.43 Ga.

Plastids

Plastids are thought to have originated from endosymbiotic cyanobacteria. The symbiosis evolved around 1.5 Ga and enabled eukaryotes to carry out oxygenic photosynthesis. Three evolutionary lineages have since emerged in which the plastids are named differently: chloroplasts in green algae and plants, rhodoplasts in red algae and cyanelles in the glaucophytes.

Sexual Reproduction and Multicellular Organisms

Evolution of Sexual Reproduction

The defining characteristics of sexual reproduction in eukaryotes are meiosis and fertilization. There is much genetic recombination in this kind of reproduction, in which offspring receive 50% of their genes from each parent, in contrast with asexual reproduction, in which there is no recombination. Bacteria also exchange DNA by bacterial conjugation, the benefits of which include resistance to antibiotics and other toxins, and the ability to utilize new metabolites. However, conjugation is not a means of reproduction, and is not limited to members of the same species – there are cases where bacteria transfer DNA to plants and animals.

On the other hand, bacterial transformation is clearly an adaptation for transfer of DNA between bacteria of the same species. Bacterial transformation is a complex process involving the products of numerous bacterial genes and can be regarded as a bacterial form of sex.This process occurs naturally in at least 67 prokaryotic species (in seven different phyla). Sexual reproduction in eukaryotes may have evolved from bacterial transformation. (Also see Evolution of sexual reproduction#Origin of sexual reproduction.)

The disadvantages of sexual reproduction are well-known: the genetic reshuffle of recombination may break up favorable combinations of genes; and since males do not directly increase the number of offspring in the next generation, an asexual population can out-breed and displace in as little as 50 generations a sexual population that is equal in every other respect. Nevertheless, the great majority of animals, plants, fungi and protists reproduce sexually. There is strong evidence that sexual reproduction arose early in the history of eukaryotes and that the genes controlling it have changed very little since then. How sexual reproduction evolved and survived is an unsolved puzzle.

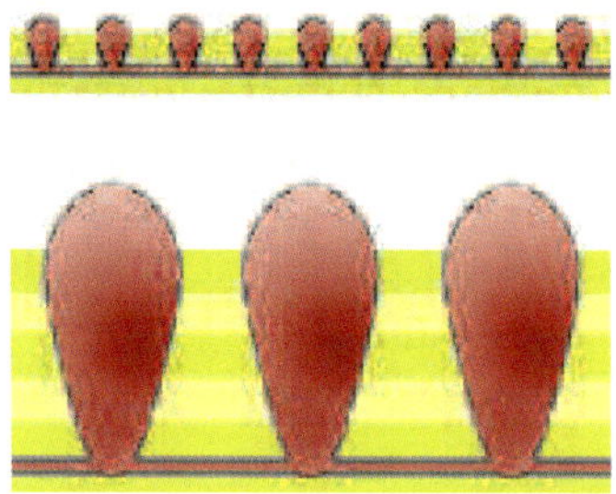

Horodyskia may have been an early metazoan,or a colonialforaminiferan. It apparently re-arranged itself into fewer but larger main masses as the sediment grew deeper round its base.

The Red Queen hypothesis suggests that sexual reproduction provides protection against parasites, because it is easier for parasites to evolve means of overcoming the defenses of genetically identical clones than those of sexual species that present moving targets, and there is some experimental evidence for this. However, there is still doubt about whether it would explain the survival of sexual species if multiple similar clone species were present, as one of the clones may survive the attacks of parasites for long enough to out-breed the sexual species.Furthermore, contrary to the expectations of the Red Queen hypothesis, Kathryn A. Hanley et al. found that the prevalence, abundance and mean intensity of mites was significantly higher in sexual geckos than in asexuals sharing the same habitat. In addition, biologist Matthew Parker, after reviewing numerous genetic studies on plant disease resistance, failed to find a single example consistent with the concept that pathogens are the primary selective agent responsible for sexual reproduction in the host.

Alexey Kondrashov›s *deterministic mutation hypothesis* (DMH) assumes that each organism has more than one harmful mutation and the combined effects of these mutations are more harmful than the sum of the harm done by each individual mutation. If so, sexual recombination of genes will reduce the harm that bad mutations do to offspring and at the same time eliminate some bad mutations from the gene pool by isolating them in individuals that perish quickly because they have an above-average number of bad mutations. However, the evidence suggests that the DMH's assumptions are shaky, because many species have on average less than one harmful mutation per individual and no species that has been investigated shows evidence of synergy between harmful mutations. (Further criticisms of this hypothesis are discussed in the article Evolution of sexual reproduction#Removal of deleterious genes)

The random nature of recombination causes the relative abundance of alternative traits to vary from one generation to another. This genetic drift is insufficient on its own to make sexual reproduction advantageous, but a combination of genetic drift and natural selection may be sufficient. When chance produces combinations of good traits, natural selection gives a large advantage to lineages in which these traits become genetically linked. On the other hand, the benefits of good traits are neutralized if they appear along with bad traits. Sexual recombination gives good traits the opportunities to become linked with other good traits, and mathematical models suggest this may be more than enough to offset the disadvantages of sexual reproduction. Other combinations of hypotheses that are inadequate on their own are also being examined.

The adaptive function of sex today remains a major unresolved issue in biology. The competing models to explain the adaptive function of sex were reviewed by John A. Birdsell and Christopher Wills. The hypotheses discussed above all depend on possible beneficial effects of random genetic variation produced by genetic recombination. An alternative view is that sex arose, and is maintained, as a process for repairing DNA damage, and that the genetic variation produced is an occasionally beneficial byproduct.

Multicellularity

The simplest definitions of "multicellular," for example "having multiple cells," could include colonial cyanobacteria like *Nostoc*. Even a technical definition such as "having the same genome but different types of cell" would still include some genera of the green algae Volvox, which have cells that specialize in reproduction. Multicellularity evolved independently in organisms as diverse as sponges and other animals, fungi, plants, brown algae, cyanobacteria, slime molds and myxobacteria. For the sake of brevity, this article focuses on the organisms that show the greatest specialization of cells and variety of cell types, although this approach to the evolution of biological complexity could be regarded as "rather anthropocentric.'

A slime mold solves a maze. The mold (yellow) explored and filled the maze (left). When the researchers placed sugar (red) at two separate points, the mold concentrated most of its mass there and left only the most efficient connection between the two points (right).

The initial advantages of multicellularity may have included: more efficient sharing of nutrients that are digested outside the cell, increased resistance to predators, many of which attacked by engulfing; the ability to resist currents by attaching to a firm surface; the ability to reach upwards to filter-feed or to obtain sunlight for photosynthesis; the ability to create an internal environment that gives protection against the external one; and even the opportunity for a group of cells to behave intelligently' by sharing information. These features would also have provided opportunities for other organisms to diversify, by creating more varied environments than flat microbial mats could.

Multicellularity with differentiated cells is beneficial to the organism as a whole but disadvantageous from the point of view of individual cells, most of which lose the opportunity to reproduce themselves. In an asexual multicellular organism, rogue cells which retain the ability to reproduce may take over and reduce the organism to a mass of undifferentiated cells. Sexual reproduction eliminates such rogue cells from the next generation and therefore appears to be a prerequisite for complex multicellularity.

The available evidence indicates that eukaryotes evolved much earlier but remained inconspicuous until a rapid diversification around 1 Ga. The only respect in which eukaryotes clearly surpass bacteria and archaea is their capacity for variety of forms, and sexual reproduction enabled eukaryotes to exploit that advantage by producing organisms with multiple cells that differed in form and function.

By comparing the composition of transcription factor families and regulatory network motifs between unicellular organisms and multicellular organisms, scientists found there are many novel transcription factor families and three novel types of regulatory network motifs in multicellular organisms, and novel family transcription factors are preferentially wired into these novel network motifs which are essential for multicullular development. These results propose a plausible mechanism for the contribution of novel-family transcription factors and novel network motifs to the origin of multicellular organisms at transcriptional regulatory level.

Fossil Evidence

The Francevillian biota fossils, dated to 2.1 Ga, are the earliest known fossil organisms that are clearly multicellular. They may have had differentiated cells. Another early multicellular fossil, *Qingshania*, dated to 1.7 Ga, appears to consist of virtually identical cells. The red algae called *Bangiomorpha*, dated at 1.2 Ga, is the earliest known organism that certainly has differentiated, specialized cells, and is also the oldest known sexually reproducing organism. The 1.43 billion-year-old fossils interpreted as fungi appear to have been multicellular with differentiated cells. The "string of beads" organism *Horodyskia*, found in rocks dated from 1.5 Ga to 900 Ma, may have been an early metazoan; however, it has also been interpreted as a colonial foraminiferan.

Emergence of Animals

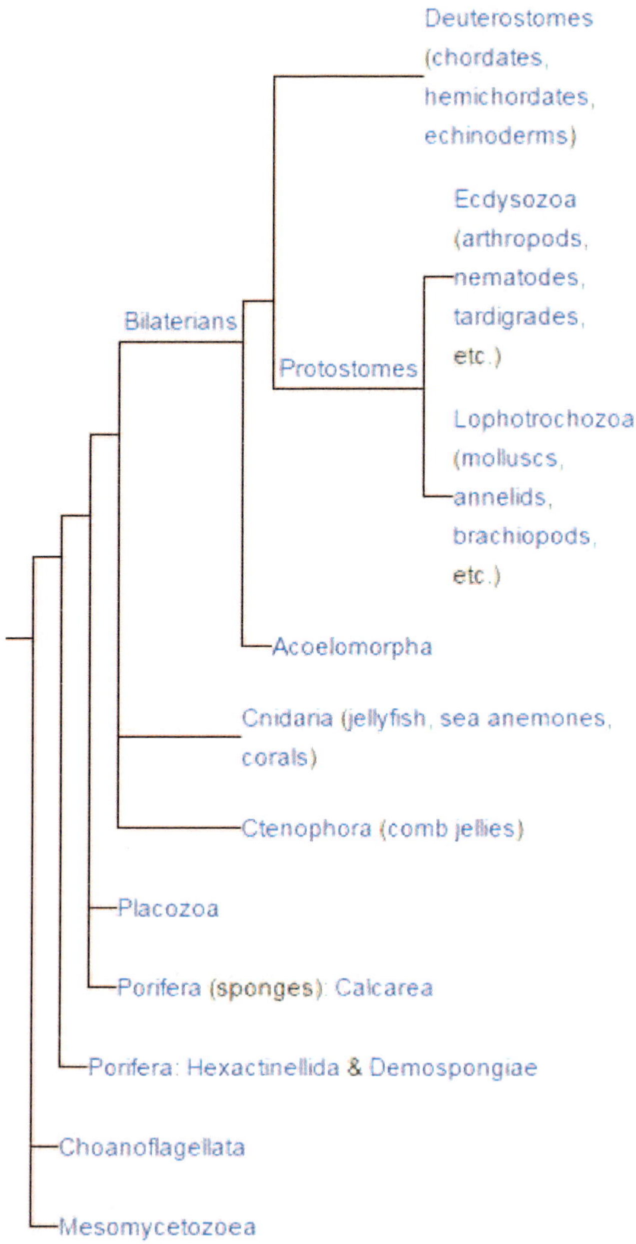

Family tree of the animals

Animals are multicellular eukaryotes, and are distinguished from plants, algae, and fungi by lacking cell walls. All animals are motile, if only at certain life stages. All animals except sponges have bodies differentiated into separate tissues, including muscles, which move parts of the animal by contracting, and nerve tissue, which transmits and processes signals.

The earliest widely accepted animal fossils are the rather modern-looking cnidarians (the group that includes jellyfish, sea anemones and *Hydra*), possibly from around 580 Ma, although fossils from the Doushantuo Formation can only be dated approximately. Their presence implies that the cnidarian and bilaterian lineages had already diverged.

The Ediacara biota, which flourished for the last 40 million years before the start of the Cambrian, were the first animals more than a very few centimetres long. Many were flat and had a quilted' appearance, and seemed so strange that there was a proposal to classify them as a separate kingdom, Vendozoa. Others, however, have been interpreted as early molluscs (*Kimberella*), echinoderms (*Arkarua*), and arthropods (*Spriggina*, *Parvancorina*). There is still debate about the classification of these specimens, mainly because the diagnostic features which allow taxonomists to classify more recent organisms, such as similarities to living organisms, are generally absent in the Ediacarans. However, there seems little doubt that *Kimberella* was at least a triploblastic bilaterian animal, in other words, an animal significantly more complex than the cnidarians.

The small shelly fauna are a very mixed collection of fossils found between the Late Ediacaran and Middle Cambrian periods. The earliest, *Cloudina*, shows signs of successful defense against predation and may indicate the start of an evolutionary arms race. Some tiny Early Cambrian shells almost certainly belonged to molluscs, while the owners of some "armor plates," *Halkieria* and *Microdictyon*, were eventually identified when more complete specimens were found in Cambrian lagerstätten that preserved soft-bodied animals.

Opabinia made the largest single contribution to modern interest in the Cambrian explosion.

In the 1970s there was already a debate about whether the emergence of the modern phyla was explosive' or gradual but hidden by the shortage of Precambrian animal fossils. A re-analysis of fossils from the Burgess Shale lagerstätte increased interest in the issue when it revealed animals, such as *Opabinia*, which did not fit into any known phylum. At the time these were interpreted as evidence that the modern phyla had evolved very rapidly in the Cambrian explosion and that the Burgess Shale's "weird wonders" showed that the Early Cambrian was a uniquely experimental period of animal evolution. Later discoveries of similar animals and the development of new theoretical approaches led to the conclusion that many of the "weird wonders" were evolutionary "aunts" or "cousins" of modern groups—for example that *Opabinia* was a member of the lobopods, a group which includes the ancestors of the arthropods, and that it may have been closely related to the modern tardigrades. Nevertheless, there is still much debate about whether the Cambrian explosion was really explosive and, if so, how and why it happened and why it appears unique in the history of animals.

Deuterostomes and the First Vertebrates

Acanthodians were among the earliest vertebrates with jaws.

Most of the animals at the heart of the Cambrian explosion debate are protostomes, one of the two main groups of complex animals. The other major group, the deuterostomes, contains invertebrates such as starfish and sea urchins (echinoderms), as well as chordates (see below). Many echinoderms have hard calcite "shells," which are fairly common from the Early Cambrian small shelly fauna onwards. Other deuterostome groups are soft-bodied, and most of the significant Cambrian deuterostome fossils come from the Chengjiang fauna, a lagerstätte in China. The chordates are another major deuterostome group: animals with a distinct dorsal nerve cord. Chordates include soft-bodied invertebrates such as tunicates as well as vertebrates—animals with a backbone. While tunicate fossils predate the Cambrian explosion, the Chengjiang fossils *Haikouichthys* and *Myllokunmingia* appear to be true vertebrates, and *Haikouichthys* had distinct vertebrae, which may have been slightly mineralized. Vertebrates with jaws, such as the acanthodians, first appeared in the Late Ordovician.

Colonization of Land

Adaptation to life on land is a major challenge: all land organisms need to avoid drying-out and all those above microscopic size must create special structures to withstand gravity; respiration and gas exchange systems have to change; reproductive systems cannot depend on water to carry eggs and sperm towards each other. Although the earliest good evidence of land plants and animals dates back to the Ordovician period (488 to 444 Ma), and a number of microorganism lineages made it onto land much earlier, modern land ecosystems only appeared in the Late Devonian, about 385 to 359 Ma.

Evolution of Terrestrial Antioxidants

Oxygen is a potent oxidant whose accumulation in terrestrial atmosphere resulted from the development of photosynthesis over 3 Ga, in cyanobacteria (blue-green algae), which were the most primitive oxygenic photosynthetic organisms. Brown algae accumulate inorganic mineral antioxidants such as rubidium, vanadium, zinc, iron, copper, molybdenum, selenium and iodine which is concentrated more than 30,000 times the concentration of this element in seawater. Protective endogenous antioxidant enzymes and exogenous dietary antioxidants helped to prevent oxidative damage. Most marine mineral antioxidants act in the cells as essential trace elements in redox and antioxidant metalloenzymes.

When plants and animals began to enter rivers and land about 500 Ma, environmental deficiency of these marine mineral antioxidants was a challenge to the evolution of terrestrial life. Terrestrial plants slowly optimized the production of "new" endogenous antioxidants such as ascorbic acid, polyphenols, flavonoids, tocopherols, etc. A few of these appeared more recently, in last 200-50 Ma, in fruits and flowers of angiosperm plants.

In fact, angiosperms (the dominant type of plant today) and most of their antioxidant pigments evolved during the Late Jurassic period. Plants employ antioxidants to defend their structures against reactive oxygen species produced during photosynthesis. Animals are exposed to the same oxidants, and they have evolved endogenous enzymatic antioxidant systems. Iodine is the most primitive and abundant electron-rich essential element in the diet of marine and terrestrial organisms, and as iodide acts as an electron donor and has this ancestral antioxidant function in all iodide-concentrating cells from primitive marine algae to more recent terrestrial vertebrates.

Evolution of Soil

Before the colonization of land, soil, a combination of mineral particles and decomposed organic matter, did not exist. Land surfaces would have been either bare rock or unstable sand produced by weathering. Water and any nutrients in it would have drained away very quickly.

Lichens growing on concrete

Films of cyanobacteria, which are not plants but use the same photosynthesis mechanisms, have been found in modern deserts, and only in areas that are unsuitable for vascular plants. This suggests that microbial mats may have been the first organisms to colonize dry land, possibly in the Precambrian. Mat-forming cyanobacteria could have gradually evolved resistance to desiccation as they spread from the seas to intertidal zones and then to land. Lichens, which are symbiotic combinations of a fungus (almost always an ascomycete) and one or more photosynthesizers (green algae or cyanobacteria),are also important colonizers of lifeless environments, and their ability to break down rocks contributes to soil formation in situations where plants cannot survive. The earliest known ascomycete fossils date from 423 to 419 Ma in the Silurian.

Soil formation would have been very slow until the appearance of burrowing animals, which mix the mineral and organic components of soil and whose feces are a major source of the organic components. Burrows have been found in Ordovician sediments, and are attributed to annelids ("worms") or arthropods.

Plants and the Late Devonian Wood Crisis

Reconstruction of *Cooksonia*, a vascular plant from the Silurian

Fossilized trees from the Middle DevonianGilboa Fossil Forest

In aquatic algae, almost all cells are capable of photosynthesis and are nearly independent. Life on land required plants to become internally more complex and specialized: photosynthesis was most efficient at the top; roots were required in order to extract water from the ground; the parts in between became supports and transport systems for water and nutrients.

Spores of land plants, possibly rather like liverworts, have been found in Middle Ordovician rocks dated to about 476 Ma. In Middle Silurian rocks 430 Ma, there are fossils of actual plants including clubmosses such as *Baragwanathia*; most were under 10 centimetres (3.9 in) high, and some appear closely related to vascular plants, the group that includes trees.

By the Late Devonian 370 Ma, trees such as *Archaeopteris* were so abundant that they changed river systems from mostly braided to mostly meandering, because their roots bound the soil firmly. In fact, they caused the Late Devonian wood crisis' because:

- They removed more carbon dioxide from the atmosphere, reducing the greenhouse effect and thus causing an ice age in the Carboniferous period. In later ecosystems the carbon dioxide "locked up" in wood is returned to the atmosphere by decomposition of dead wood. However, the earliest fossil evidence of fungi that can decompose wood also comes from the Late Devonian.
- The increasing depth of plants› roots led to more washing of nutrients into rivers and seas by rain. This caused algal blooms whose high consumption of oxygen caused anoxic events in deeper waters, increasing the extinction rate among deep-water animals.

Land Invertebrates

Animals had to change their feeding and excretory systems, and most land animals developed internal fertilization of their eggs. The difference in refractive index between water and air required changes in their eyes. On the other hand, in some ways movement and breathing became easier, and the better transmission of high-frequency sounds in air encouraged the development of hearing.

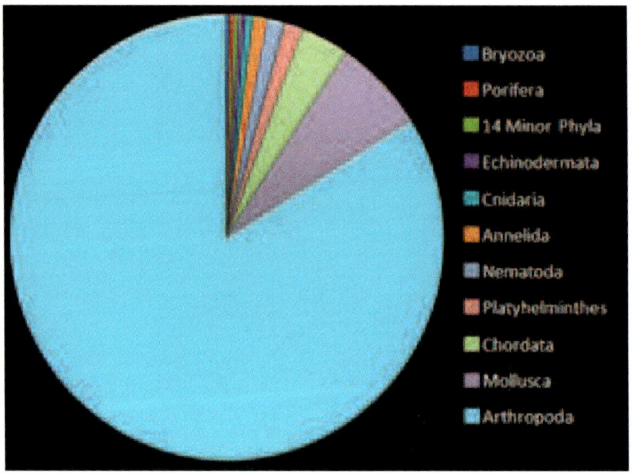

The relative number of species contributed to the total by each phylum of animals. Nematoda is the phylum with the most individual organisms while arthropod has the most species.

The oldest known air-breathing animal is *Pneumodesmus*, an archipolypodan millipede from the Middle Silurian, about 428 Ma. Its air-breathing, terrestrial nature is evidenced by the presence of spiracles, the openings to tracheal systems. However, some earlier trace fossils from the Cambrian-Ordovician boundary about 490 Ma are interpreted as the tracks of large amphibious arthropods on coastal sand dunes, and may have been made by euthycarcinoids, which are thought to be evolutionary aunts' of myriapods. Other trace fossils from the Late Ordovician a little over 445 Ma probably represent land invertebrates, and there is clear evidence of numerous arthropods on coasts and alluvial plains shortly before the Silurian-Devonian boundary, about 415 Ma, including signs that some arthropods ate plants. Arthropods were well pre-adapted to colonise land, because their existing jointed exoskeletons provided protection against desiccation, support against gravity and a means of locomotion that was not dependent on water.

The fossil record of other major invertebrate groups on land is poor: none at all for non-parasitic flatworms, nematodes or nemerteans; some parasitic nematodes have been fossilized in amber; annelid worm fossils are known from the Carboniferous, but they may still have been aquatic animals; the earliest fossils of gastropods on land date from the Late Carboniferous, and this group may have had to wait until leaf litter became abundant enough to provide the moist conditions they need.

The earliest confirmed fossils of flying insects date from the Late Carboniferous, but it is thought that insects developed the ability to fly in the Early Carboniferous or even Late Devonian. This gave them a wider range of ecological niches for feeding and breeding, and a means of escape from predators and from unfavorable changes in the environment. About 99% of modern insect species fly or are descendants of flying species.

Early land Vertebrates

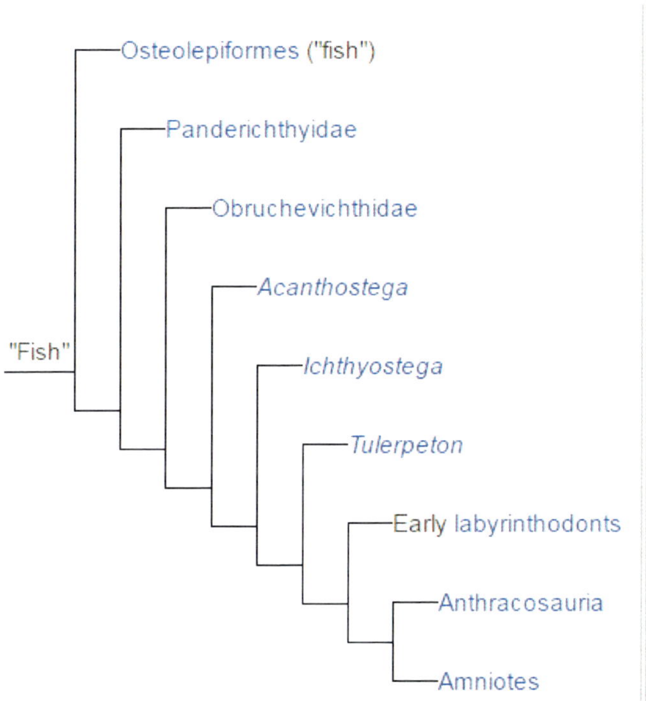

Acanthostega changed views about the early evolution of tetrapods.

Family Tree of Tetrapods

Tetrapods, vertebrates with four limbs, evolved from other rhipidistian fish over a relatively short timespan during the Late Devonian (370 to 360 Ma). The early groups are grouped together as Labyrinthodontia. They retained aquatic, fry-like tadpoles, a system still seen in modern amphibians.

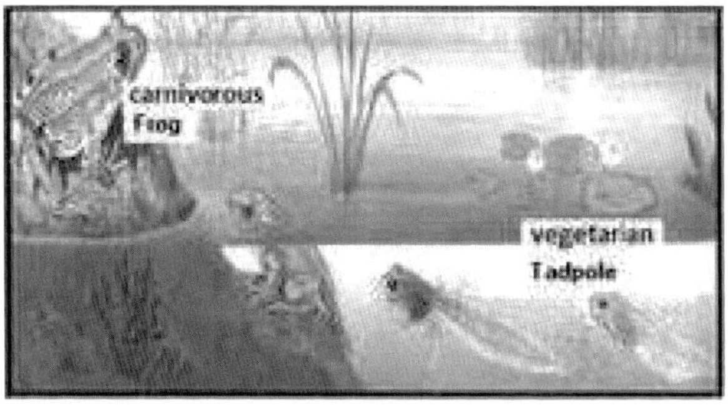

Amphibian Metamorphosis

Iodine and T4/T3 stimulate the amphibian metamorphosis and the evolution of nervous systems transforming the aquatic, vegetarian tadpole into a "more evoluted" terrestrial, carnivorous frog with better neurological, visuospatial, olfactory and cognitive abilities for hunting. The new hormonal action of T3 was made possible by the formation of T3-receptors in the cells of vertebrates. Firstly, about 600-500 million years ago, in primitive Chordata appeared the alpha T3-receptors with a metamorphosing action and then, about 250-150 million years ago, in the Birds and Mammalia appeared the beta T3-receptors with metabolic and thermogenetic actions.

From the 1950s to the early 1980s it was thought that tetrapods evolved from fish that had already acquired the ability to crawl on land, possibly in order to go from a pool that was drying out to one that was deeper. However, in 1987, nearly complete fossils of *Acanthostega* from about 363 Ma showed that this Late Devonian transitional animal had legs and both lungs and gills, but could never have survived on land: its limbs and its wrist and ankle joints were too weak to bear its weight; its ribs were too short to prevent its lungs from being squeezed flat by its weight; its fish-like tail fin would have been damaged by dragging on the ground. The current hypothesis is that *Acanthostega*, which was about 1 metre (3.3 ft) long, was a wholly aquatic predator that hunted in shallow water. Its skeleton differed from that of most fish, in ways that enabled it to raise its head to breathe air while its body remained submerged, including: its jaws show modifications that would have enabled it to gulp air; the bones at the back of its skull are locked together, providing strong attachment points for muscles that raised its head; the head is not joined to the shoulder girdle and it has a distinct neck.

The Devonian proliferation of land plants may help to explain why air breathing would have been an advantage: leaves falling into streams and rivers would have encouraged the growth of aquatic vegetation; this would have attracted grazing invertebrates and small fish that preyed on them; they would have been attractive prey but the environment was unsuitable for the big marine predatory fish; air-breathing would have been necessary because these waters would have been short of oxygen, since warm water holds less dissolved oxygen than cooler marine water and since the decomposition of vegetation would have used some of the oxygen.

Later discoveries revealed earlier transitional forms between *Acanthostega* and completely fish-like animals. Unfortunately, there is then a gap (Romer›s gap) of about 30 Ma between the fossils of ancestral tetrapods and Middle Carboniferous fossils of vertebrates that look well-adapted for life on land. Some of these look like early relatives of modern amphibians, most of which need to keep their skins moist and to lay their eggs in water, while others are accepted as early relatives of the amniotes, whose waterproof skin and egg membranes enable them to live and breed far from water.

Dinosaurs, Birds and Mammals

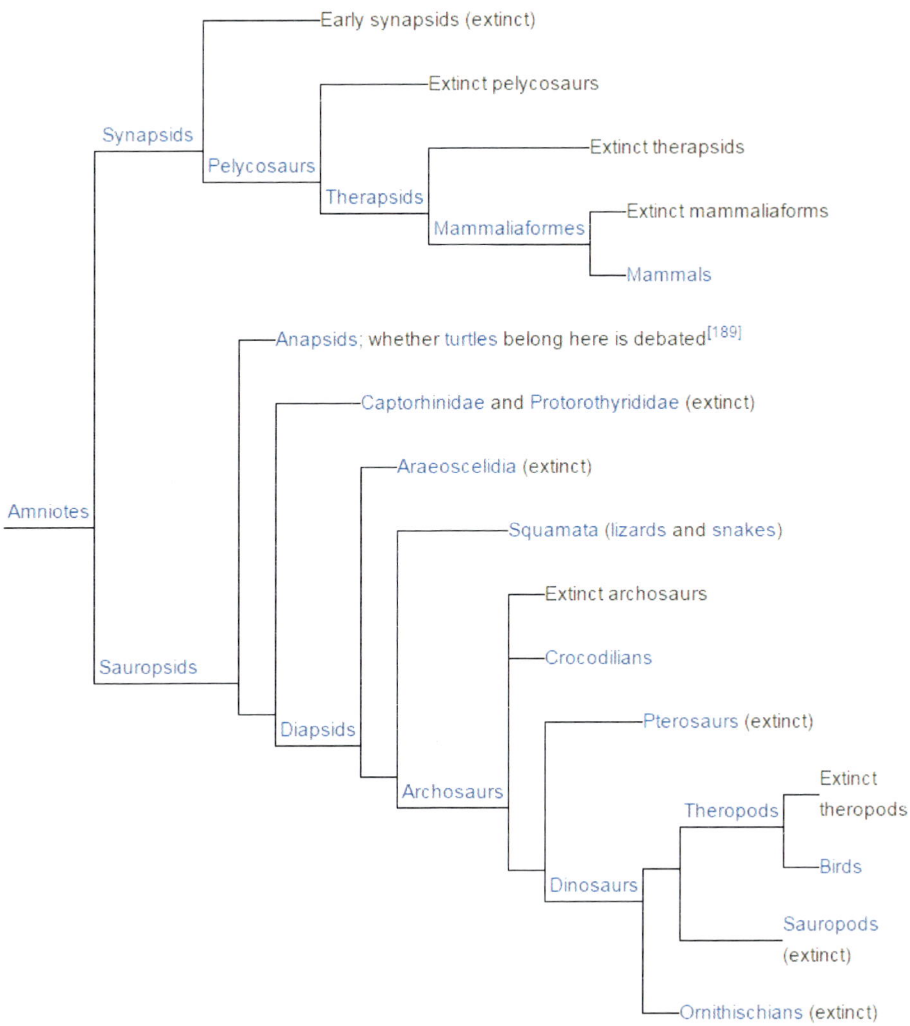

Possible family tree of dinosaurs, birds and mammals

Amniotes, whose eggs can survive in dry environments, probably evolved in the Late Carboniferous period (330 to 298.9 Ma). The earliest fossils of the two surviving amniote groups, synapsids and sauropsids, date from around 313 Ma. The synapsid pelycosaurs and their descendants the therapsids are the most common land vertebrates in the best-known Permian (298.9 to 252.17 Ma) fossil beds. However, at the time these were all in temperate zones at middle latitudes, and there is evidence that hotter, drier environments nearer the Equator were dominated by sauropsids and amphibians.

The Permian–Triassic extinction event wiped out almost all land vertebrates, as well as the great majority of other life. During the slow recovery from this catastrophe, estimated to have taken 30 million years, a previously obscure sauropsid group became the most abundant and diverse terrestrial vertebrates: a few fossils of archosauriformes ("ruling lizard forms") have been found in Late Permian rocks, but, by the Middle Triassic, archosaurs were the dominant land vertebrates. Dinosaurs distinguished themselves from other archosaurs in the Late Triassic, and became the dominant land vertebrates of the Jurassic and Cretaceous periods (201.3 to 66 Ma).

During the Late Jurassic, birds evolved from small, predatory theropod dinosaurs. The first birds inherited teeth and long, bony tails from their dinosaur ancestors, but some had developed horny, toothless beaks by the very Late Jurassic and short pygostyle tails by the Early Cretaceous.

While the archosaurs and dinosaurs were becoming more dominant in the Triassic, the mammaliaform successors of the therapsids evolved into small, mainly nocturnal insectivores. This ecological role may have promoted the evolution of mammals, for example nocturnal life may have accelerated the development of endothermy ("warm-bloodedness") and hair or fur. By 195 Ma in the Early Jurassic there were animals that were very like today's mammals in a number of respects. Unfortunately, there is a gap in the fossil record throughout the Middle Jurassic.However, fossil teeth discovered in Madagascar indicate that the split between the lineage leading to monotremes and the one leading to other living mammals had occurred by 167 Ma. After dominating land vertebrate niches for about 150 Ma, the non-avian dinosaurs perished in the Cretaceous–Paleogene extinction event (66 Ma) along with many other groups of organisms. Mammals throughout the time of the dinosaurs had been restricted to a narrow range of taxa, sizes and shapes, but increased rapidly in size and diversity after the extinction, with bats taking to the air within 13 million years, and cetaceans to the sea within 15 million years.

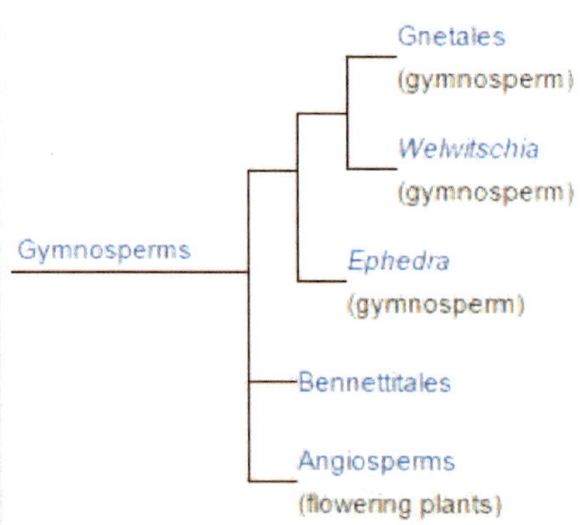

One possible family tree of flowering plants[210]

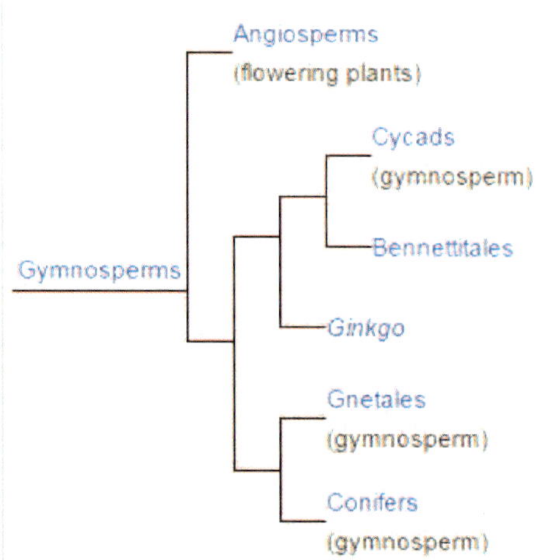

Another possible family tree[211]

Flowering plants

The first flowering plants appeared around 130 Ma. The 250,000 to 400,000 species of flowering plants outnumber all other ground plants combined, and are the dominant vegetation in most terrestrial ecosystems. There is fossil evidence that flowering plants diversified rapidly in the Early Cretaceous, from 130 to 90 Ma, and that their rise was associated with that of pollinating insects. Among modern flowering plants *Magnolia* are thought to be close to the common ancestor of the group.However, paleontologists have not succeeded in identifying the earliest stages in the evolution of flowering plants.

Social Insects

These termite mounds have survived a bush fire.

The social insects are remarkable because the great majority of individuals in each colony are sterile. This appears contrary to basic concepts of evolution such as natural selection and the selfish gene. In fact, there are very few eusocial insect species: only 15 out of approximately 2,600 living families of insects contain eusocial species, and it seems that eusociality has evolved independently only 12 times among arthropods, although some eusocial lineages have diversified into several families. Nevertheless, social insects have been spectacularly successful; for example although ants and termites account for only about 2% of known insect species, they form over 50% of the total mass of insects. Their ability to control a territory appears to be the foundation of their success.

The sacrifice of breeding opportunities by most individuals has long been explained as a consequence of these species› unusual haplodiploid method of sex determination, which has the paradoxical consequence that two sterile worker daughters of the same queen share more genes with each other than they would with their offspring if they could breed. However, E. O. Wilson and Bert Hölldobler argue that this explanation is faulty: for example, it is based on kin selection, but there is no evidence of nepotism in colonies that have multiple queens. Instead, they write, eusociality evolves only in species that are under strong pressure from predators and competitors, but in environments where it is possible to build fortresses'; after colonies have established this security, they gain other advantages through co-operative foraging. In support of this explanation they cite the appearance of eusociality in bathyergid mole rats, which are not haplodiploid.

The earliest fossils of insects have been found in Early Devonian rocks from about 400 Ma, which preserve only a few varieties of flightless insect. The Mazon Creek lagerstätten from the Late Carboniferous, about 300 Ma, include about 200 species, some gigantic by modern standards, and indicate that insects had occupied their main modern ecological niches as herbivores, detritivores and insectivores. Social termites and ants first appear in the Early Cretaceous, and advanced social bees have been found in Late Cretaceous rocks but did not become abundant until the Middle Cenozoic.

Humans

The idea that, along with other life forms, modern-day humans evolved from an ancient, common ancestor was proposed by Robert Chambers in 1844 and taken up by Charles Darwin in 1871. Modern humans evolved from a lineage of upright-walking apes that has been traced back over 6 Ma to *Sahelanthropus*. The first known stone tools were made about 2.5 Ma, apparently by *Australopithecus garhi*, and were found near animal bones that bear scratches made by these tools. The earliest hominines had chimpanzee-sized brains, but there has been a fourfold increase in the last 3 Ma; a statistical analysis suggests that hominine brain sizes depend almost completely on the date of the fossils, while the species to which they are assigned has only slight influence. There is a long-running debate about whether modern humans evolved all over the world simultaneously from existing advanced hominines or are descendants of a single small population in Africa, which then migrated all over the world less than 200,000 years ago and replaced previous hominine species. There is also debate about whether anatomically modern humans had an intellectual, cultural and technological Great Leap Forward' under 100,000 years ago and, if so, whether this was due to neurological changes that are not visible in fossils.

Mass Extinctions

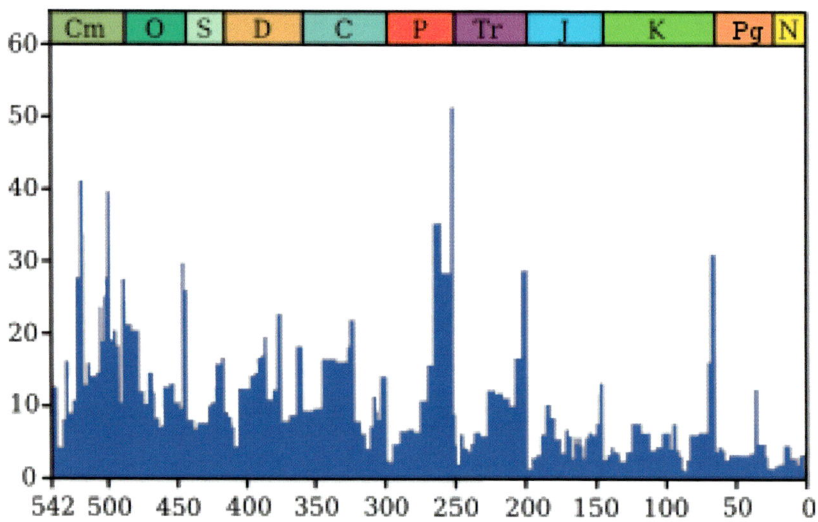

Marine extinction intensity during the Phanerozoic %Millions of years ago

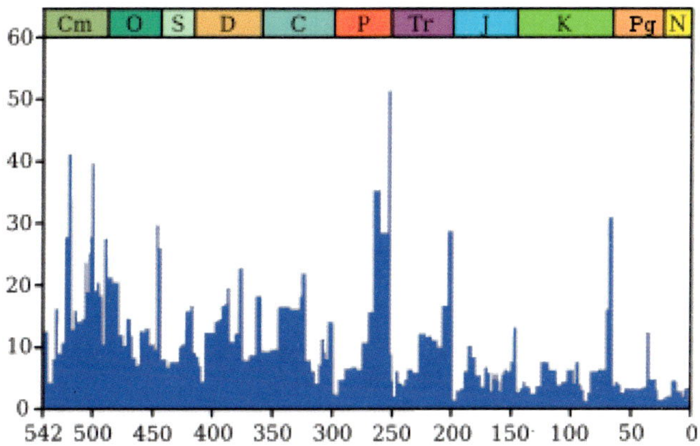

Apparent extinction intensity, i.e. the fraction of genera going extinct at any given time, as reconstructed from the fossil record. (Graph is not meant to include the recent, ongoing Holocene extinction event).

Life on Earth has suffered occasional mass extinctions at least since 542 Ma. Although they were disasters at the time, mass extinctions have sometimes accelerated the evolution of life on Earth. When dominance of particular ecological niches passes from one group of organisms to another, it is rarely because the new dominant group is "superior" to the old and usually because an extinction event eliminates the old dominant group and makes way for the new one.

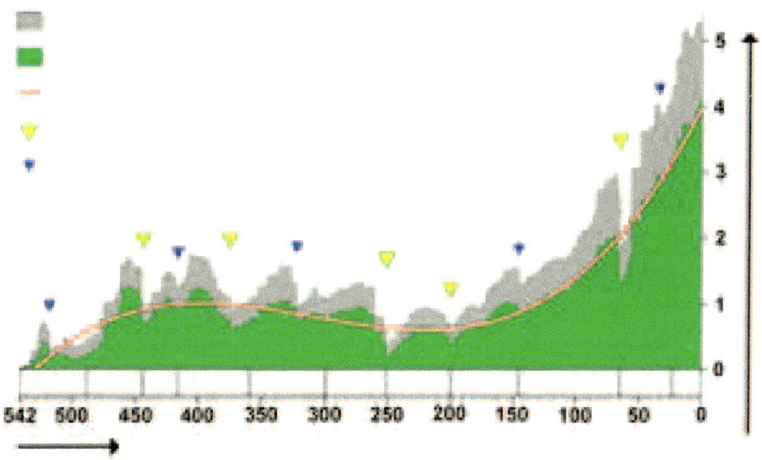

All genera

"Well-defined" genera

Trend line

"Big Five" mass extinctions

Other mass extinctions

Million years ago

Thousands of genera

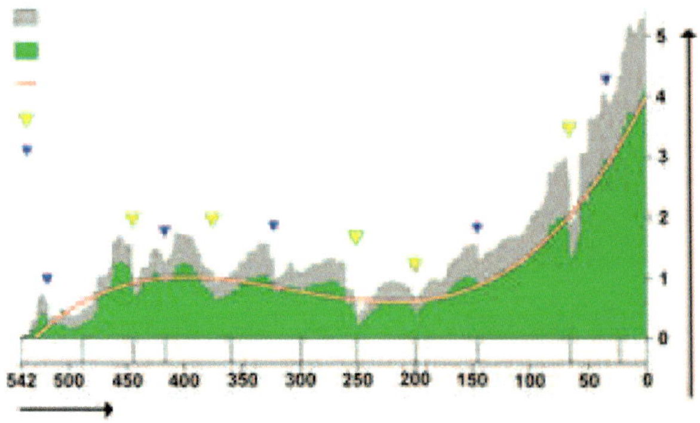

Phanerozoic biodiversity as shown by the fossil record

The fossil record appears to show that the gaps between mass extinctions are becoming longer and the average and background rates of extinction are decreasing. Both of these phenomena could be explained in one or more ways:

- The oceans may have become more hospitable to life over the last 500 Ma and less vulnerable to mass extinctions: dissolved oxygen became more widespread and penetrated to greater depths; the development of life on land reduced the run-off of nutrients and hence the risk of eutrophication and anoxic events; and marine ecosystems became more diversified so that food chains were less likely to be disrupted.
- Reasonably complete fossils are very rare, most extinct organisms are represented only by partial fossils, and complete fossils are rarest in the oldest rocks. So paleontologists have mistakenly assigned parts of the same organism to different genera, which were often defined solely to accommodate these finds—the story of *Anomalocaris* is an example of this. The risk of this mistake is higher for older fossils because these are often both unlike parts of any living organism and poorly conserved. Many of the "superfluous" genera are represented by fragments which are not found again and the "superfluous" genera appear to become extinct very quickly.

Biodiversity in the fossil record, which is "...the number of distinct genera alive at any given time; that is, those whose first occurrence predates and whose last occurrence postdates that time" shows a different trend: a fairly swift rise from 542 to 400 Ma; a slight decline from 400 to 200 Ma, in which the devastating Permian–Triassic extinction event is an important factor; and a swift rise from 200 Ma to the present.

REFERENCES

1. scientists thought published by "Phys.org" in October 19, 2015 (Potentially biogenic carbon preserved in a 4.1 billion-year-old zircon, Proceedings of the National Academy of Sciences, www.pnas.org/cgi/doi/10.1073/pnas.1517557112)
2. Futuyma 2005
3. McKinney 1997, p. 110
4. Stearns & Stearns 1999, p. x

5. Novacek, Michael J. (November 8, 2014). Prehistory›s Brilliant Future'. The New York Times. New York: The New York Times Company. ISSN 0362-4331. Retrieved 2014-12-25.

6. Mora, Camilo; Tittensor, Derek P.; Adl, Sina; et al. (August 23, 2011). How Many Species Are There on Earth and in the Ocean?'. *PLOS Biology. San Francisco, CA: Public Library of Science.* **9** *(8): e1001127. doi:10.1371/journal.pbio.1001127. ISSN 1544-9173. PMC 3160336* . PMID 21886479.

7. Miller & Spoolman 2012, p. 62

8. Chapman, Arthur D. (2009). Numbers of Living Species in Australia and the World. 2nd edition. Canberra: Australian Biological Resources Study. ISBN 978-0-642-56860-1. Retrieved 2016-11-06.

9. Catalogue of Life: 2016 Annual Checklist'. 2016. Retrieved 2016-11-06.

10. Ohtomo, Yoko; Kakegawa, Takeshi; Ishida, Akizumi; et al. (January 2014). Evidence for biogenic graphite in early Archaean Isua metasedimentary rocks'. Nature Geoscience. London: Nature Publishing Group. **7** *(1): 25–28. Bibcode*:2014NatGe...7...25O. doi:10.1038/ngeo2025. ISSN 1752-0894.

11. Borenstein, Seth (November 13, 2013). Oldest fossil found: Meet your microbial mom'. Excite. Yonkers, NY: Mindspark Interactive Network. Associated Press. Retrieved 2015-05-30.

12. Noffke, Nora; Christian, Daniel; Wacey, David; Hazen, Robert M. (December 16, 2013). Microbially Induced Sedimentary Structures Recording an Ancient Ecosystem in the ca. 3.48 Billion-Year-Old Dresser Formation, Pilbara, Western Australia'. Astrobiology. New Rochelle, NY: Mary Ann Liebert, Inc. **13** *(12):* 1103–1124. Bibcode:2013AsBio..13.1103N. doi:10.1089/ast.2013.1030. ISSN 1531-1074. PMC 3870916 . PMID 24205812. Retrieved 2013-11-15.

13. Borenstein, Seth (19 October 2015). Hints of life on what was thought to be desolate early Earth'. Excite. Yonkers, NY: Mindspark Interactive Network. Associated Press. Retrieved 2015-10-20.

14. Bell, Elizabeth A.; Boehnike, Patrick; Harrison, T. Mark; et al. (19 October 2015). Potentially biogenic carbon preserved in a 4.1 billion-year-old zircon' (PDF). Proc. Natl. Acad. Sci. U.S.A. Washington, D.C.: National Academy of Sciences. **112**: 14518–21. doi:10.1073/pnas.1517557112. ISSN 1091-6490. PMC 4664351 . PMID 26483481. Retrieved 2015-10-20. Early edition, published online before print.

15. ᶜ Nisbet, Euan G.; Fowler, C. M. R. (December 7, 1999). Archaean metabolic evolution of microbial mats'. Proceedings of the Royal Society B. London: Royal Society. **266** *(1436): 2375–2382. doi:10.1098/rspb.1999.0934. ISSN 0962-8452. PMC 1690475* .

16. Anbar, Ariel D.; Yun Duan; Lyons, Timothy W.; et al. (September 28, 2007). A Whiff of Oxygen Before the Great Oxidation Event?'. *Science. Washington, D.C.: American Association for the Advancement of Science.* **317** *(5846): 1903–1906. Bibcode*:2007Sci...317.1903A. doi:10.1126/science.1140325. ISSN 0036-8075. PMID 17901330.

17. Knoll, Andrew H.; Javaux, Emmanuelle J.; Hewitt, David; Cohen, Phoebe (June 29, 2006). Eukaryotic organisms in Proterozoic oceans'. Philosophical Transactions of the

Royal Society B. London: Royal Society. **361** *(1470): 1023–1038. doi:10.1098/rstb.2006.1843. ISSN 0962-8436. PMC 1578724* . PMID 16754612.

18. Fedonkin, Mikhail A. (March 31, 2003). The origin of the Metazoa in the light of the Proterozoic fossil record' (PDF). Paleontological Research. Tokyo: Palaeontological Society of Japan. **7** *(1): 9–41. doi:10.2517/prpsj.7.9. ISSN 1880-0068. Retrieved 2008-09-02.*
19. [c] Bonner, John Tyler (1998). The origins of multicellularity'. Integrative Biology. London: Royal Society of Chemistry. **1** *(1): 27–36.* doi:10.1002/(SICI)1520-6602(1998)1:1<27::AID-INBI4>3.0.CO;2-6. ISSN 1757-9694. Retrieved 2008-09-03.
20. Otto, Sarah P.; Lenormand, Thomas (April 2002). Evolution of sex: Resolving the paradox of sex and recombination'. Nature Reviews Genetics. London: Nature Publishing Group. **3** *(4): 252–261.* doi:10.1038/nrg761. ISSN 1471-0056. PMID 11967550.
21. Letunic, Ivica; Bork, Peer. iTOL: Interactive Tree of Life'. Heidelberg, Germany: European Molecular Biology Laboratory. Retrieved 2015-07-21.
22. Letunic, Ivica; Bork, Peer (January 1, 2007). Interactive Tree Of Life (iTOL): an online tool for phylogenetic tree display and annotation' (PDF). Bioinformatics. Cambridge: Oxford University Press. **23** *(1): 127–128. doi:10.1093/bioinformatics/btl529. ISSN 1367-4803. PMID 17050570. Retrieved 2015-07-21.*
23. Letunic, Ivica; Bork, Peer (July 1, 2011). Interactive Tree Of Life v2: online annotation and display of phylogenetic trees made easy'(PDF). Nucleic Acids Research. Oxford: Oxford University Press. **39** *(Suppl. 2): W475–W478. doi:10.1093/*nar/gkr201. ISSN 1362-4962. PMC 3125724 . PMID 21470960. Retrieved 2015-07-21.
24. Fedonkin, Mikhail A.; Simonetta, Alberto; Ivantsov, Andrei Yu. (2007). New data on Kimberella, the Vendian mollusc-like organism (White Sea region, Russia): palaeoecological and evolutionary implications'(PDF). Geological Society Special Publication. Oxford; Boston: Published for the Geological Society of London by Blackwell Scientific Publications. **286**: 157–179. Bibcode:2007GSLSP.286..157F. doi:10.1144/SP286.12. ISSN 0305-8719. Retrieved 2013-05-16.
25. Ciesielski, Paul F. Transition of Plants to Land'. Gainesville, FL: University of Florida. Archived from the original on 2013-11-02. Retrieved 2015-01-22. The oldest fossils reveal evolution of non-vascular plants by the middle to late Ordovician Period (~450-440 m.y.a.) on the basis of fossil spores.
26. Beraldi-Campesi, Hugo (February 2013). Early life on land and the first terrestrial ecosystems'. Ecological Processes. Heidelberg: Springer-Verlag. **2** *(1).* doi:10.1186/2192-1709-2-1. ISSN 2192-1709. Retrieved 2015-01-21.
27. [c] Algeo, Thomas J.; Scheckler, Stephen E. (January 29, 1998). Terrestrial-marine teleconnections in the Devonian: links between the evolution of land plants, weathering processes, and marine anoxic events'. Philosophical Transactions of the Royal Society B. London: Royal Society. **353** *(1365): 113–130. doi:10.1098/rstb.1998.0195. ISSN 0962-8436. PMC 1692181* .
28. Jun-Yuan Chen; Oliveri, Paola; Chia-Wei Li; et al. (April 25, 2000). Precambrian animal diversity: Putative phosphatized embryos from the Doushantuo Formation of China'. Proc. Natl. Acad. Sci. U.S.A. Washington, D.C.: National Academy of Sciences. **97** *(9): 4457–4462. Bibcode:*2000PNAS...97.4457C. doi:10.1073/pnas.97.9.4457. ISSN 0027-8424. PMC 18256 . PMID 10781044. Retrieved 2009-04-30.

29. D-G. Shu; H-L. Luo; Conway Morris, Simon; et al. (November 4, 1999). Lower Cambrian vertebrates from south China' (PDF). Nature. London: Nature Publishing Group. **402** *(6757): 42–46. Bibcode*:1999Natur.402...42S. doi:10.1038/46965. ISSN 0028-0836. Archived from the original (PDF) on 2009-02-26. Retrieved 2015-01-22.

30. Hoyt, Donald F. (February 17, 1997). Synapsid Reptiles'. ZOO 138 Vertebrate Zoology (Lecture). Pomona, CA: California State Polytechnic University, Pomona. Archived from the original on 2009-05-20. Retrieved 2015-01-22.

31. Barry, Patrick L. (January 28, 2002). Phillips, Tony, ed. The Great Dying'. Science@ NASA. Science and Technology Directorate, Marshall Space Flight Center. Retrieved 2015-01-22.

32. Tanner, Lawrence H.; Lucas, Spencer G.; Chapman, Mary G. (March 2004). Assessing the record and causes of Late Triassic extinctions'(PDF). Earth-Science Reviews. Amsterdam, the Netherlands: Elsevier. **65** *(1–2): 103–139. Bibcode*:2004ESRv...65..103T. doi:10.1016/S0012-8252(03)00082-5. ISSN 0012-8252. Archived from the original (PDF) on 2007-10-25. Retrieved 2007-10-22.

33. Benton 1997

34. Fastovsky, David E.; Sheehan, Peter M. (March 2005). The Extinction of the Dinosaurs in North America' (PDF). GSA Today. Boulder, CO: Geological Society of America. **15** *(3): 4–10.* doi:10.1130/1052-5173(2005)015<4:TEOTDI>2.0.CO;2. ISSN 1052-5173. Retrieved 2015-01-23.

35. Roach, John (June 20, 2007). Dinosaur Extinction Spurred Rise of Modern Mammals'. National Geographic News. Washington, D.C.: National Geographic Society. Retrieved 2009-03-08.

36. Van Valkenburgh, Blaire (May 1999). Major patterns in the history of carnivorous mammals'. Annual Review of Earth and Planetary Sciences. Palo Alto, CA: Annual Reviews. **27**: 463–493. Bibcode:1999AREPS..27..463V. doi:10.1146/annurev.earth.27.1.463. ISSN 1545-4495. Retrieved 2015-01-23.

37. Erwin, Douglas H. (9 November 2015). Early metazoan life: divergence, environment and ecology'. Phil. Trans. R. Soc. B. **370** *(20150036). doi:10.1098/rstb.2015.0036. Retrieved 7 January 2016.*

38. El Albani, Abderrazak; Bengtson, Stefan; Canfield, Donald E.; et al. (July 1, 2010). Large colonial organisms with coordinated growth in oxygenated environments 2.1 Gyr ago'. Nature. London: Nature Publishing Group. **466** *(7302): 100–104. Bibcode*:2010Natur.466..100A. doi:10.1038/nature09166. ISSN 0028-0836. PMID 20596019.

39. Dalrymple 1991

40. Newman 2007

41. Dalrymple, G. Brent (2001). The age of the Earth in the twentieth century: a problem (mostly) solved'. Geological Society Special Publication. Oxford; Boston: Published for the Geological Society of London by Blackwell Scientific Publications. **190**: 205–221. Bibcode:2001GSLSP.190..205D. doi:10.1144/GSL.SP.2001.190.01.14. ISSN 0305-8719. Retrieved 2015-01-23.

42. Galimov, Erik M.; Krivtsov, Anton M. (December 2005). Origin of the Earth—Moon system' (PDF). Journal of Earth System Science. Springer Science+Business Media on behalf

of the Indian Academy of Sciences. **114** *(6): 593–600. Bibcode*:2005JESS..114..593G. doi:10.1007/BF02715942. ISSN 0253-4126. Retrieved 2015-01-23.

43. Thompson, Andrea (September 25, 2008). Oldest Rocks on Earth Found'. LiveScience. Watsonville, CA: Imaginova. Retrieved 2015-01-23.
44. Dalrymple 1991
45. Newman 2007
46. Dalrymple, G. Brent (2001). The age of the Earth in the twentieth century: a problem (mostly) solved'. Geological Society Special Publication. Oxford; Boston: Published for the Geological Society of London by Blackwell Scientific Publications. **190**: 205–221. Bibcode:2001GSLSP.190..205D. doi:10.1144/GSL.SP.2001.190.01.14. ISSN 0305-8719. Retrieved 2015-01-23.
47. Cohen, Barbara A.; Swindle, Timothy D.; Kring, David A. (December 1, 2000). Support for the Lunar Cataclysm Hypothesis from Lunar Meteorite Impact Melt Ages'. Science. Washington, D.C.: American Association for the Advancement of Science. **290** *(5497): 1754–1756. Bibcode*:2000Sci...290.1754C. doi:10.1126/science.290.5497.1754. ISSN 0036-8075. PMID 11099411. Retrieved 2015-01-23.
48. Early Earth Likely Had Continents And Was Habitable' (Press release). Boulder, CO: University of Colorado. November 17, 2005. Retrieved 2015-01-23.
49. Cavosie, Aaron J.; Valley, John W.; Wilde, Simon A.; Edinburgh Ion Microprobe Facility (July 15, 2005). Magmatic $\delta^{18}O$ in 4400-3900 Ma detrital zircons: A record of the alteration and recycling of crust in the Early Archean". Earth and Planetary Science Letters. Amsterdam, the Netherlands: Elsevier. **235** *(3–4): 663–681. Bibcode*:2005E&PSL.235..663C. doi:10.1016/j.epsl.2005.04.028. ISSN 0012-821X.
50. [c] Garwood, Russell J. (2012). Patterns In Palaeontology: The first 3 billion years of evolution'. Palaeontology Online. Godalming, UK: Palaeontological Association (sponsor). **2** *(11): 1–14. Retrieved 2015-06-25.*
51. Britt, Robert Roy (July 24, 2002). Evidence for Ancient Bombardment of Earth'. Space.com. New York: Space Holding Corp. Archived from the original on 2006-04-15. Retrieved 2015-01-23.
52. Valley, John W.; Peck, William H.; King, Elizabeth M.; Wilde, Simon A. (April 2002). A cool early Earth' (PDF). Geology. Boulder, CO: Geological Society of America. **30** *(4): 351–354. Bibcode*:2002Geo....30..351V. doi:10.1130/0091-7613(2002)030<0351:ACEE>2.0.CO;2. ISSN 0091-7613. Retrieved 2008-09-13.
53. Dauphas, Nicolas; Robert, François; Marty, Bernard (December 2000). The Late Asteroidal and Cometary Bombardment of Earth as Recorded in Water Deuterium to Protium Ratio'. Icarus. Amsterdam, the Netherlands: Elsevier. **148** *(2): 508–512. Bibcode*:2000Icar..148..508D. doi:10.1006/icar.2000.6489. ISSN 0019-1035.
54. Scalice, Daniella (May 20, 2009). Fletcher, Julie, ed. Microbial Habitability During the Late Heavy Bombardment'. Astrobiology. Mountain View, CA: NASA Astrobiology Program. Retrieved 2015-01-23.
55. Brasier, Martin; McLoughlin, Nicola; Green, Owen; Wacey, David (June 2006). A fresh look at the fossil evidence for early Archaean cellular life' (PDF). Philosophical Transactions of the Royal Society B. London: Royal Society. **361** *(1470): 887–902. doi:10.1098/rstb.2006.1835. ISSN 0962-8436. PMC 1578727* . PMID 16754605.

Retrieved 2008-08-30.

56. Schopf, J. William (April 30, 1993). Microfossils of the Early Archean Apex Chert: New Evidence of the Antiquity of Life'. Science. Washington, D.C.: American Association for the Advancement of Science. **260** *(5108): 640–646. Bibcode*:1993Sci...260..640S. doi:10.1126/science.260.5108.640. ISSN 0036-8075. PMID 11539831. Retrieved 2015-01-24.

57. Altermann, Wladyslaw; Kazmierczak, Józef (November 2003). Archean microfossils: a reappraisal of early life on Earth'. Research in Microbiology. Amsterdam, the Netherlands: Elsevier for the Pasteur Institute. **154** *(9): 611–617. doi:10.1016/j.resmic.2003.08.006. ISSN 0923-2508. PMID 14596897.*

58. Mojzsis, Stephen J.; Arrhenius, Gustaf; McKeegan, Kevin D.; et al. (November 1996). Evidence for life on Earth before 3,800 million years ago'. Nature. London: Nature Publishing Group. **384** *(6604): 55–59. Bibcode*:1996Natur.384...55M. doi:10.1038/384055a0. ISSN 0028-0836. PMID 8900275. Retrieved 2008-08-30.

59. Grotzinger, John P.; Rothman, Daniel H. (October 3, 1996). An abiotic model for stromatolite morphogenesis'. Nature. London: Nature Publishing Group. **383** *(6599): 423–425. Bibcode*:1996Natur.383..423G. doi:10.1038/383423a0. ISSN 0028-0836.

60. Fedo, Christopher M.; Whitehouse, Martin J. (May 24, 2002). Metasomatic Origin of Quartz-Pyroxene Rock, Akilia, Greenland, and Implications for Earth›s Earliest Life'. Science. Washington, D.C.: American Association for the Advancement of Science. **296** *(5572): 1448–1452. Bibcode*:2002Sci...296.1448F. doi:10.1126/science.1070336. ISSN 0036-8075. PMID 12029129. Retrieved 2015-01-24.

61. Lepland, Aivo; van Zuilen, Mark A.; Arrhenius, Gustaf; et al. (January 2005). Questioning the evidence for Earth›s earliest life—Akilia revisited'. Geology. Boulder, CO: Geological Society of America. **33** *(1): 77–79. Bibcode*:2005Geo....33...77L. doi:10.1130/G20890.1. ISSN 0091-7613. Retrieved 2015-01-24.

62. Schopf, J. William (June 29, 2006). Fossil evidence of Archaean life'. Philosophical Transactions of the Royal Society B. London: Royal Society. **361** *(1470): 869–885. doi:10.1098/rstb.2006.1834. ISSN 0962-8436. PMC 1578735* . PMID 16754604.

63. Ciccarelli, Francesca D.; Doerks, Tobias; von Mering, Christian; et al. (March 3, 2006). Toward Automatic Reconstruction of a Highly Resolved Tree of Life'. Science. Washington, D.C.: American Association for the Advancement of Science. **311** *(5765): 1283–1287. Bibcode*:2006Sci...311.1283C. doi:10.1126/science.1123061. ISSN 0036-8075. PMID 16513982.

64. Mason, Stephen F. (1984). Origins of biomolecular handedness'. Nature. London: Nature Publishing Group. **311** *(5981): 19–23. Bibcode*:1984Natur.311...19M. doi:10.1038/311019a0. ISSN 0028-0836. PMID 6472461.

65. Orgel, Leslie E. (October 1994). The Origin of Life on the Earth'. Scientific American. Stuttgart: Georg von Holtzbrinck Publishing Group. **271** *(4): 76–83.* doi:10.1038/scientificamerican1094-76. ISSN 0036-8733. PMID 7524147. Archived from the original (PDF) on 2008. Retrieved 2008-08-30.

66. O'Leary 2008

67. Arrhenius 1980, p. 32

68. Hoyle, Fred; Wickramasinghe, Nalin C. (November 1979). On the Nature of Interstellar

Grains'. Astrophysics and Space Science. Dordrecht, the Netherlands: Kluwer Academic Publishers. **66** *(1): 77–90. Bibcode*:1979Ap&SS..66...77H. doi:10.1007/BF00648361. ISSN 0004-640X.

69. Crick, Francis H.; Orgel, Leslie E (July 1973). Directed Panspermia'. Icarus. Amsterdam, the Netherlands: Elsevier. **19** *(3): 341–348. Bibcode*:1973Icar...19..341C. doi:10.1016/0019-1035(73)90110-3. ISSN 0019-1035.

70. [c] Warmflash, David; Weiss, Benjamin (November 2005). Did Life Come From Another World?'. *Scientific American. Stuttgart: Georg von Holtzbrinck Publishing Group.* **293** *(5): 64–71.* doi:10.1038/scientificamerican1105-64. ISSN 0036-8733. Retrieved 2015-01-25.

71. Wickramasinghe, Nalin C.; Wickramasinghe, Janaki T. (September 2008). On the possibility of microbiota transfer from Venus to Earth'. Astrophysics and Space Science. Dordrecht, the Netherlands: Springer Netherlands. **317** *(1–2): 133–137. Bibcode*:2008Ap&SS.317..133W. doi:10.1007/s10509-008-9851-2. ISSN 0004-640X.

72. Clancy, Brack & Horneck 2005

73. Horneck, Gerda; Klaus, David M.; Mancinelli, Rocco L. (March 2010). Space Microbiology'. Microbiology and Molecular Biology Reviews. Washington, D.C.: American Society for Microbiology. **74** *(1): 121–156. doi:10.1128/mmbr.00016-09. ISSN 1092-2172. Retrieved 2013-07-29.*

74. Than, Ker (August 23, 2007). Claim of Martian Life Called ‹Bogus›'. Space.com. Watsonville, CA: Imaginova. Retrieved 2015-01-25.

75. Bennett 2008, pp. 82–85

76. Schulze-Makuch, Dirk; Irwin, Louis N. (April 2006). The prospect of alien life in exotic forms on other worlds'. Naturwissenschaften. Heidelberg: Springer Science+Business Media. **93** *(4): 155–72. Bibcode*:2006NW.....93..155S. doi:10.1007/s00114-005-0078-6. ISSN 0028-1042. PMID 16525788.

77. Peretó, Juli (2005). Controversies on the origin of life' (PDF). International Microbiology. Barcelona: Spanish Society for Microbiology. **8** *(1): 23–31. ISSN 1139-6709. PMID 15906258. Retrieved 2007-10-07.*

78. Szathmáry, Eörs *(February 3, 2005). "Life: In search of the simplest cell". Nature. London: Nature Publishing Group.* **433** *(7025): 469–470. Bibcode*:2005Natur.433..469S. doi:10.1038/433469a. ISSN 0028-0836. PMID 15690023. Retrieved 2008-09-01.

79. Luisi, Pier Luigi; Ferri, Francesca; Stano, Pasquale (January 2006). Approaches to semi-synthetic minimal cells: a review'. Naturwissenschaften. Heidelberg: Springer Science+Business Media. **93** *(1): 1–13. Bibcode*:2006NW.....93....1L. doi:10.1007/s00114-005-0056-z. ISSN 0028-1042. PMID 16292523.

80. Joyce, Gerald F. (July 11, 2002). The antiquity of RNA-based evolution'. Nature. London: Nature Publishing Group. **418** *(6894): 214–221. Bibcode*:2002Natur.418..214J. doi:10.1038/418214a. ISSN 0028-0836. PMID 12110897.

81. Hoenigsberg, Hugo (December 30, 2003). Evolution without speciation but with selection: LUCA, the Last Universal Common Ancestor in Gilbert›s RNA world'. Genetic and Molecular Research. Ribeirão Preto, Brazil: Fundação de Pesquisas Científicas de Ribeirão Preto. **2** *(4): 366–375. ISSN 1676-5680. PMID 15011140. Retrieved 2008-08-30.*

82. Trevors, Jack T.; Abel, David L. (November 2004). Chance and necessity do not explain the origin of life'. Cell Biology International. London: Portland Press for the International Federation for Cell Biology. **28** *(11): 729–739. doi:10.1016/j.cellbi.2004.06.006. ISSN 1065-6995. PMID 15563395.*

83. Forterre, Patrick; Benachenhou-Lahfa, Nadia; Confalonieri, Fabrice; et al. (1992). The nature of the last universal ancestor and the root of the tree of life, still open questions'. BioSystems. Amsterdam, the Netherlands: Elsevier. **28** *(1–3): 15–32.* doi:10.1016/0303-2647(92)90004-I. ISSN 0303-2647. PMID 1337989.

84. Cech, Thomas R. (August 11, 2000). The Ribosome Is a Ribozyme'. Science. Washington, D.C.: American Association for the Advancement of Science. **289** *(5481): 878–879. doi:10.1126/science.289.5481.878. ISSN 0036-8075. PMID 10960319. Retrieved 2015-01-26.*

85. Johnston, Wendy K.; Unrau, Peter J.; Lawrence, Michael S.; et al. (May 18, 2001). RNA-Catalyzed RNA Polymerization: Accurate and General RNA-Templated Primer Extension'. Science. Washington, D.C.: American Association for the Advancement of Science. **292** *(5520): 1319–1325. Bibcode*:2001Sci...292.1319J. doi:10.1126/science.1060786. ISSN 0036-8075. PMID 11358999.

86. Levy, Matthew; Miller, Stanley L. (July 7, 1998). The stability of the RNA bases: Implications for the origin of life'. Proc. Natl. Acad. Sci. U.S.A. Washington, D.C.: National Academy of Sciences. **95** *(14): 7933–7938. Bibcode*:1998PNAS...95.7933L. doi:10.1073/pnas.95.14.7933. ISSN 0027-8424. PMC 20907 . PMID 9653118. Retrieved 2015-01-26.

87. Larralde, Rosa; Robertson, Michael P.; Miller, Stanley L. (August 29, 1995). "Rates of decomposition of ribose and other sugars: Implications for chemical evolution". Proc. Natl. Acad. Sci. U.S.A. Washington, D.C.: National Academy of Sciences. **92** *(18): 8158–8160. Bibcode*:1995PNAS...92.8158L. doi:10.1073/pnas.92.18.8158. ISSN 0036-8075. PMC 41115 . PMID 7667262. Retrieved 2015-01-26.

88. Lindahl, Tomas (April 22, 1993). "Instability and decay of the primary structure of DNA". Nature. London: Nature Publishing Group. **362** *(6422): 709–715. Bibcode*:1993Natur.362..709L. doi:10.1038/362709a0. ISSN 0028-0836. PMID 8469282.

89. Orgel, Leslie (November 17, 2000). A Simpler Nucleic Acid'. Science. Washington, D.C.: American Association for the Advancement of Science. **290** *(5495): 1306–1307. doi:10.1126/science.290.5495.1306. ISSN 0036-8075. PMID 11185405.*

90. Nelson, Kevin E.; Levy, Matthew; Miller, Stanley L. (April 11, 2000). Peptide nucleic acids rather than RNA may have been the first genetic molecule'. Proc. Natl. Acad. Sci. U.S.A. Washington, D.C. **97** *(8): 3868–3871. Bibcode*:2000PNAS...97.3868N. doi:10.1073/pnas.97.8.3868. ISSN 0036-8075. PMC 18108 . PMID 10760258. Retrieved 2015-01-26.

91. Martin, William; Russell, Michael J. (January 29, 2003). On the origins of cells: a hypothesis for the evolutionary transitions from abiotic geochemistry to chemoautotrophic prokaryotes, and from prokaryotes to nucleated cells'. Philosophical Transactions of the Royal Society B. London: Royal Society. **358** *(1429): 59–85. doi:10.1098/rstb.2002.1183. ISSN 0962-8436. PMC 1693102* . PMID 12594918.

92. Wächtershäuser, Günter (August 25, 2000). Life as We Don›t Know It'. Science. Washington, D.C.: American Association for the Advancement of Science. **289** *(5483):*

1307–1308. doi:10.1126/science.289.5483.1307. ISSN 0036-8075. PMID 10979855.

93. Trevors, Jack T.; Psenner, Roland (December 2001). From self-assembly of life to present-day bacteria: a possible role for nanocells'. FEMS Microbiology Reviews. Amsterdam, the Netherlands: Elsevier on behalf of the Federation of European Microbiological Societies. **25** *(5): 573–582. doi:10.1111/j.1574-6976.2001.tb00592.x. ISSN 1574-6976. PMID 11742692.*

94. Segré, Daniel; Ben-Eli, Dafna; Deamer, David W.; Lancet, Doron (February 2001). The Lipid World' (PDF). Origins of Life and Evolution of Biospheres. Dordrecht, the Netherlands: Kluwer Academic Publishers. **31** *(1–2): 119–145.* doi:10.1023/A:1006746807104. ISSN 0169-6149. PMID 11296516. Retrieved 2015-01-26.

95. Cairns-Smith 1968, pp. 57–66

96. Ferris, James P. (June 1999). Prebiotic Synthesis on Minerals: Bridging the Prebiotic and RNA Worlds'. The Biological Bulletin. Woods Hole, MA: Marine Biological Laboratory. **196** *(3): 311–314.* doi:10.2307/1542957. ISSN 0006-3185. JSTOR 1542957. PMID 10390828. "This paper was originally presented at a workshop titled Evolution: A Molecular Point of View.'

97. Hanczyc, Martin M.; Fujikawa, Shelly M.; Szostak, Jack W. (October 24, 2003). Experimental Models of Primitive Cellular Compartments: Encapsulation, Growth, and Division'. Science. Washington, D.C.: American Association for the Advancement of Science. **302** *(5645): 618–622. Bibcode*:2003Sci...302..618H. doi:10.1126/science.1089904. ISSN 0036-8075. PMID 14576428. Retrieved 2015-01-26.

98. Hartman, Hyman (October 1998). Photosynthesis and the Origin of Life'. Origins of Life and Evolution of Biospheres. Dordrecht, the Netherlands: Kluwer Academic Publishers. **28** *(4–6): 512–521. Bibcode*:1998OLEB...28..515H. doi:10.1023/A:1006548904157. ISSN 0169-6149. PMID 11536891. Retrieved 2015-01-27.

99. Krumbein et al. 2003, pp. 1–28

100. Risatti, J. Bruno; Capman, William C.; Stahl, David A. (October 11, 1994). Community structure of a microbial mat: The phylogenetic dimension' (PDF). Proc. Natl. Acad. Sci. U.S.A. Washington, D.C.: National Academy of Sciences. **91** *(21): 10173–10177. Bibcode*:1994PNAS...9110173R. doi:10.1073/pnas.91.21.10173. ISSN 0036-8075. PMC 44980 . PMID 7937858. Retrieved 2008-07-09.

101. Biodiversity rocks'. Nature (Editor›s summary). London: Nature Publishing Group. **441** *(7094). June 8, 2006. ISSN 0028-0836. Retrieved 2009-01-10.*

102. Allwood, Abigail C.; Walter, Malcolm R.; Kamber, Balz S.; et al. (June 8, 2006). Stromatolite reef from the Early Archaean era of Australia'. Nature. London: Nature Publishing Group. **441** *(7094): 714–718. Bibcode*:2006Natur.441..714A. doi:10.1038/nature04764. ISSN 0028-0836. PMID 16760969. Retrieved 2008-08-31.

103. Blankenship, Robert E. (January 1, 2001). Molecular evidence for the evolution of photosynthesis'. Trends in Plant Science. Cambridge, MA: Cell Press. **6** *(1): 4–6.* doi:10.1016/S1360-1385(00)01831-8. ISSN 1360-1385. PMID 11164357. Retrieved 2015-01-28.

104. Hoehler, Tori M.; Bebout, Brad M.; Des Marais, David J. (July 19, 2001). The role of microbial mats in the production of reduced gases on the early Earth'. Nature. London:

Nature Publishing Group. **412** *(6844): 324–327.* doi:10.1038/35085554. ISSN 0028-0836. PMID 11460161. Retrieved 2008-07-14.

105. Abele, Doris (November 7, 2002). Toxic oxygen: The radical life-giver'. Nature. London: Nature Publishing Group. **420** *(6911): 27. Bibcode*:2002Natur.420...27A. doi:10.1038/420027a. ISSN 0028-0836. PMID 12422197. Retrieved 2008-07-14.

106. Westerdahl, Becky B. (2007). Introduction to Aerobic Respiration'. Biological Science 10V (Lecture). Davis, CA: University of California, Davis. Archived from the original on 2007-10-29. Retrieved 2008-07-14.

107. Textbook used for lecture: Biology Today and Tomorrow With Physiology (2007), ISBN 0-495-01654-3.

108. Goldblatt, Colin; Lenton, Timothy M.; Watson, Andrew J. (2006). The Great Oxidation at ~2.4 Ga as a bistability in atmospheric oxygen due to UV shielding by ozone' (PDF). Geophysical Research Abstracts. Katlenburg-Lindau, Germany: Copernicus Publications on behalf of the European Geosciences Union. **8** *(00770). ISSN 1029-7006.* SRef-ID: 1607-7962/gra/EGU06-A-00770. Retrieved 2008-09-01.

109. Glansdorff, Nicolas; Ying Xu; Labedan, Bernard (July 9, 2008). The Last Universal Common Ancestor: emergence, constitution and genetic legacy of an elusive forerunner'. Biology Direct. London: BioMed Central. **3** *(29): 29.* doi:10.1186/1745-6150-3-29. ISSN 1745-6150. PMC 2478661 . PMID 18613974.

110. Brocks, Jochen J.; Logan, Graham A.; Buick, Roger; Summons, Roger E. (August 13, 1999). Archean Molecular Fossils and the Early Rise of Eukaryotes'. Science. Washington, D.C.: American Association for the Advancement of Science. **285** *(5430): 1033–1036. doi:10.1126/science.285.5430.1033. ISSN 0036-8075. PMID 10446042. Retrieved 2015-01-29.*

111. [c] Hedges, S. Blair; Blair, Jaime E.; Venturi, Maria L.; Shoe, Jason L. (January 28, 2004). A molecular timescale of eukaryote evolution and the rise of complex multicellular life'. BMC Evolutionary Biology. London: BioMed Central. **4**: 2. doi:10.1186/1471-2148-4-2. ISSN 1471-2148. PMC 341452 . PMID 15005799. Retrieved 2008-07-14.

112. Adl, Sina M.; et al. (September 2012). The revised classification of eukaryotes' (PDF). Journal of Eukaryotic Microbiology. **59** *(5): 429–514. doi:10.1111/j.1550-7408.2012.00644.x. PMC 3483872* . PMID 23020233.

113. Burki, F. (2014). The eukaryotic tree of life from a global phylogenomic perspective'. Cold Spring Harbor Perspectives in Biology. **6**: 1–17. doi:10.1101/cshperspect.a016147.

114. Margulis 1981

115. Vellai, Tibor; Vida, Gábor (August 7, 1999). The origin of eukaryotes: the difference between prokaryotic and eukaryotic cells'. Proceedings of the Royal Society B. London: Royal Society. **266** *(1428): 1571–1577. doi:10.1098/rspb.1999.0817. ISSN 0962-8452. PMC 1690172* . PMID 10467746.

116. Selosse, Marc-André; Abert, Béatrice; Godelle, Bernard (March 1, 2001). Reducing the genome size of organelles favours gene transfer to the nucleus'. Trends in Ecology & Evolution. Cambridge, MA: Cell Press. **16** *(3): 135–141.* doi:10.1016/S0169-5347(00)02084-X. ISSN 0169-5347. Retrieved 2015-01-29.

117. Pisani, Davide; Cotton, James A.; McInerney, James O. (August 2007). Supertrees Disentangle the Chimerical Origin of Eukaryotic Genomes'. Molecular Biology and

Evolution. Oxford: Oxford University Press on behalf of the Society for Molecular Biology and Evolution. **24** *(8): 1752–1760. doi:10.1093/molbev/msm095. ISSN 0737-4038. PMID 17504772.*

118. Gray, Michael W.; Burger, Gertraud; Lang, B. Franz (March 5, 1999). Mitochondrial Evolution'. Science. Washington, D.C.: American Association for the Advancement of Science. **283** *(5407): 1476–1481. Bibcode*:1999Sci...283.1476G. doi:10.1126/science.283.5407.1476. ISSN 0036-8075. PMID 10066161. Retrieved 2015-01-29.

119. Rasmussen, Birger; Fletcher, Ian R.; Brocks, Jochen J.; Kilburn, Matt R. (October 23, 2008). Reassessing the first appearance of eukaryotes and cyanobacteria'. Nature. London: Nature Publishing Group. **455** *(7216): 1101–1104. Bibcode*:2008Natur.455.1101R. doi:10.1038/nature07381. ISSN 0028-0836. PMID 18948954.

120. Tsu-Ming Han; Runnegar, Bruce (July 10, 1992). Megascopic eukaryotic algae from the 2.1-billion-year-old negaunee iron-formation, Michigan'. Science. Washington, D.C.: American Association for the Advancement of Science. **257** *(5067): 232–235. Bibcode*:1992Sci...257..232H. doi:10.1126/science.1631544. ISSN 0036-8075. PMID 1631544. Retrieved 2015-01-30.

121. Javaux, Emmanuelle J.; Knoll, Andrew H.; Walter, Malcolm R. (July 2004). TEM evidence for eukaryotic diversity in mid-Proterozoic oceans'. Geobiology. Hoboken, NJ: Wiley-Blackwell. **2** *(3): 121–132. doi:10.1111/j.1472-4677.2004.00027.x. ISSN 1472-4677. Retrieved 2015-01-30.*

122. Butterfield, Nicholas J. (Winter 2005). Probable Proterozoic fungi'. Paleobiology. Paleontological Society. **31** *(1): 165–182.* doi:10.1666/0094-8373(2005)031<0165:PPF>2.0.CO;2. ISSN 0094-8373. Retrieved 2015-01-30.

123. c d e Neiman, Maurine; Jokela, Jukka (2010). Sex: Advantage'. Encyclopedia of Life Sciences. Hoboken, NJ: John Wiley & Sons. doi:10.1002/9780470015902.a0001716.pub2. ISBN 0-470-01617-5. Retrieved 2015-01-21.

124. Holmes & Jobling 1996

125. Christie, Peter J. (April 2001). Type IV secretion: intercellular transfer of macromolecules by systems ancestrally related to conjugation machines'. Molecular Microbiology. Hoboken, NJ: Wiley-Blackwell. **40** *(2): 294–305. doi:10.1046/j.1365-2958.2001.02302.x. ISSN 0950-382X. PMID 11309113.*

126. Michod, Richard E.; Bernstein, Harris; Nedelcu, Aurora M. (May 2008). Adaptive value of sex in microbial pathogens' (PDF). Infection, Genetics and Evolution. Amsterdam, the Netherlands: Elsevier. **8** *(3): 267–285. doi:10.1016/j.meegid.2008.01.002. ISSN 1567-1348. PMID 18295550.*

127. Bernstein, Harris; Bernstein, Carol (July 2010). Evolutionary Origin of Recombination during Meiosis'. BioScience. Oxford: Oxford University Press on behalf of the American Institute of Biological Sciences. **60** *(7): 498–505. doi:10.1525/bio.2010.60.7.5. ISSN 0006-3568.*

128. Johnsborg, Ola; Eldholm, Vegard; Håvarstein, Leiv Sigve (December 2007). Natural genetic transformation: prevalence, mechanisms and function'. Research in Microbiology. Amsterdam, the Netherlands: Elsevier for the Pasteur Institute. **158** *(10): 767–778. doi:10.1016/j.resmic.2007.09.004. ISSN 0923-2508. PMID 17997281.*

129. Bernstein, Bernstein & Michod 2012, pp. 1–50

130. Ramesh, Marilee A.; Malik, Shehre-Banoo; Logsdon, John M., Jr. (January 26, 2005). A Phylogenomic Inventory of Meiotic Genes: Evidence for Sex in Giardia and an Early Eukaryotic Origin of Meiosis' (PDF). Current Biology. Cambridge, MA: Cell Press. **15** *(2): 185–191. doi:10.1016/j.cub.2005.01.003. ISSN 0960-9822. PMID 15668177. Retrieved 2008-12-22.*

131. Otto, Sarah P.; Gerstein, Aleeza C. (August 2006). Why have sex? The population genetics of sex and recombination'. Biochemical Society Transactions. London: Portland Press on behalf of the Biochemical Society. **34** *(Pt 4): 519–522.* doi:10.1042/BST0340519. ISSN 0300-5127. PMID 16856849. Retrieved 2008-12-22.

132. Lin Dong; Shuhai Xiao; Bing Shen; Chuanming Zhou (January 2008). Silicified Horodyskia *and* Palaeopascichnus *from upper Ediacaran* cherts in South China: tentative phylogenetic interpretation and implications for evolutionary stasis'. Journal of the Geological Society. London: Geological Society of London. **165** *(1): 367–378.* doi:10.1144/0016-76492007-074. ISSN 0016-7649. Retrieved 2015-02-01.

133. Hanley, Kathryn A.; Fisher, Robert N.; Case, Ted J. (June 1995). Lower Mite Infestations in an Asexual Gecko Compared With Its Sexual Ancestors'. Evolution. Hoboken, NJ: John Wiley & Sons on behalf of the Society for the Study of Evolution. **49** *(3): 418–426.* doi:10.2307/2410266. ISSN 0014-3820.

134. Parker, Matthew A. (September 1994). Pathogens and sex in plants'. Evolutionary Ecology. Dordrecht, the Netherlands: Kluwer Academic Publishers. **8** *(5): 560–584.* doi:10.1007/bf01238258. ISSN 0269-7653.

135. Birdsell & Wills 2003, pp. 27–137

136. Bernstein, Hopf & Michod 1987, pp. 323–370

137. Bell, Graham; Mooers, Arne O. (1997). Size and complexity among multicellular organisms' (PDF). Biological Journal of the Linnean Society. Hoboken, NJ: John Wiley & Sons on behalf of the Linnean Society of London. **60** *(3): 345–363. doi:10.1111/j.1095-8312.1997.tb01500.x. ISSN 0024-4074. Retrieved 2015-02-02.*

138. Kaiser, Dale (December 2001). Building a multicellular organism'. Annual Review of Genetics. Palo Alto, CA: Annual Reviews. **35**: 103–123. doi:10.1146/annurev.genet.35.102401.090145. ISSN 0066-4197. PMID 11700279.

139. Nakagaki, Toshiyuki; Yamada, Hiroyasu; Tóth, *Ágota* (September 28, 2000). Intelligence: Maze-solving by an amoeboid organism'. Nature. London: Nature Publishing Group. **407** *(6803): 470.* doi:10.1038/35035159. ISSN 0028-0836. PMID 11028990. Retrieved 2008-09-03.

140. Koschwanez, John H.; Foster, Kevin R.; Murray, Andrew W. (August 9, 2011). Sucrose Utilization in Budding Yeast as a Model for the Origin of Undifferentiated Multicellularity'. PLOS Biology. San Francisco, CA: Public Library of Science. **9** *(8): e1001122. doi:10.1371/journal.pbio.1001122. ISSN 1545-7885. Retrieved 2015-02-02.*

141. Butterfield, Nicholas J. (Summer 2000). Bangiomorpha pubescens *n. gen., n. sp.: implications for the evolution of sex, multicellularity, and the Mesoproterozoic/Neoproterozoic radiation of eukaryotes".* Paleobiology. Paleontological Society. **26** *(3): 386–404.* doi:10.1666/0094-8373(2000)026<0386:BPNGNS>2.0.CO;2. ISSN 0094-8373. Retrieved 2015-02-01.

142. Jin JP; et al. (July 2015). An Arabidopsis transcriptional regulatory map reveals distinct functional and evolutionary features of novel transcription factors'. Molecular Biology and Evolution. **32** *(7). doi:10.1093/*molbev/msv058.

143. Dickey, Gwyneth (July 31, 2010). Evidence for earlier multicellular life' (PNG). Science News. Washington, D.C.: Society for Science & the Public. **178** *(3): 17. doi:10.1002/scin.5591780322. ISSN 0036-8423. Retrieved 2015-02-02.*

144. Gaidos, Eric J.; Dubuc, Timothy; Dunford, Mike; et al. (December 2007). The Precambrian emergence of animal life: a geobiological perspective'. Geobiology. Hoboken, NJ: Wiley-Blackwell. **5** *(4): 351–373. doi:10.1111/j.1472-4669.2007.00125.x. ISSN 1472-4677.*

145. Davidson, Michael W. (May 26, 2005). Animal Cell Structure'. Molecular Expressions. Tallahassee, FL: Florida State University. Retrieved 2008-09-03.

146. Saupe, Stephen G. (January 3, 2004). Concepts of Biology'. Concepts of Biology (BIOL116) (Lecture). St. Joseph, MN: College of Saint Benedict and Saint John›s University. Retrieved 2008-09-03.

147. Hinde 2001, pp. 28–57

148. Jun-Yuan Chen; Oliveri, Paola; Feng Gao; et al. (August 1, 2002). Precambrian Animal Life: Probable Developmental and Adult Cnidarian Forms from Southwest China' (PDF). Developmental Biology. Amsterdam, the Netherlands: Elsevier. **248** *(1): 182–196. doi:10.1006/dbio.2002.0714. ISSN 0012-1606. PMID 12142030. Retrieved 2015-02-04.*

149. Grazhdankin, Dima (June 2004). Patterns of distribution in the Ediacaran biotas: facies versus biogeography and evolution'. Paleobiology. Paleontological Society. **30** *(2): 203–221.* doi:10.1666/0094-8373(2004)030<0203:PODITE>2.0.CO;2. ISSN 0094-8373.

150. Seilacher, Adolf (August 1992). Vendobionta and Psammocorallia: lost constructions of Precambrian evolution'. Journal of the Geological Society. London: Geological Society of London. **149** *(4): 607–613. doi:10.1144/gsjgs.149.4.0607. ISSN 0016-7649. Retrieved 2015-02-04.*

151. Martin, Mark W.; Grazhdankin, Dmitriy V.; Bowring, Samuel A.; et al. (May 5, 2000). Age of Neoproterozoic Bilaterian Body and Trace Fossils, White Sea, Russia: Implications for Metazoan Evolution'. Science. Washington, D.C.: American Association for the Advancement of Science. **288** *(5467): 841–845. Bibcode:*2000Sci...288..841M. doi:10.1126/science.288.5467.841. ISSN 0036-8075. PMID 10797002. Retrieved 2015-02-05.

152. Fedonkin, Mikhail A.; Waggoner, Benjamin M. (August 28, 1997). The late Precambrian fossil Kimberella *is a* mollusc-like bilaterian organism'. Nature. London: Nature Publishing Group. **388** *(6645): 868–871. Bibcode:*1997Natur.388..868F. doi:10.1038/42242. ISSN 0028-0836. Retrieved 2008-07-03.

153. Mooi, Rich; David, Bruno (December 1998). Evolution Within a Bizarre Phylum: Homologies of the First Echinoderms' (PDF). American Zoologist. Society for Integrative and Comparative Biology. **38** *(6): 965–974. doi:10.1093/*icb/38.6.965. ISSN 1540-7063. Retrieved 2015-02-05.

154. McMenamin, Mark A. S. (September 2003). Spriggina *is a* trilobitoid ecdysozoan. Geoscience Horizons Seattle 2003. Abstracts with Programs. **35**. Boulder, CO: Geological Society of America. p. 105. OCLC 249088612. Retrieved 2007-11-24. Paper No.

40-2 presented at the Geological Society of America's 2003 Seattle Annual Meeting (November 2–5, 2003) on November 2, 2003, at the Washington State Convention Center.

155. Jih-Pai Lin; Gon, Samuel M., III; Gehling, James G.; et al. (2006). A Parvancorina-like arthropod from the Cambrian of South China'. Historical Biology: An International Journal of Paleobiology. Taylor & Francis. **18** *(1): 33–45.* doi:10.1080/08912960500508689. ISSN 1029-2381.

156. Butterfield, Nicholas J. (December 2006). Hooking some stem-group ‹worms›: fossil lophotrochozoans in the Burgess Shale'. BioEssays. Hoboken, NJ: John Wiley & Sons. **28** *(12): 1161–1166. doi:10.1002/bies.20507. ISSN 0265-9247. PMID 17120226.*

157. [c] Bengtson 2004, pp. 67–78

158. Gould 1989, pp. 124–136

159. Gould 1989

160. Budd, Graham E. (February 2003). The Cambrian Fossil Record and the Origin of the Phyla' (PDF). Integrative and Comparative Biology. Oxford: Oxford University Press for the Society for Integrative and Comparative Biology. **43** *(1): 157–165. doi:10.1093/icb/43.1.157. ISSN 1557-7023. PMID 21680420. Retrieved 2015-02-06.*

161. Budd, Graham E. (March 1996). The morphology of Opabinia regalis and the reconstruction of the arthropod stem-group'. Lethaia. Hoboken, NJ: Wiley-Blackwell. **29** *(1): 1–14. doi:10.1111/j.1502-3931.1996.tb01831.x. ISSN 0024-1164.*

162. Marshall, Charles R. (May 2006). Explaining the Cambrian ‹Explosion› of Animals'. Annual Review of Earth and Planetary Sciences. Palo Alto, CA: Annual Reviews. **34**: 355–384. Bibcode:2006AREPS..34..355M. doi:10.1146/annurev.earth.33.031504.103001. ISSN 1545-4495. Retrieved 2015-02-06.

163. Janvier, Philippe (2001). Vertebrata (Vertebrates)'. Encyclopedia of Life Sciences. Hoboken, NJ: John Wiley & Sons. doi:10.1038/npg.els.0001531. ISBN 0-470-01617-5. Retrieved 2015-01-21.

164. Conway Morris, Simon (August 2, 2003). Once we were worms'. New Scientist. London: Reed Business Information. **179** *(2406): 34. ISSN 0262-4079. Archived from the original on 2008-07-25. Retrieved 2008-09-05.*

165. Jun-Yuan Chen; Di-Ying Huang; Qing-Qing Peng; et al. (July 8, 2003). The first tunicate from the Early Cambrian of South China'. Proc. Natl. Acad. Sci. U.S.A. Washington, D.C.: National Academy of Sciences. **100** *(14): 8314–8318. Bibcode*:2003PNAS..100.8314C. doi:10.1073/pnas.1431177100. ISSN 0036-8075. PMC 166226 . PMID 12835415.

166. D.-G. Shu; Conway Morris, Simon; J. Han; et al. (January 30, 2003). Head and backbone of the Early Cambrian vertebrateHaikouichthys'. Nature. London: Nature Publishing Group. **421** *(6922): 526–529. Bibcode*:2003Natur.421..526S. doi:10.1038/nature01264. ISSN 0028-0836. PMID 12556891. Retrieved 2008-09-05.

167. Sansom, Smith & Smith 2001, pp. 156–171

168. Cowen 2000, pp. 120–122

169. [c] Selden 2001, "Terrestrialization of Animals,' pp. 71–74

170. [c] Garwood, Russell J.; Edgecombe, Gregory D. (September 2011). Early Terrestrial Animals, Evolution, and Uncertainty'. Evolution: Education and Outreach. New York:

Springer Science+Business Media. **4** *(3): 489–501.* doi:10.1007/s12052-011-0357-y. ISSN 1936-6426. Retrieved 2015-07-21.

171. Battistuzzi, Fabia U.; Feijao, Andreia; Hedges, S. Blair (November 9, 2004). A genomic timescale of prokaryote evolution: insights into the origin of methanogenesis, phototrophy, and the colonization of land'. BMC Evolutionary Biology. London: BioMed Central. **4**: 44. doi:10.1186/1471-2148-4-44. ISSN 1471-2148. PMC 533871 . PMID 15535883.

172. [g] Shear 2000, "The Early Development of Terrestrial Ecosystems," pp. 169–184

173. Venturi, Sebastiano (September 2011). Evolutionary Significance of Iodine'. Current Chemical Biology. Sharjah, U.A.E.: Bentham Science Publishers. **5** *(3): 155–162.* doi:10.2174/187231311796765012. ISSN 1872-3136.

174. Crockford, Susan J. (August 2009). Evolutionary roots of iodine and thyroid hormones in cell-cell signaling'. Integrative and Comparative Biology. Oxford: Oxford University Press for the Society for Integrative and Comparative Biology. **49** *(2): 155–166. doi:10.1093/*icb/icp053. ISSN 1557-7023. PMID 21669854.

175. Venturi, Sebastiano; Donati, Francesco M.; Venturi, Alessandro; Venturi, Mattia (August 2000). Environmental Iodine Deficiency: A Challenge to the Evolution of Terrestrial Life?'. *Thyroid. New Rochelle, NY: Mary Ann Liebert, Inc.* **10** *(8): 727–729.* doi:10.1089/10507250050137851. ISSN 1050-7256. PMID 11014322.

176. Küpper, Frithjof C.; Carpenter, Lucy J.; McFiggans, Gordon B.; et al. (May 13, 2008). Iodide accumulation provides kelp with an inorganic antioxidant impacting atmospheric chemistry'. Proc. Natl. Acad. Sci. U.S.A. Washington, D.C.: National Academy of Sciences. **105** *(19): 6954–6958. Bibcode*:2008PNAS..105.6954K. doi:10.1073/pnas.0709959105. ISSN 0036-8075. PMC 2383960 . PMID 18458346.

177. Hawksworth, David L. (2002). Lichens'. Encyclopedia of Life Sciences. Hoboken, NJ: John Wiley & Sons. doi:10.1038/npg.els.0000368. ISBN 0-470-01617-5. Retrieved 2015-01-21.

178. Retallack, Gregory J.; Feakes, Carolyn R. (January 2, 1987). Trace Fossil Evidence for Late Ordovician Animals on Land'. Science. Washington, D.C.: American Association for the Advancement of Science. **235** *(4784): 61–63. Bibcode*:1987Sci...235...61R. doi:10.1126/science.235.4784.61. ISSN 0036-8075. PMID 17769314.

179. Kenrick, Paul; Crane, Peter R. (September 4, 1997). The origin and early evolution of plants on land' (PDF). Nature. London: Nature Publishing Group. **389** *(6646): 33–39. Bibcode*:1997Natur.389...33K. doi:10.1038/37918. ISSN 0028-0836. Retrieved 2015-02-10.

180. Scheckler 2001, Afforestation—the First Forests,' pp. 67–70

181. The phrase "Late Devonian wood crisis" is used at Tetrapoda:Acanthostega'. Palaeos. Retrieved 2015-02-10.

182. Taylor, Thomas N.; Osborn, Jeffrey M. (February 1996). The importance of fungi in shaping the paleoecosystem'. Review of Palaeobotany and Palynology. Amsterdam, the Netherlands: Elsevier. **90** *(3–4): 249–262.* doi:10.1016/0034-6667(95)00086-0. ISSN 0034-6667. Retrieved 2015-02-10.

183. Wilson, Heather M.; Anderson, Lyall I. (January 2004). Morphology and taxonomy of Paleozoic millipedes (Diplopoda: Chilognatha: Archipolypoda) from Scotland'. Journal of Paleontology. Paleontological Society. **78** *(1): 169–184.* doi:10.1666/0022-

3360(2004)078<0169:MATOPM>2.0.CO;2. ISSN 0022-3360.

184. Selden, Paul; Read, Helen J. (2008). The Oldest Land Animals: Silurian Millipedes from Scotland' (PDF). Bulletin of the British Myriapod & Isopod Group. London: British Myriapod and Isopod Group. **23**: 36–37. ISSN 1475-1739. Retrieved 2015-07-21.

185. Shear, William A.; Edgecombe, Gregory D. (March–May 2010). The geological record and phylogeny of the Myriapoda'. Arthropod Structure & Development. Amsterdam, the Netherlands: Elsevier. **39** *(2–3): 174–190. doi:10.1016/j.asd.2009.11.002. ISSN 1467-8039. PMID 19944188.*

186. MacNaughton, Robert B.; Cole, Jennifer M.; Dalrymple, Robert W.; et al. (May 2002). First steps on land: Arthropod trackways in Cambrian-Ordovician eolian sandstone, southeastern Ontario, Canada'. Geology. Boulder, CO: Geological Society of America. **30** *(5): 391–394. Bibcode*:2002Geo....30..391M. doi:10.1130/0091-7613(2002)030<0391:FSOLAT>2.0.CO;2. ISSN 0091-7613. Retrieved 2015-02-11.

187. Vaccari, N. Emilio; Edgecombe, Gregory D.; Escudero, C. (July 29, 2004). Cambrian origins and affinities of an enigmatic fossil group of arthropods'. Nature. London: Nature Publishing Group. **430** *(6999): 554–557. Bibcode*:2004Natur.430..554V. doi:10.1038/nature02705. ISSN 0028-0836. PMID 15282604.

188. Buatois, Luis A.; Mangano, M. Gabriela; Genise, Jorge F.; Taylor, Thomas N. (June 1998). The Ichnologic Record of the Continental Invertebrate Invasion: Evolutionary Trends in Environmental Expansion, Ecospace Utilization, and Behavioral Complexity'. PALAIOS. Tulsa, OK: SEPM Society for Sedimentary Geology. **13** *(3): 217–240.* doi:10.2307/3515447. ISSN 0883-1351. JSTOR 3515447. Retrieved 2015-02-11.

189. Cowen 2000, p. 126

190. Grimaldi & Engel 2005, pp. 155–160

191. Grimaldi & Engel 2005, p. 12

192. Clack, Jennifer A. (December 2005). Getting a Leg Up on Land'. Scientific American. Stuttgart: Georg von Holtzbrinck Publishing Group. **293** *(6): 100–107.* doi:10.1038/scientificamerican1205-100. ISSN 0036-8733. PMID 16323697. Retrieved 2015-02-12.

193. Ahlberg, Per E.; Milner, Andrew R. (April 7, 1994). The origin and early diversification of tetrapods'. Nature. London: Nature Publishing Group. **368** *(6471): 507–514. Bibcode*:1994Natur.368..507A. doi:10.1038/368507a0. ISSN 0028-0836. Retrieved 2008-09-06.

194. Gordon, Malcolm S.; Graham, Jeffrey B.; Wang, Tobias (September–October 2004). Introduction to the Special Collection: Revisiting the Vertebrate Invasion of the Land'. Physiological and Biochemical Zoology. Chicago, IL: University of Chicago Press on behalf of the Society for Integrative and Comparative Biology. **77** *(5): 697–699.* doi:10.1086/425182. ISSN 1522-2152.

195. Venturi, Sebastiano (2011). Evolutionary Significance of Iodine'. Current Chemical Biology-. **5** *(3): 155–162.* doi:10.2174/187231311796765012. ISSN 1872-3136.

196. Venturi, S; Bégin ME (2010). Thyroid Hormone, Iodine and Human Brain Evolution'. In Cunnane S; Stewart K. Environmental Influences on Human Brain Evolution. John Wiley & Sons. pp. 105–124. ISBN 978-0-470-45268-4.

197. Daeschler, Edward B.; Shubin, Neil H.; Jenkins, Farish A., Jr. (April 6, 2006). A Devonian tetrapod-like fish and the evolution of the tetrapod body plan' (PDF). Nature.

London: Nature Publishing Group. **440** *(7085): 757–763. Bibcode*:2006Natur.440..757D. doi:10.1038/nature04639. ISSN 0028-0836. PMID 16598249. Retrieved 2015-02-12.

198. deBraga, Michael; Rieppel, Olivier (July 1997). Reptile phylogeny and the interrelationships of turtles'. Zoological Journal of the Linnean Society. Hoboken, NJ: Wiley-Blackwell. **120** *(3): 281–354. doi:10.1111/j.1096-3642.1997.tb01280.x. ISSN 1096-3642. Retrieved 2015-02-12.*

199. Benton, Michael J.; Donoghue, Philip C. J. (January 2007). Paleontological Evidence to Date the Tree of Life'. Molecular Biology and Evolution. Oxford: Oxford University Press on behalf of the Society for Molecular Biology and Evolution. **24** *(1): 26–53. doi:10.1093/*molbev/msl150. ISSN 0737-4038. PMID 17047029. Retrieved 2015-02-12.

200. Benton, Michael J. (May 1990). Phylogeny of the Major Tetrapod Groups: Morphological Data and Divergence Dates'. Journal of Molecular Evolution. New York: Springer-Verlag. **30** *(5): 409–424.* doi:10.1007/BF02101113. ISSN 0022-2844. PMID 2111854. Retrieved 2015-02-12.

201. Sidor, Christian A.; O›Keefe, F. Robin; Damiani, Ross; et al. (April 14, 2005). Permian tetrapods from the Sahara show climate-controlled endemism in Pangaea'. Nature. London: Nature Publishing Group. **434** *(7035): 886–889. Bibcode*:2005Natur.434..886S. doi:10.1038/nature03393. ISSN 0028-0836. PMID 15829962. Retrieved 2008-09-08.

202. Smith, Roger; Botha, Jennifer (September–October 2005). The recovery of terrestrial vertebrate diversity in the South African Karoo Basin after the end-Permian extinction'. Comptes Rendus Palevol. New York: Elsevier on behalf of the French Academy of Sciences. **4** *(6–7): 623–636. doi:10.1016/j.crpv.2005.07.005. ISSN 1631-0683. Retrieved 2015-02-13.*

203. Benton 2005

204. Sahney, Sarda; Benton, Michael J. (April 7, 2008). Recovery from the most profound mass extinction of all time' (PDF). Proceedings of the Royal Society B. London: Royal Society. **275** *(1636): 759–765. doi:10.1098/rspb.2007.1370. ISSN 0962-8452. PMC 2596898* . PMID 18198148. Retrieved 2015-02-13.

205. Gauthier et al. 1989, p. 345

206. Benton, Michael J. (March 1983). Dinosaur Success in the Triassic: A Noncompetitive Ecological Model' (PDF). The Quarterly Review of Biology. Stony Brook Foundation. **58** *(1): 29–55.* doi:10.1086/413056. ISSN 0033-5770. JSTOR 2828101. Retrieved 2008-09-08.

207. Padian 2004, pp. 210–231

208. Lian-hai Hou; Zhonghe Zhou; Martin, Larry D.; Feduccia, Alan (October 19, 2002). A beaked bird from the Jurassic of China'. Nature. London: Nature Publishing Group. **377** *(6550): 616–618. Bibcode*:1995Natur.377..616H. doi:10.1038/377616a0. ISSN 0028-0836. Retrieved 2008-09-08.

209. Clarke, Julia A.; Zhonghe Zhou; Fucheng Zhang (March 2006). Insight into the evolution of avian flight from a new clade of Early Cretaceous ornithurines from China and the morphology of Yixianornis grabaui'. Journal of Anatomy. Hoboken, NJ: John Wiley & Sons. **208** *(3): 287–308. doi:10.1111/j.1469-7580.2006.00534.x. ISSN 1469-7580. PMC 2100246* . PMID 16533313. Retrieved 2015-02-15.

210. Ruben, John; Jones, Terry D. (August 2000). Selective Factors Associated with the

Origin of Fur and Feathers'. American Zoologist. Society for Integrative and Comparative Biology. **40** *(4): 585–596. doi:10.1093/*icb/40.4.585. ISSN 1540-7063. Retrieved 2015-02-16.

211. Zhe-Xi Luo; Crompton, Alfred W.; Ai-Lin Sun (May 25, 2001). A New Mammaliaform from the Early Jurassic and Evolution of Mammalian Characteristics'. Science. Washington, D.C.: American Association for the Advancement of Science. **292** *(5521): 1535–1540. Bibcode*:2001Sci...292.1535L. doi:10.1126/science.1058476. ISSN 0036-8075. PMID 11375489. Retrieved 2015-02-16.

212. Cifelli, Richard L. (November 2001). Early mammalian radiations'. Journal of Paleontology. Paleontological Society. **75** *(6): 1214–1226.* doi:10.1666/0022-3360(2001)075<1214:EMR>2.0.CO;2. ISSN 0022-3360. Retrieved 2015-02-16.

213. Flynn, John; Parrish, J. Michael; Rakotosamimanana, Berthe; et al. (September 2, 1999). A Middle Jurassic mammal from Madagascar'. Nature. London: Nature Publishing Group. **401** *(6748): 57–60. Bibcode*:1999Natur.401...57F. doi:10.1038/43420. ISSN 0028-0836. Retrieved 2008-09-08.

214. MacLeod, Norman; Rawson, Peter F.; Forey, Peter L.; et al. (April 1997). The Cretaceous–Tertiary biotic transition'. Journal of the Geological Society. London: Geological Society of London. **154** *(2): 265–292. doi:10.1144/gsjgs.154.2.0265. ISSN 0016-7649. Retrieved 2015-02-16.*

215. Alroy, John (March 1999). The Fossil Record of North American Mammals: Evidence for a Paleocene Evolutionary Radiation'. Systematic Biology. Oxford: Oxford University Press on behalf of the Society of Systematic Biologists. **48** *(1): 107–118.* doi:10.1080/106351599260472. ISSN 1063-5157. PMID 12078635.

216. Archibald, J. David; Deutschman, Douglas H. (June 2001). Quantitative Analysis of the Timing of the Origin and Diversification of Extant Placental Orders' (PDF). Journal of Mammalian Evolution. New York: Kluwer Academic Publishers/Plenum Publishers on behalf of the Society for Study of Mammalian Evolution. **8** *(2): 107–124.* doi:10.1023/A:1011317930838. ISSN 1064-7554. Retrieved 2015-02-16.

217. Simmons, Nancy B.; Seymour, Kevin L.; Habersetzer, Jörg; Gunnell, Gregg F. (February 14, 2008). Primitive Early Eocene bat from Wyoming and the evolution of flight and echolocation'. Nature. London: Nature Publishing Group. **451** *(7180): 818–821. Bibcode*:2008Natur.451..818S. doi:10.1038/nature06549. ISSN 0028-0836. PMID 18270539.

218. Thewissen, Madar & Hussain 1996

219. Crane, Friis and Pedersen 2000, The Origin and Early Diversification of Angiosperms,' pp. 233–250

220. Crepet, William L. (November 21, 2000). Progress in understanding angiosperm history, success, and relationships: Darwin›s abominably ‹perplexing phenomenon›' (PDF). Proc. Natl. Acad. Sci. U.S.A. Washington, D.C.: National Academy of Sciences. **97** *(24): 12939–12941. Bibcode*:2000PNAS...9712939C. doi:10.1073/pnas.97.24.12939. ISSN 0036-8075. PMC 34068 . PMID 11087846. Retrieved 2008-09-09.

221. evolution: plant timeline'. Encyclopædia Britannica Online. Encyclopædia Britannica, Inc. 1996. OCLC 42796406. Retrieved 2015-02-16.

222. Wilson, Edward O.; Hölldobler, Bert (September 20, 2005). Eusociality: Origin and

consequences' (PDF). Proc. Natl. Acad. Sci. U.S.A. Washington, D.C.: National Academy of Sciences. **102** *(38): 13367–13371. Bibcode*:2005PNAS..10213367W. doi:10.1073/pnas.0505858102. ISSN 0036-8075. PMC 1224642 . PMID 16157878. Retrieved 2008-09-07.

223. Hughes, William O. H.; Oldroyd, Benjamin P.; Beekman, Madeleine; Ratnieks, Francis L. W. (May 30, 2008). Ancestral Monogamy Shows Kin Selection Is Key to the Evolution of Eusociality'. Science. Washington, D.C.: American Association for the Advancement of Science. **320** *(5880): 1213–1216. Bibcode*:2008Sci...320.1213H. doi:10.1126/science.1156108. ISSN 0036-8075. PMID 18511689. Retrieved 2015-02-17.

224. Lovegrove, Barry G. (January 1991). The evolution of eusociality in molerats (Bathyergidae): a question of risks, numbers, and costs'. Behavioral Ecology and Sociobiology. Berlin: Springer-Verlag. **28** *(1): 37–45.* doi:10.1007/BF00172137. ISSN 0340-5443.

225. Labandeira & Eble 1999

226. Montgomery, Stephen (2009). Darwin & the Descent of Man'. Charles Darwin & Evolution. Cambridge: Christ›s College. Retrieved 2015-02-17.

227. Brunet, Michel; Guy, Franck; Pilbeam, David; et al. (July 11, 2002). A new hominid from the Upper Miocene of Chad, Central Africa'. Nature. London: Nature Publishing Group. **418** *(6894): 145–151.* doi:10.1038/nature00879. ISSN 0028-0836. PMID 12110880. Retrieved 2008-09-09.

228. de Heinzelin, Jean; Clark, J. Desmond; White, Tim; et al. (April 23, 1999). Environment and Behavior of 2.5-Million-Year-Old Bouri Hominids'. Science. Washington, D.C.: American Association for the Advancement of Science. **284** *(5414): 625–629. doi:10.1126/science.284.5414.625. ISSN 0036-8075. PMID 10213682. Retrieved 2015-02-19.*

229. De Miguel, Carmen; Henneberg, Maciej (2001). Variation in hominid brain size: How much is due to method?'. *HOMO - Journal of Comparative Human Biology. Elsevier.* **52** *(1): 3–58.* doi:10.1078/0018-442X-00019. ISSN 1618-1301.

230. Leakey 1994, pp. 87–89

231. Mellars, Paul (June 20, 2006). Why did modern human populations disperse from Africa ca. 60,000 years ago? A new model' (PDF). Proc. Natl. Acad. Sci. U.S.A. Washington, D.C.: National Academy of Sciences. **103** *(25): 9381–9386. Bibcode*:2006PNAS..103.9381M. doi:10.1073/pnas.0510792103. ISSN 0027-8424. PMC 1480416 . PMID 16772383. Retrieved 2015-02-20.

232. Benton 2005a, Chapter 6: "Tetrapods of the Triassic'

233. MacLeod, Norman (January 6, 2001). Extinction!'. *FirstScience.com. Retrieved 2015-02-20.*

234. Martin, Ronald E. (June 1995). Cyclic and secular variation in microfossil biomineralization: clues to the biogeochemical evolution of Phanerozoic oceans'. Global and Planetary Change. Amsterdam, the Netherlands: Elsevier. **11** *(1–2): 1–23. Bibcode*:1995GPC....11....1M. doi:10.1016/0921-8181(94)00011-2. ISSN 0921-8181.

235. Martin, Ronald E. (June 1996). Secular Increase in Nutrient Levels through the Phanerozoic: Implications for Productivity, Biomass, and Diversity of the Marine Biosphere'. PALAIOS. Tulsa, OK: SEPM Society for Sedimentary Geology. **11** *(3):*

209–219. doi:10.2307/3515230. ISSN 0883-1351. JSTOR 3515230.

236. Rohde, Robert A.; Muller, Richard A. (March 10, 2005). Cycles in fossil diversity' (PDF). Nature. London: Nature Publishing Group. **434** *(7030): 208–210. Bibcode*:2005Natur.434..208R. doi:10.1038/nature03339. ISSN 0028-0836. PMID 15758998. Retrieved 2008-09-22.

CHAPTER 2

Cell Theory

Chapter Contents

Cell theory in the science world is an idea supported by many tests that have been repeated over and over to prove the same result. In this case, cell theory is a collection of ideas and conclusions from many different scientists over time that describes cells and how cells operate.

What we know about cells has continuously evolved over time as new technologies have emerged and new information has been collected by scientists. Popular knowledge throughout history, such as that cells can grow spontaneously, has been disproved, and cell theory has been revised to reflect the new information.

INTRODUCTION

Human cancer cells with nuclei (specifically the DNA) stained blue. The central and rightmost cell are in interphase, so the entire nuclei are labeled. The cell on the left is going through mitosis and its DNA has condensed.

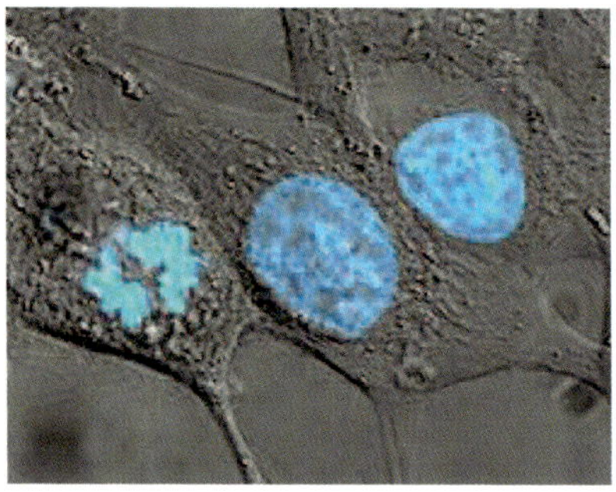

In biology, **cell theory** is a scientific theory which describes the properties of cells. These cells are the basic unit of structure in all organisms and also the basic unit of reproduction. With continual improvements made to microscopes over time, magnification technology advanced enough to discover cells in the 17th century. This discovery is largely attributed to Robert Hooke, and began the scientific study of cells, also known as cell biology. Over a century later, many debates about cells began amongst scientists. Most of these debates involved the nature of cellular regeneration, and the idea of cells as a fundamental unit of life. Cell theory was eventually formulated in 1839. This is usually credited to Matthias Schleiden and Theodor Schwann. However, many other scientists like Rudolf Virchow contributed to the theory. Cell theory has become the foundation of biology and is the most widely accepted explanation of the function of cells.

The three tenets to the cell theory are as described below:

- All living organisms are composed of one or more cells. (However, this is considered a controversy because non-cellular life such as viruses are disputed as a life form. See Non-cellular life.)
- The cell is the basic unit of structure and organization in organisms.
- Cells arise from pre-existing cells.

MICROSCOPES

Anton van Leeuwenhoek's microscope from the 17th century with a magnification of 270x.

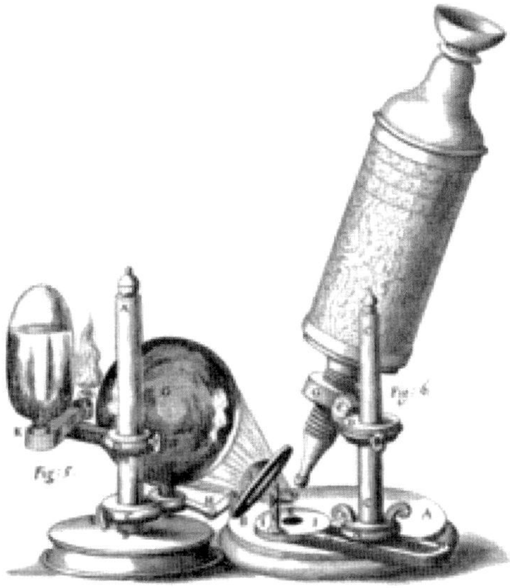

Robert Hooke's Microscope

The discovery of the cell was made possible through the invention of the microscope. In the first century BC, Romans were able to make glass, discovering that objects appeared to be larger under the glass. In Italy during the 12th century, Salvino D'Armate made a piece of glass fit over one eye, allowing for a magnification effect to that eye. It was not until the 1590s when a Dutch spectacle maker Zacharias Jansen began to test lenses that progress had been made to microscopes. Jansen was able to obtain about 9x magnification, but the objects appeared to be blurry. In 1595, Jansen and his father built the first compound microscope. While simple glasses were able to magnify objects, they were not considered to be a microscope. A compound microscope was defined by having two or more lenses in a hollow tube. In 1665, Robert Hooke used a microscope about six inches long with two convex lenses inside and examined specimens under reflected light for the observations in his book *Micrographia*. Hooke also used a simpler microscope with a single lens for examining specimens with directly transmitted light, because this allowed for a clearer image.

However, the first real invention and use of the microscope were done by Anton van Leeuwenhoek. He was a Dutch draper that took the interest in microscopes after seeing one while on an apprenticeship in Amsterdam in 1648. At some point in his life before 1668, he was able to learn how to grind lenses. This eventually led to Leeuwenhoek making his own unique microscope. His were instead simple powerful magnifying glasses, rather than a compound microscope. This was because he was able to use a single lens that was a small glass sphere but allowed for a magnification of 270x. This was a large progression since the magnification before was only a maximum of 50x. After Leeuwenhoek, there was not much progress for the microscopes until the 1850s, two hundred years later. Carl Zeiss, a German engineer who manufactured microscopes, began to make changes to the lenses used. But the optical quality did not improve until the 1880s when he hired Otto Schott and eventually Ernst Abbe.

These microscopes could focus on objects the size of a wavelength or larger, giving restrictions still to advancement in discoveries with objects smaller than a wavelength. Later in the 1920s, the electron microscope was developed, making it possible to view objects that are smaller than a wavelength, once again, changing the possibilities in science.

DISCOVERY OF CELLS

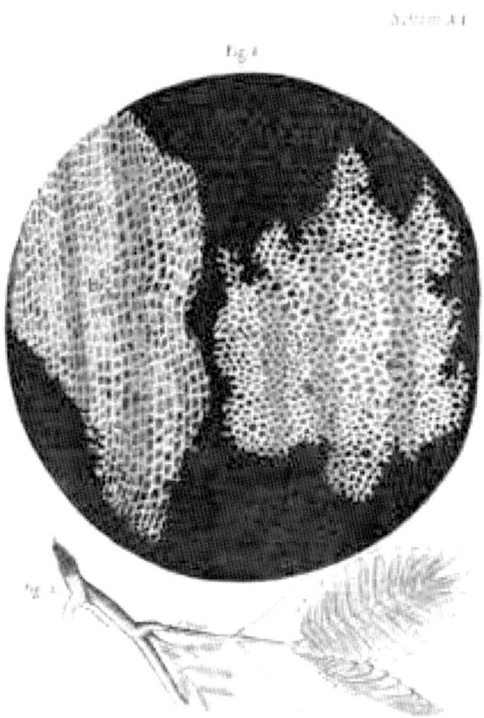

Drawing of the structure of cork by Robert Hooke that appeared in *Micrographia*.

The cell was first discovered by Robert Hooke in 1665, which can be found to be described in his book Micrographia. In this book, he gave 60 'observations' in detail of various objects under a coarse, compound microscope. One observation was from very thin slices of bottle cork. Hooke discovered a multitude of tiny pores that he named cells. This came from the Latin word Cella, meaning 'a small room' like monks lived in and also Cellulae, which meant the six sided cell of a honeycomb. However, Hooke did not know their real structure or function. What Hooke had thought were cells, were actually empty cell walls of plant tissues. With microscopes during this time having a low magnification, Hooke was unable to see that there were other internal components to the cells he was observing. Therefore, he did not think the cellulae were alive. His cell observations gave no indication of the nucleus and other organelles found in most living cells. In Micrographia, Hooke also observed mould, bluish in color, found on leather. After studying it under his microscope, he was unable to observe "seeds" that would have indicated how the mould was multiplying in quantity. This led to Hooke suggesting that spontaneous generation, from either natural or artificial heat, was the cause. Since this was an old Aristotelian theory still accepted at the time, others did not reject it and was not disproved until Leeuwenhoek later discovers generation is achieved otherwise.

Anton van Leeuwenhoek is another scientist who saw these cells soon after Hooke did. He made use of a microscope containing improved lenses that could magnify objects almost 300-fold, or 270x. Under these microscopes, Leeuwenhoek found motile objects. In a letter to The Royal Society on October 9, 1676, he states that motility is a quality of life therefore these were living organisms. Over time, he wrote many more papers in which described many specific forms of microorganisms. Leeuwenhoek named these "animalcules," which included protozoa and other unicellular organisms, like bacteria. Though he did not have much formal education, he was able to identify the first accurate description of red blood cells and discovered bacteria after gaining interest in the sense of taste that resulted in Leeuwenhoek to observe the tongue of an ox, then leading him to study pepper water in 1676. He also found for the first time the sperm cells of animals and humans. Once discovering these types of cells, Leeuwenhoek saw that the fertilization process requires the sperm cell to enter the egg cell. This put an end to the previous theory of spontaneous generation. After reading letters by Leeuwenhoek, Hooke was the first to confirm his observations that were thought to be unlikely by other contemporaries.

The cells in animal tissues were observed after plants were because the tissues were so fragile and susceptible to tearing, it was difficult for such thin slices to be prepared for studying. Biologists believed that there was a fundamental unit to life, but were unsure what this was. It would not be until over a hundred years later that this fundamental unit was connected to cellular structure and existence of cells in animals or plants. This conclusion was not made until Henri Dutrochet. Besides stating "the cell is the fundamental element of organization", Dutrochet also claimed that cells were not just a structural unit, but also a physiological unit.

In 1804, Karl Rudolphi and J.H.F. Link were awarded the prize for "solving the problem of the nature of cells", meaning they were the first to prove that cells had independent cell walls by the Königliche Societät der Wissenschaft (Royal Society of Science), Göttingen. Before, it had been thought that cells shared walls and the fluid passed between them this way.

CELL THEORY

Matthias Jakob Schleiden (1804–1881)

Theodor Schwann (1810–1882)

Credit for developing cell theory is usually given to two scientists: Theodor Schwann and Matthias Jakob Schleiden. While Rudolf Virchow contributed to the theory, he is not as credited for his attributions toward it. In 1839, Schleiden suggested that every structural part of a plant was made up of cells or the result of cells. He also suggested that cells were made by a crystallization process either within other cells or from the outside. However, this was not an original idea of Schlieden. He claimed this theory as his own, though Barthelemy Dumortier had stated it years before him. This crystallization process is no longer accepted with modern cell theory. In 1839, Theodor Schwann states that along with plants, animals are composed of cells or the product of cells in their structures. This was a major advancement in the field of biology since little was known about animal structure up to this point compared to plants. From these conclusions about plants and animals, two of the three tenets of cell theory were postulated.

- All living organisms are composed of one or more cells
- The cell is the most basic unit of life

Schleiden's theory of free cell formation through crystallization was refuted in the 1850s by Robert Remak, Rudolf Virchow, and Albert Kolliker. In 1855, Rudolf Virchow added the third tenet to cell theory. In Latin, this tenet states *Omnis cellula e cellula*. This translated to:

- All cells arise only from pre-existing cells

However, the idea that all cells come from pre-existing cells had in fact already been proposed by Robert Remak; it has been suggested that Virchow plagiarized Remak and did not give him credit. Remak published observations in 1852 on cell division, claiming Schleiden and Schawnn were incorrect about generation schemes. He instead said that binary fission, which was first introduced by Dumortier, was how reproduction of new animal cells were made. Once this tenet was added, the classical cell theory was complete.

MODERN INTERPRETATION

The generally accepted parts of modern cell theory include:

- All known living things are made up of one or more cells
- All living cells arise from pre-existing cells by division.
- The cell is the fundamental unit of structure and function in all living organisms.
- The activity of an organism depends on the total activity of independent cells.
- Energy flow (metabolism and biochemistry) occurs within cells.
- Cells contain DNA which is found specifically in the chromosome and RNA found in the cell nucleus and cytoplasm.
- All cells are basically the same in chemical composition in organisms of similar species .

THE MODERN VERSION OF THE CELL THEORY

The modern version of the cell theory includes the ideas that:

- Energy flow occurs within cells.
- Heredity information (DNA) is passed on from cell to cell.
- All cells have the same basic chemical composition.

Opposing concepts in cell theory: history and background

The cell was first discovered by Robert Hooke in 1665 using a microscope. The first cell theory is credited to the work of Theodor Schwann and Matthias Jakob Schleiden in the 1830s. In this theory the internal contents of cells were called protoplasm and described as a jelly-like substance, sometimes called living jelly. At about the same time, colloidal chemistry began its development, and the concepts of bound water emerged. A colloid being something between a solution and a suspension, where Brownian motion is sufficient to prevent sedimentation. The idea of a semipermeable membrane, a barrier that is permeable to solvent but impermeable to solute molecules was developed at about the same time. The term osmosis originated in 1827 and its importance to physiological phenomena realized, but it wasn't until 1877, when the botanist Pfeffer proposed the membrane theory of cell physiology. In this view, the cell was seen to be enclosed by a thin surface, the plasma membrane, and cell water and solutes such as a potassium ion existed in a physical state like that of a dilute solution. In 1889 Hamburger used hemolysis of erythrocytes to determine the permeability of various solutes. By measuring the time required for the cells to swell past their elastic limit, the rate at which solutes entered the cells could be estimated by the accompanying change in cell volume. He also found that there was an apparent nonsolvent volume of about 50% in red blood cells and later showed that this includes water of hydration in addition to the protein and other nonsolvent components of the cells.

Evolution of the Membrane and Bulk Phase Theories

Two opposing concepts developed within the context of studies on osmosis, permeability, and electrical properties of cells. The first held that these properties all belonged to the plasma membrane whereas the other predominant view was that the protoplasm was responsible for these properties. The membrane theory developed as a succession of ad-hoc additions and changes to the theory to overcome experimental hurdles. Overton (a distant cousin of Charles Darwin) first proposed the concept of a lipid (oil) plasma membrane in 1899. The major weakness of the lipid membrane was the lack of an explanation of the high permeability to water, so Nathansohn (1904) proposed the mosaic theory. In this view, the membrane is not a pure lipid layer, but a mosaic of areas with lipid and areas with semipermeable gel. Ruhland refined the mosaic theory to include pores to allow additional passage of small molecules. Since membranes are generally less permeable to anions, Leonor Michaelis concluded that ions are adsorbed to the walls of the pores, changing the permeability of the pores to ions by electrostatic repulsion. Michaelis demonstrated the membrane potential (1926) and proposed that it was related to the distribution of ions across the membrane. Harvey and Danielli (1939) proposed a lipid bilayer membrane covered on each side with a layer of protein to account for measurements of surface tension. In 1941 Boyle & Conway showed that the membrane of frog muscle was permeable to both K+ and Cl-, but apparently not to Na+, so the idea of electrical charges in the pores was unnecessary since a single critical pore size would explain the permeability to K+, H+, and Cl- as well as the impermeability to Na+, Ca+, and Mg++. Over the same time period, it was shown (Procter & Wilson, 1916) that gels, which do not have a semipermeable membrane, would swell in dilute solutions. Loeb (1920) also studied gelatin extensively, with and without a membrane, showing that more of the properties attributed to the plasma membrane could be duplicated in gels without a membrane. In particular, he found that an electrical potential difference between the gelatin and the outside medium could be developed, based on the H+ concentration. Some criticisms of the membrane theory developed in the 1930s, based on observations such as the ability of some cells to swell and increase their surface area by a factor of 1000. A lipid layer cannot stretch to that extent without becoming a patchwork (thereby losing its barrier properties. Such criticisms stimulated continued studies on protoplasm as the principal agent determining cell permeability properties. In 1938, Fischer and Suer proposed that water in the protoplasm is not free but in a chemically combined form—the protoplasm represents a combination of protein, salt and water—and demonstrated the basic similarity between swelling in living tissues and the swelling of gelatin and fibrin gels. Dimitri Nasonov (1944) viewed proteins as the central components responsible for many properties of the cell, including electrical properties. By the 1940s, the bulk phase theories were not as well developed as the membrane theories. In 1941, Brooks & Brooks published a monograph The Permeability of Living Cells, which rejects the bulk phase theories.

The Emergence of the steady-state Membrane Pump Concept

With the development of radioactive tracers, it was shown that cells are not impermeable to Na^{+}. This was difficult to explain with the membrane barrier theory, so the sodium pump was proposed to continually remove Na+ as it permeates cells. This drove the concept that cells are in a state of dynamic equilibrium, constantly using energy to maintain ion gradients. In 1935, Karl Lohmann discovered ATP and its role as a source of energy for cells, so the concept of a metabolically-driven sodium pump was proposed. The tremendous success of Hodgkin, Huxley, and Katz in the development of the membrane theory of cellular membrane potentials, with differential equations that modeled the phenomena correctly, provided even more support for the membrane pump hypothesis.

The modern view of the plasma membrane is of a fluid lipid bilayer that has protein components embedded within it. The structure of the membrane is now known in great detail, including 3D models of many of the hundreds of different proteins that are bound to the membrane. These major developments in cell physiology placed the membrane theory in a position of dominance and stimulated the imagination of most physiologists, who now apparently accept the theory as fact—there are, however, a few dissenters.

The Reemergence of the Bulk Phase Theories

In 1956, Afanasy S. Troshin published a book, *The Problems of Cell Permeability*, in Russian (1958 in German, 1961 in Chinese, 1966 in English) in which he found that permeability was of secondary importance in determination of the patterns of equilibrium between the cell and its environment. Troshin showed that cell water decreased in solutions of galactose or urea although these compounds did slowly permeate cells. Since the membrane theory requires an impermanent solute to sustain cell shrinkage, these experiments cast doubt on the theory. Others questioned whether the cell has enough energy to sustain the sodium/potassium pump. Such questions became even more urgent as dozens of new metabolic pumps were added as new chemical gradients were discovered.

In 1962, Gilbert Ling became the champion of the bulk phase theories and proposed his association-induction hypothesis of living cells.

Types of Cells

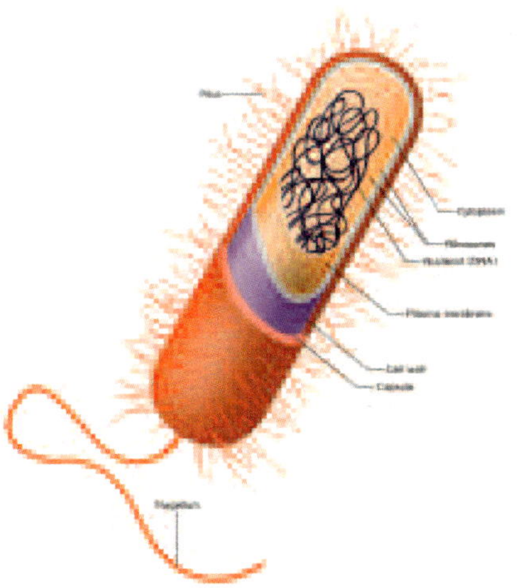

Prokaryote cell.

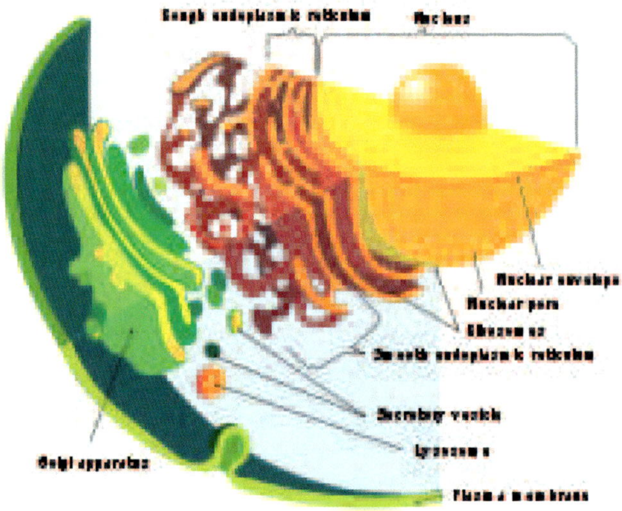

Eukaryote cell.

Cells can be subdivided into the following subcategories:

- *Prokaryotes*: Prokaryotes are relatively small cells surrounded by the plasma membrane, with a characteristic cell wall that may differ in composition depending on the particular organism.Prokaryotes lack a nucleus (although they do have circular or linear DNA) and other membrane-bound organelles (though they do contain ribosomes). The protoplasm of a prokaryote contains the chromosomal region that appears as fibrous deposits under the microscope, and the cytoplasm.Bacteria and Archaea are the two domains of prokaryotes.
- *Eukaryotes*: Eukaryotic cells are also surrounded by the plasma membrane, but on the other hand, they have distinct nuclei bound by a nuclear membrane or envelope. Eukaryotic cells also contain membrane-bound organelles, such as (mitochondria, chloroplasts, lysosomes, rough and smooth endoplasmic reticulum, vacuoles). In addition, they possess organized chromosomes which store genetic material.

Animals have evolved a greater diversity of cell types in a multicellular body (100–150 different cell types), compared with 10–20 in plants, fungi, and protoctists.

CELL ADHESION

Cell adhesion is the process cells interact and attach to a surface, substrate or another cell, mediated by interactions between molecules of the cell surface. Cell adhesion occurs from the action of transmembrane glycoproteins, called cell adhesion molecules. Examples of these proteins include selectins, integrins, syndecans, and cadherins. Cellular adhesion is essential in maintaining multicellular structure. Cellular adhesion can link cells in different ways and can be involved in signal transduction. Cell adhesion is also essential for the pathogenesis of infectious organisms.

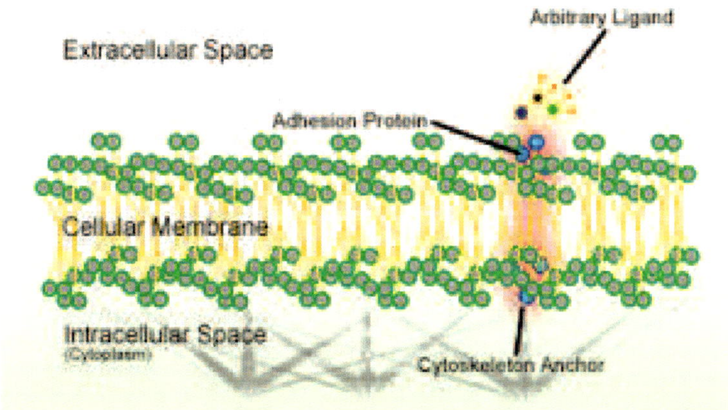

Schematic of cell adhesion

Eukaryotes

Protozoans express multiple adhesion molecules. An example of a pathogenic protozoan is the malarial parasite (*Plasmodium falciparum*), which uses one adhesion molecule called the circumsporozoite protein to bind to liver cells, and another adhesion molecule called the merozoite surface protein to bind red blood cells.

Mammalian cells express many different types of adhesion molecules. The major classes are named integrins, Ig superfamily members, cadherins, and selectins. Each of these adhesion molecules has a different function and recognizes different ligands. Defects in cell adhesion are usually attributable to defects in expression of adhesion molecules.

Cell junctions allow cells to adhere to one another in multicellular organisms. There are 4 types of cell junctions:

- Anchoring junctions (Adherens junctions, Desmosomes and Hemidesmosomes), which transmit stress through tethering to cytoskeleton are cell–cell or cell–matrix
- Occluding Junctions (Tight junctions), which seal gaps between cells, making an impermeable barrier
- Channel forming junctions (Gap junctions and plasmodesmata), which links cytoplasm of adjacent cells
- Signal relaying junction, synapses in nervous systems

Cell-cell adhesions in Anchoring junctions are mediated by cadherins. The cadherin subunit, β-catenin plays a role in cell-cell adhesion by controlling cadherin-mediated cell adhesion at the plasma membrane. Cell–matrix adhesions however, are usually mediated by integrins.

Selective cell–cell adhesion enables vertebrate cells to assemble into organised tissues. homophilic attachment allows selective recognition resulting. Cells of a similar type stick together whereas cells of a different type stay segregated.

Fungal adhesion allows fungi to bind to host tissues.

Human Genetic Diseases

There are human genetic diseases caused by an inability to express a specific adhesion molecule. An example is leukocyte adhesion deficiency-I (LAD-I), where patients do not express the β2-integrin subunit precursor. This integrin is required for leukocytes to adhere to the blood vessel wall during inflammation in order to fight infection. The leukocytes from LAD-I patients fail to adhere and patients exhibit serious episodes of infection that can be life-threatening.

Tumor Adhesion

Cancer metastasis tumors that spread through the circulatory system use mechanisms of cell adhesion to establish new tumors in the body. Release of epoxyeicosatrienoic acid is believed to increase this propensity.

Prokaryotes

Prokaryotes have adhesion molecules usually termed "adhesins. Adhesins may occur on pili (fimbriae), flagellae, or the cell surface. Adhesion of bacteria is the first step in colonization and regulates tropism (tissue- or cell-specific interactions).

Viruses

Viruses also have adhesion molecules required for viral binding to host cells. For example, influenza virus has a hemagglutinin on its surface that is required for recognition of the sugar sialic acid on host cell surface molecules. HIV has an adhesion molecule termed gp120 that binds to its ligand CD4, which is expressed on lymphocytes.

Differential Adhesion Hypothesis

The differential adhesion hypothesis (sometimes called the "thermodynamic hypothesis") is a theory of cell adhesion advanced by Malcolm Steinberg in 1964 to explain the mechanism by which heterotypic cells in mixed aggregates sort out into isotypic territories. The DAH postulates that tissues are viscoelastic liquids, and as such possess measurable tissue surface tensions. These surface tensions have been determined for a variety of tissues, including embryonic tissues and cell lines. The surface tensions correspond to the mutual sorting behavior: the tissue type with the higher surface tension will occupy an internal position relative to a tissue with a lower surface tension (if these tissues can interact with each other through their adhesion machinery).

Quantitative differences in homo and heterotypic adhesion are supposed to be sufficient to account for the phenomenon without the need to postulate cell type specific adhesion systems: fairly generally accepted, although some tissue specific cell adhesion molecules are now known to exist.

Clinical Implications

Dysfunction of cell-adhesion and cell-migration occurs during cancer metastasis. Cellular adhesion and traction can allow cells to migrate. Cells can form integrin mediated attachments sites called focal adhesions. Focal adhesions at the leading edge provide the necessary traction allowing the cell to pull itself forward.

Pemphigus is the result of auto-antibodies which target desmosomal cadherins, resulting in loss of cell adhesion.

Measurement and Applications

Cell adhesion forces of mammalian cells could specifically be measured down to the single receptor-ligand level. Researchers at Japan's Nagoya University have developed a test to measure the adhesion of a single cell and potentially determine cellular viability.

ROLE OF CELL ADHESIONS IN NEURAL DEVELOPMENT

Cellular adhesions can be defined as proteins or protein aggregates that form mechanical and chemical linkages between the intracellular and extracellular space. Adhesions serve several critical processes including cell migration, signal transduction, tissue development and repair. Due to this functionality, adhesions and adhesion molecules have been a topic of study within the scientific community. Specifically, it has been found that adhesions are involved in tissue development, plasticity, and memory formation within the central nervous system (CNS), and may prove vital in the generation of CNS-specific therapeutics.

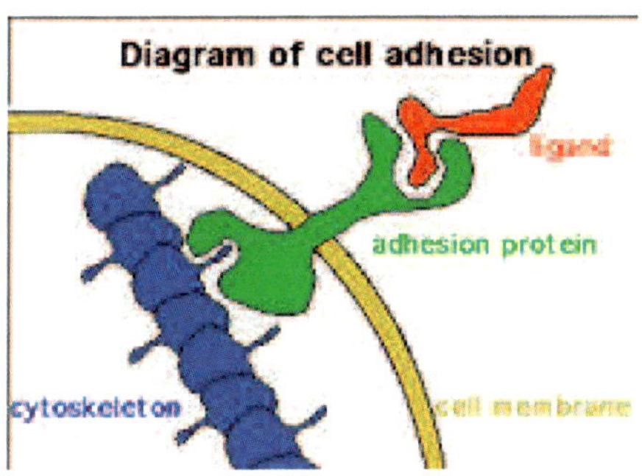

Image courtesy of Wikipedia user JWSchmidt under the GNU Free Documentation License

Adhesion Classifications

- **Cell-cell adhesions** provide chemical and mechanical connections between adjacent cells. Of special importance to neuronal tissue development are the subcategory n-cadherins. These cadherin molecules have been shown to be important in formation of the CNS structure, as well as neuronal migration along glial fibers.

- **Cell-'***Extra-Cellular Matrix'* (ECM) **adhesions** also form mechanical and chemical linkages, but the connection occurs between cellular-matrix and the extra-cellular matrix through a plethora of adhesive proteins that form cohesive functional units. These focal adhesion plaques are highly dynamic in nature and undergo a maturation process through which they have altered functionality and protein content. Maturation stages are summarized in the following table:

Adhesion Classification	Approximate Size
Nascent	0.25 μm
Focal Complex	0.5 μm
Focal Adhesion	1-5 μm
Fibrillar Adhesion	>5 μm

Adhesions role in cell migration

During early development, cell migration plays a crucial role in neuronal tissue organization. Although still largely under investigation, networks of highly ordered neurons are known to be a vital component of the nervous systems communication with the body. A major mechanism of cellular migration is the translation of internal force, to the external environment. Force transmission can occur through a variety of mechanisms, though adhesion complexes between cell-cell and cell-extracellular matrix (ECM) are a known to be chief mechanisms of this activity. Cell migration is generally classified with four cell processes:

- Leading edge protrusion
- Adhesion formation
- Cell body translation
- Trailing edge adhesion detachment

The coordination of these processes allows for the efficient migration of cells through their environment.

Cadherin Dependent Migration

Scaffold cell-dependent migration, in which neuronal cadherin (N-cadherin) adhesive molecules are tightly regulated, provides one mode of motility in developing neuron tissue. During cell migration, N-cadherin binds the neuron to a glial fiber, and allows for transfer of force, generated by an intracellular actin network treadmilling, to the glial fiber. Force transmission across the cell-glial fiber interface sums over many individual N-cadherin/glial-fiber interactions, allowing required levels of traction force essential for migration. It has also been shown that these adhesive cadherin molecules are internalized, and recycled by the migratory neuron. This cadherin recycling mechanism is thought to be substantial in the neural adhesion-based migratory pathway. Cadherin based migration is essential to tissue organization in the central nervous system, specifically in cortical layer formation.

It has also been sugguested that the N-cadherin pathway may be crucial in neuron differentiation, as knockdown of the N-cadherin pathway leads to premature neuron differentiation.

Integrin Dependent Migration

Integrin dependent cell migration can be described as protein plaques that form the mechanical linkage between the intracellular and extracellular environments. One major components of this classification of cell migration, integrin, is a trans-membrenal protein dimer, which binds ECM components on its external domains and actin cytoskeletal components on its intra-cellular domains. These adhesions couple forces between the intracellular and extracellular space through both actin retrograde flow mechanisms (which have been described as a molecular clutch), and through actin-myosin protein contraction machinery. It is thought that these adhesions are involved in mechanosensing, that is, they respond both physically and chemically when exposed to various physical environments.

ADHESION-RELATED MECHANISMS INVOLVED IN NEURONAL TISSUE DEVELOPMENT

Growth Cone Extensions

Growth cones function as structural and chemically sensitive axon directing cellular organelles. Growth cones are highly dynamic in nature and contain a dynamic actin cytoskeleton in their peripheral region undergoing a constant retrograde flow. This retrograde force provides a mechanism for the growth cone to respond to direction cue, thereby directing neuronal axons. Growth cones are known to respond to various mechanical cues, which may be vital in proper nervous system development as growth cones experience a wide variety of mechanical environments as they navigate the extracellular space. Interestingly, research suggest that growth cones from different regions of the brain may respond to mechanical cues differently. It has been demonstrated that neural cells located in the hippocampus aren't sensitive to varying mechanical stiffness as it related to outgrowth, where cells originating from the dorsal root ganglion show maximal outgrowth on surfaces of approximately 1 kPa. Both hippocampal and dorsal root ganglion neural growth cones show increased traction force generation on increased stiffness substrates. Growth cones utilize integrin migratory machinery such as integrins, but are not a class of cell migration.

Thy-1 Adhesion Protein

Thy-1 (or CD90.2) is a membrane bound glycoprotein that has been shown to be involved in the axon guidance pathway. This protein has been shown to be highly mobile, as it contains a GPI membrane anchor. Although much of the details are elusive, it is known that thy-1 interacts with the protein dimer integrin found on astrocytes, forming aggregates that can inhibit neurite outgrowth and extension. Thy-1 has also been shown to have involvement in the src-family kinase pathway. This astrocyte-neuron feedback has been proposed as a mechanism involved in CNS tissue repair post-injury, as a down regulation of thy-1 may lead to enhanced neurite outgrowth. Additional research has shown that thy-1 expression in post natal humans is elevated for several weeks. This suggests that in addition to tissue repair, thy-1 might have roles in early CNS tissue development and organization.

L1 Family Protein

The L1 family of proteins are involved in neuronal migration, as well as in axon growth and proper synapse formation, and include L1CAM, CHL1, NrCAM and neurofascin. L1-Cell Adhesion Molecule (L1CAM) was first discovered to be important in neuron-related tissue development in the mid-1980s, and is a trans-membranal glycoprotein of approximately 200-220 kDa. On its extracellular domain, the L1CAM protein includes IgG-like and fibronectin-III (FN-III) repeats which allow for interaction with integrins and ECM proteins. Similarly to integrin, F1CAM expresses domains intracellularly that interact with the actin cytoskeleton. Supporting the claim that L1-family proteins are involved in CNS development is the finding that L1CAM is highly expressed in neuronal tissue during its early stages of growth, especially at the ends of axons. Some areas of the brain, such as the hippocampus, have been found to highly express L1CAM into adulthood, though the exact reason for this has not been elucidated.

Due to its involvement in neuronal development and axon guidance, it has been proposed that L1CAM and L1-family proteins may be useful therapeutics to treat tissue damage in the CNS. Some have even proposed that L1CAM expression is elevated in vivo during tissue repair, which would support the notion that it yields benefit during CNS tissue repair.

Mechanosensing in Neurons

Mechanosensing is a process by which cells alter their bio-physical properties in response to mechanical cues present in the environment. It is well known that a wide-variety of cell types change their behavior to mechanical environmental signals.

In addition to providing force transmission to the ECM for neuron extension and development, Integrin mediated adhesions are also functional in these mechanosensing processes in neurons. Sensing of the external environments mechanical properties *in vivo* can determine cell behaviors such as differentiation and branching. It has been experimentally determined that increasing substrate stiffness (~2-80kPa) can result in sequestered neurite branching and branch length.

Relevant Neurological Conditions

Several debilitating diseases are brought about from errors in neural development due in part to problems involving neural cell adhesions and adhesion mechanisms.

- CRASH syndrome (or L1 syndrome) is brought about by a mutation in the L1CAM gene on the x-chromosome, resulting in a malfunctioning L1CAM protein. CRASH (acronym) syndrome include the conditions:

Summary of CRASH conditions

Year	Comments
Corpus callosum hypoplasia	Incomplete corpus-callosum development
Retardation	Impaired cognitive function
Adducted thumbs	Abnormal thumb development

Spastic paraplegia	Stiffening and contraction in the lower limbs
Hydrocephalus	Abnormal accumulations of Cerebrospinal fluid within skull

- Additionally, studies have shown that alterations in the expression of the protein thy-1 may be partially responsible for the abnormal neuronal outgrowth observed in Alzheimer's patients. It was found that abnormal neural outgrowth and thy-1 presence were correlated spatially, though mechanistic work is still needed to better understand thy-1's involvement in this condition.

Cell Biology

Cell biology (formerly called cytology, from the Greek κυτος, *kytos*, "vessel") is a branch of biology that studies the different structures and functions of the cell and focuses mainly on the idea of the cell as the basic unit of life. Cell biology explains the structure, organization of the organelles they contain, their physiological properties, metabolic processes, signaling pathways , life cycle, and interactions with their environment. This is done both on a microscopic and molecular level as it encompasses prokaryotic cells and eukaryotic cells. Knowing the components of cells and how cells work is fundamental to all biological sciences; it is also essential for research in bio-medical fields such as cancer, and other diseases. Research in cell biology is closely related to genetics, biochemistry, molecular biology, immunology, and developmental biology

Internal Cellular Structures

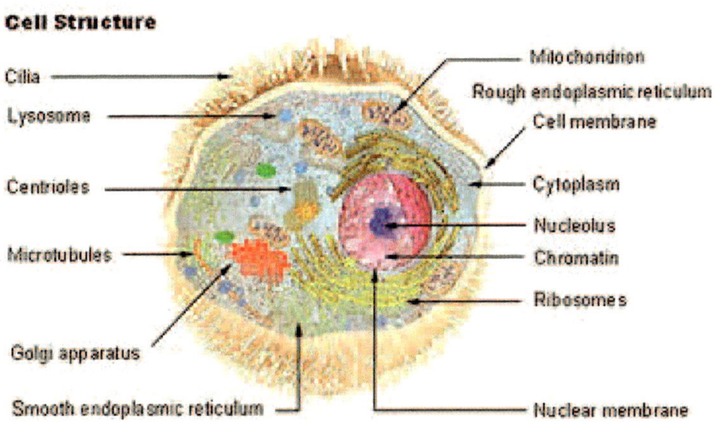

The generalized structure and molecular components of a cell

Chemical and Molecular Environment

The study of the cell is done on a molecular level; however, most of the processes within the cell are made up of a mixture of small organic molecules, inorganic ions, hormones, and water.

Approximately 75-85% of the cell's volume is due to water making it an indispensable solvent as a result of its polarity and structure. These molecules within the cell, which operate as substrates, provide a suitable environment for the cell to carry out metabolic reactions and signalling. The cell shape varies among the different types of organisms, and are thus then classified into two categories: eukaryotes and prokaryotes. In the case of eukaryotic cells - which are made up of animal, plant, fungi, and protozoa cells - the shapes are generally round and spherical, while for prokaryotic cells – which are composed of bacteria and archaea - the shapes are: spherical (cocci), rods (bacillus), curved (vibrio), and spirals (*spirochetes)*.

Cell biology focuses more on the study of eukaryotic cells, and their signalling pathways, rather than on prokaryotes which is covered under microbiology. The main constituents of the general molecular composition of the cell includes: proteins and lipids which are either free flowing or membrane bound, along with different internal compartments known as organelles. This environment of the cell is made up of hydrophilic and hydrophobic regions which allows for the exchange of the above-mentioned molecules and ions. The hydrophilic regions of the cell are mainly on the inside and outside of the cell, while the hydrophobic regions are within the phospholipid bilayer of the cell membrane. The cell membrane consists of lipids and proteins which accounts for its hydrophobicity as a result of being non-polar substances. Therefore, in order for these molecules to participate in reactions, within the cell, they need to be able to cross this membrane layer to get into the cell. They accomplish this process of gaining access to the cell via: osmotic pressure, diffusion, concentration gradients, and membrane channels. Inside of the cell are extensive internal sub-cellular membrane-bounded compartments called organelles.

Organelles

- Centrosome - an associated pair of cylindrical shaped protein structures (centrioles) that organize microtubules and aid in forming the mitotic spindle during cell division in eukaryotes
- Cell membrane (plasma membrane) - the part of the cell which separates the cells from the outside environment and protects the cell, as well as regulating what goes in and out of the cell
- Cell wall - extra layer of protection and gives structural support (only found in plant cells)
- Chloroplast - key organelle for photosynthesis (only found in plant cells)
- Cilium - motile structure of eukaryotes having a cytoskeleton, the axoneme.
- Cytoplasm - contents of the main fluid-filled space inside cells, chemical reactions also happen in this jelly-like substance.
- Cytoskeleton - protein filaments inside cells (microfilaments, microtubules, and intermediate filaments)
- Endoplasmic reticulum (rough) - major site of membrane protein synthesis
- Endoplasmic reticulum (smooth) - major site of lipid synthesis
- Endosomes - vesicles that traffic membrane and intra and extra cellular contents for recycling or degradation by lysosomes
- Flagellum - motile structure of bacteria, archaea and eukaryotes

- Golgi apparatus - site of protein glycosylation in the endomembrane system
- Lipid bilayer - fundamental organizational structure of cell membranes
- Lysosome - acidic organelle that breaks down cellular waste products and debris into simple compounds (only found in animal cells)
- Microvilli - increases surface area for absorption of nutrients from surrounding medium
- Mitochondrion - major energy-producing organelle by releasing energy in the form of ATP
- Nucleus - contains chromosomes composed of DNA, the building block of life. Nuclear Architecture is important for dictating nuclear function.
- Organelle - term used for major subcellular structures
- Peroxisomes - a very small organelle that uses oxygen to breakdown and detoxify long fatty acids and other molecules
- Pili - also called fimbria is used for conjugation and sometimes movement
- Ribosome - RNA and protein complex required for protein synthesis in cells
- Starch grain - found in the cytoplasm of a typical plant cell, it stores chemical energy of the plant.
- Vacuole - contain cell sap or other storage material.
- Vesicle - small membrane-bounded spheres inside cells which transport substances.

cell surface membrane protects the cell

Processes

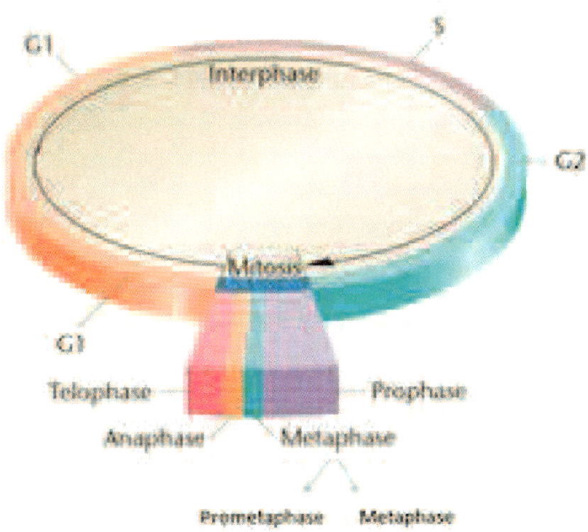

An Image of the Cell Cycle Of Mitosis

Growth and development

The growth process of the cell does not refer to the size of the cell, but instead the density of the number of cells present in the organism at a given time. Cell growth pertains to the increase in the number of cells present in an organism as it grows and develops; as the organism gets larger so too does the number of cells present. Cells are the foundation of all organisms, they are the fundamental unit of life. The growth and development of the cell are essential for the maintenance of the host, and survival of the organisms. For this process the cell goes through the steps of the cell cycle and development which involves cell growth, DNA replication, cell division, regeneration, specialization, and cell death. The cell cycle is divided into four distinct phases, G1, S, G2, and M. The G phases – which is the cell growth phase - makes up approximately 95% of the cycle. The proliferation of cells is instigated by progenitors, the cells then differentiate to become specialized, where specialized cells of the same type aggregate to form tissues, then organs and ultimately systems. The G phases along with the S phase – DNA replication, damage and repair - are considered to be the interphase portion of the cycle. While the M phase (mitosis and cytokinesis) is the cell division portion of the cycle. The cell cycle is regulated by a series of signalling factors and complexes such as CDK›s, kinases, and p53. to name a few. When the cell has completed its growth process, and if it is found to be damaged or altered it undergoes cell death, either by apoptosis or necrosis, to eliminate the threat it cause to the organism's survival.

Other Cellular Processes

- Active transport and Passive transport - Movement of molecules into and out of cells.
- Autophagy - The process whereby cells "eat" their own internal components or microbial invaders.
- Adhesion - Holding together cells and tissues.
- Cell movement - Chemotaxis, contraction, cilia and flagella.
- Cell signaling - Regulation of cell behavior by signals from outside.
- Division - By which cells reproduce either by mitosis (to produce clones of the parent cell) or Meiosis (to produce haploid gametes)
- DNA repair - Cell death and cell senescence.
- Metabolism - Glycolysis, respiration, photosynthesis, and chemosynthesis.
- Signalling - The process by which the activities in the cell are regulated
- Transcription and mRNA splicing - Gene expression.

Techniques used to Study Cells

Electron micrograph of blood cells clotting.

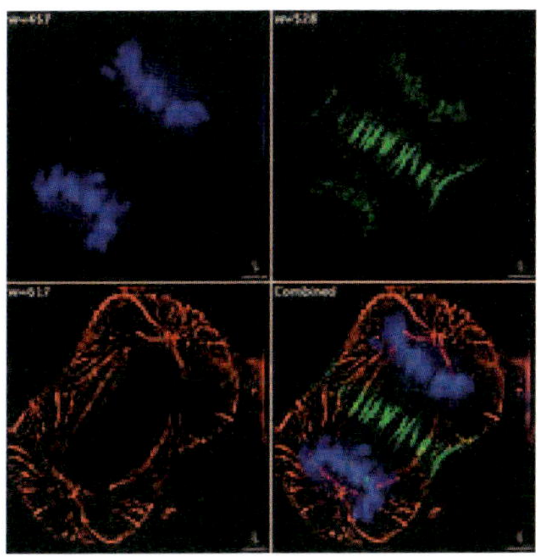

Cell division studied using fluorescence to stain specific structures

Cells may be observed under the microscope, using several different techniques; these include optical microscopy, transmission electron microscopy, scanning electron microscopy, fluorescence microscopy, and confocal microscopy.

There are several different methods used in the study of cells:

- Cell culture is the basic technique of growing cells in a laboratory independent of an organism.
- Immunostaining, also known as immunohistochemistry, is a specialized histological method used to localize proteins in cells or tissue slices. Unlike regular histology, which uses stains to identify cells, cellular components or protein classes, immunostaining requires the reaction of an antibody directed against the protein of interest within the tissue or cell. Through the use of proper controls and published protocols (need to add reference links here), specificity of the antibody-antigen reaction can be achieved. Once this complex is formed, it is identified via either a "tag" attached directly to the antibody, or added in an additional technical

step. Commonly used "tags" include fluorophores or enzymes. In the case of the former, detection of the location of the immuno-stained protein occurs via fluorescence microscopy. With an enzymatic tag, such as horse radish peroxidase, a chemical reaction is carried out that results in a dark color in the location of the protein of interest. This darkened pattern is then detected using light microscopy.

- Computational genomics is used to find patterns in genomic information
- DNA microarrays identify changes in transcript levels between different experimental conditions.
- Gene knockdown mutates a selected gene.
- In situ hybridization shows which cells are expressing a particular RNA transcript.
- PCR can be used to determine how many copies of a gene are present in a cell.
- Transfection introduces a new gene into a cell, usually an expression construct

Purification of cells and their parts Purification may be performed using the following methods:

- Cell fractionation
  - Release of cellular organelles by disruption of cells.
  - Separation of different organelles by centrifugation.
- Flow cytometry
- Immunoprecipitation
  - The binding of an antibody to a target protein
  - Collection of the target protein through elution
- Proteins extracted from cell membranes by detergents and salts or other kinds of chemicals

CELLULAR DIFFERENTIATION

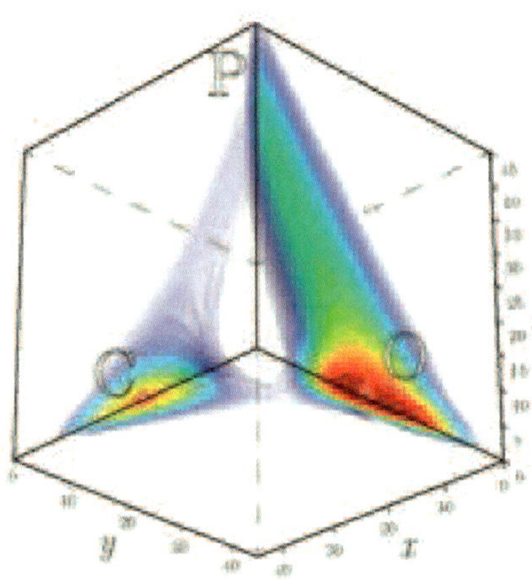

Cell-count distribution featuring cellular differentiation for three types of cells (progenitor x, osteoblast y, and chondrocyte z exposed to pro-osteoblast stimulus.

In developmental biology, cellular differentiation is the process where a cell changes from one cell type to another. Most commonly the cell changes to a more specialized type. Differentiation occurs numerous times during the development of a multicellular organism as it changes from a simple zygote to a complex system of tissues and cell types. Differentiation continues in adulthood as adult stem cells divide and create fully differentiated daughter cells during tissue repair and during normal cell turnover. Some differentiation occurs in response to antigen exposure. Differentiation dramatically changes a cell's size, shape, membrane potential, metabolic activity, and responsiveness to signals. These changes are largely due to highly controlled modifications in gene expression and are the study of epigenetics. With a few exceptions, cellular differentiation almost never involves a change in the DNA sequence itself. Thus, different cells can have very different physical characteristics despite having the same genome.

A cell that can differentiate into all cell types of the adult organism is known as *pluripotent*. Such cells are called embryonic stem cells in animals and meristematic cells in higher plants. A cell that can differentiate into all cell types, including the placental tissue, is known as *totipotent*. In mammals, only the zygote and subsequent blastomeres are totipotent, while in plants many differentiated cells can become totipotent with simple laboratory techniques. In cytopathology, the level of cellular differentiation is used as a measure of cancer progression. "Grade is a marker of how differentiated a cell in a tumor is.

Mammalian Cell Types

Three basic categories of cells make up the mammalian body: germ cells, somatic cells, and stem cells. Each of the approximately 100 trillion (10^{14}) cells in an adult human has its own copy or copies of the genome except certain cell types, such as red blood cells, that lack nuclei in their fully differentiated state. Most cells are diploid; they have two copies of each chromosome. Such cells, called somatic cells, make up most of the human body, such as skin and muscle cells. Cells differentiate to specialize for different functions.

Germ line cells are any line of cells that give rise to gametes—eggs and sperm—and thus are continuous through the generations. Stem cells, on the other hand, have the ability to divide for indefinite periods and to give rise to specialized cells. They are best described in the context of normal human development.

Development begins when a sperm fertilizes an egg and creates a single cell that has the potential to form an entire organism. In the first hours after fertilization, this cell divides into identical cells. In humans, approximately four days after fertilization and after several cycles of cell division, these cells begin to specialize, forming a hollow sphere of cells, called a blastocyst. The blastocyst has an outer layer of cells, and inside this hollow sphere, there is a cluster of cells called the inner cell mass. The cells of the inner cell mass go on to form virtually all of the tissues of the human body. Although the cells of the inner cell mass can form virtually every type of cell found in the human body, they cannot form an organism. These cells are referred to as pluripotent.

Pluripotent stem cells undergo further specialization into multipotent progenitor cells that then give rise to functional cells. Examples of stem and progenitor cells include:

- *Radial glial cells* (embryonic neural stem cells) that give rise to excitatory neurons in the fetal brain through the process of neurogenesis.

- *Hematopoietic stem cells* (adult stem cells) from the bone marrow that give rise to red blood cells, white blood cells, and platelets
- *Mesenchymal stem cells* (adult stem cells) from the bone marrow that give rise to stromal cells, fat cells, and types of bone cells
- *Epithelial stem cells* (progenitor cells) that give rise to the various types of skin cells
- *Muscle satellite cells* (progenitor cells) that contribute to differentiated muscle tissue.

A pathway that is guided by the cell adhesion molecules consisting of four amino acids, arginine, glycine, asparagine, and serine, is created as the cellular blastomere differentiates from the single-layered blastula to the three primary layers of germ cells in mammals, namely the ectoderm, mesoderm and endoderm (listed from most distal (exterior) to proximal (interior)). The ectoderm ends up forming the skin and the nervous system, the mesoderm forms the bones and muscular tissue, and the endoderm forms the internal organ tissues.

Dedifferentiation

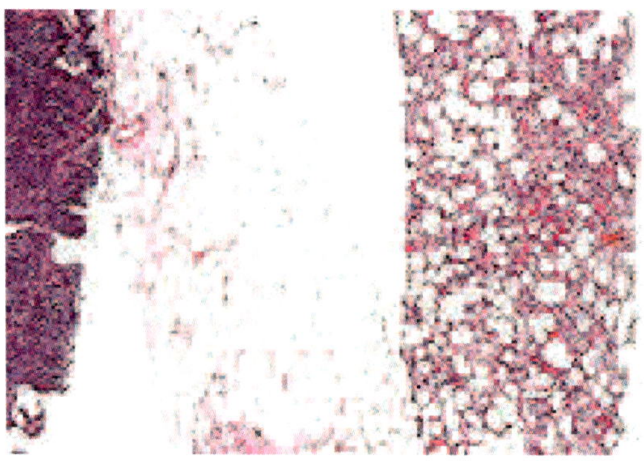

Micrograph of a liposarcoma with some dedifferentiation, that is not identifiable as a liposarcoma, (left edge of image) and a differentiated component (with lipoblasts and increased vascularity (right of image)). Fully differentiated (morphologically benign) adipose tissue (center of the image) has few blood vessels. H&E stain.

Dedifferentiation, or integration is a cellular process often seen in more basal life forms such as worms and amphibians in which a partially or terminally differentiated cell reverts to an earlier developmental stage, usually as part of a regenerative process. Dedifferentiation also occurs in plants. Cells in cell culture can lose properties they originally had, such as protein expression, or change shape. This process is also termed dedifferentiation.

Some believe dedifferentiation is an aberration of the normal development cycle that results in cancer,whereas others believe it to be a natural part of the immune response lost by humans at some point as a result of evolution.

A small molecule dubbed reversine, a purine analog, has been discovered that has proven to induce dedifferentiation in myotubes. These dedifferentiated cells could then redifferentiate into osteoblasts and adipocytes.

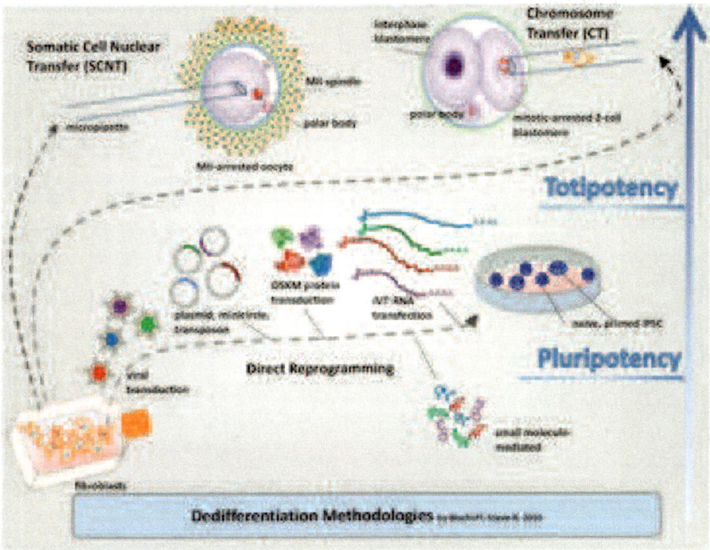

Diagram exposing several methods used to revert adult somatic cells to totipotency or pluripotency.

Mechanisms

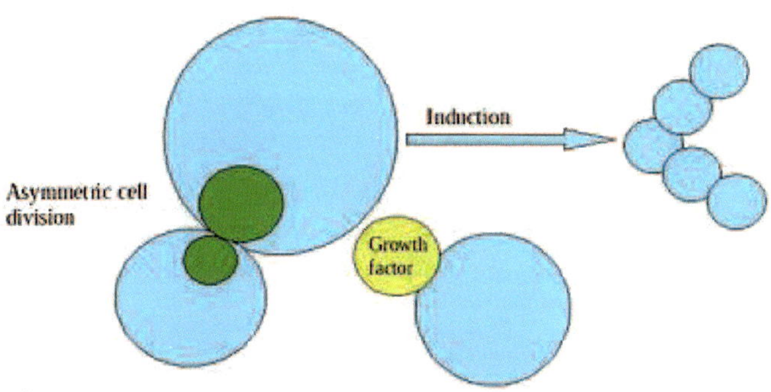

Mechanisms of cellular differentiation.

Each specialized cell type in an organism expresses a subset of all the genes that constitute the genome of that species. Each cell type is defined by its particular pattern of regulated gene expression. Cell differentiation is thus a transition of a cell from one cell type to another and it involves a switch from one pattern of gene expression to another. Cellular differentiation during development can be understood as the result of a gene regulatory network. A regulatory gene and its cis-regulatory modules are nodes in a gene regulatory network; they receive input and create output elsewhere in the network. The systems biology approach to developmental biology emphasizes the importance of investigating how developmental mechanisms interact to produce predictable patterns (morphogenesis). (However, an alternative view has been proposed recently. Based on stochastic gene expression, cellular differentiation is the result of a Darwinian selective process occurring among cells. In this frame, protein and gene networks are the result of cellular processes and not their cause. See: Cellular Darwinism)

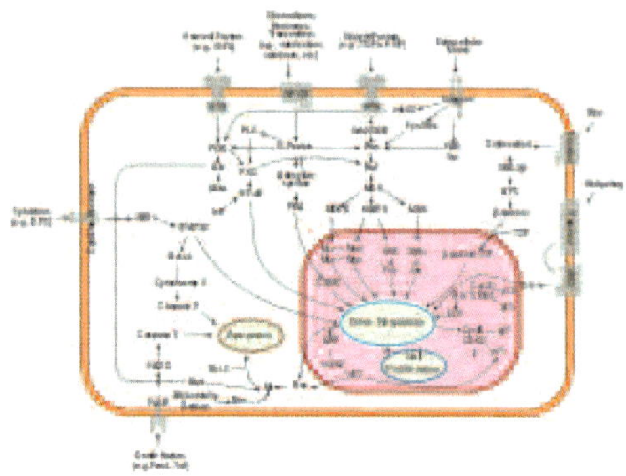

An overview of major signal transduction pathways.

A few evolutionarily conserved types of molecular processes are often involved in the cellular mechanisms that control these switches. The major types of molecular processes that control cellular differentiation involve cell signaling. Many of the signal molecules that convey information from cell to cell during the control of cellular differentiation are called growth factors. Although the details of specific signal transduction pathways vary, these pathways often share the following general steps. A ligand produced by one cell binds to a receptor in the extracellular region of another cell, inducing a conformational change in the receptor. The shape of the cytoplasmic domain of the receptor changes, and the receptor acquires enzymatic activity. The receptor then catalyzes reactions that phosphorylate other proteins, activating them. A cascade of phosphorylation reactions eventually activates a dormant transcription factor or cytoskeletal protein, thus contributing to the differentiation process in the target cell. Cells and tissues can vary in competence, their ability to respond to external signals.

Signal induction refers to cascades of signaling events, during which a cell or tissue signals to another cell or tissue to influence its developmental fate. Yamamoto and Jeffery investigated the role of the lens in eye formation in cave- and surface-dwelling fish, a striking example of induction. Through reciprocal transplants, Yamamoto and Jeffery found that the lens vesicle of surface fish can induce other parts of the eye to develop in cave- and surface-dwelling fish, while the lens vesicle of the cave-dwelling fish cannot.

Other important mechanisms fall under the category of asymmetric cell divisions, divisions that give rise to daughter cells with distinct developmental fates. Asymmetric cell divisions can occur because of asymmetrically expressed maternal cytoplasmic determinants or because of signaling. In the former mechanism, distinct daughter cells are created during cytokinesis because of an uneven distribution of regulatory molecules in the parent cell; the distinct cytoplasm that each daughter cell inherits results in a distinct pattern of differentiation for each daughter cell. A well-studied example of pattern formation by asymmetric divisions is body axis patterning in Drosophila. RNA molecules are an important type of intracellular differentiation control signal. The molecular and genetic basis of asymmetric cell divisions has also been studied in green algae of the genus *Volvox*, a model system for studying how unicellular organisms can evolve into multicellular organisms. In *Volvox carteri*, the 16 cells in the anterior hemisphere of a 32-cell embryo divide asymmetrically, each producing one large and one small daughter cell. The size of the cell at the end of all cell divisions determines whether it becomes a specialized germ or somatic cell.

Epigenetic Control of Cellular Differentiation

Since each cell, regardless of cell type, possesses the same genome, determination of cell type must occur at the level of gene expression. While the regulation of gene expression can occur through cis- and trans-regulatory elements including a gene's promoter and enhancers, the problem arises as to how this expression pattern is maintained over numerous generations of cell division. As it turns out, epigenetic processes play a crucial role in regulating the decision to adopt a stem, progenitor, or mature cell fate. This section will focus primarily on mammalian stem cells.

In systems biology and mathematical modeling of gene regulatory networks, cell-fate determination is predicted to exhibit certain dynamics, such as attractor-convergence (the attractor can be an equilibrium point, limit cycle or strange attractor) or oscillatory.

Importance of Epigenetic Control

The first question that can be asked is the extent and complexity of the role of epigenetic processes in the determination of cell fate. A clear answer to this question can be seen in the 2011 paper by Lister R, *et al.* on aberrant epigenomic programming in human induced pluripotent stem cells. As induced pluripotent stem cells (iPSCs) are thought to mimic embryonic stem cells in their pluripotent properties, few epigenetic differences should exist between them. To test this prediction, the authors conducted whole-genome profiling of DNA methylation patterns in several human embryonic stem cell (ESC), iPSC, and progenitor cell lines.

Female adipose cells, lung fibroblasts, and foreskin fibroblasts were reprogrammed into induced pluripotent state with the OCT4, SOX2, KLF4, and MYC genes. Patterns of DNA methylation in ESCs, iPSCs, somatic cells were compared. Lister R, *et al.* observed significant resemblance in methylation levels between embryonic and induced pluripotent cells. Around 80% of CG dinucleotides in ESCs and iPSCs were methylated, the same was true of only 60% of CG dinucleotides in somatic cells. In addition, somatic cells possessed minimal levels of cytosine methylation in non-CG dinucleotides, while induced pluripotent cells possessed similar levels of methylation as embryonic stem cells, between 0.5 and 1.5%. Thus, consistent with their respective transcriptional activities, DNA methylation patterns, at least on the genomic level, are similar between ESCs and iPSCs.

However, upon examining methylation patterns more closely, the authors discovered 1175 regions of differential CG dinucleotide methylation between at least one ES or iPS cell line. By comparing these regions of differential methylation with regions of cytosine methylation in the original somatic cells, 44-49% of differentially methylated regions reflected methylation patterns of the respective progenitor somatic cells, while 51-56% of these regions were dissimilar to both the progenitor and embryonic cell lines. In vitro-induced differentiation of iPSC lines saw transmission of 88% and 46% of hyper and hypo-methylated differentially methylated regions, respectively.

Two conclusions are readily apparent from this study. First, epigenetic processes are heavily involved in cell fate determination, as seen from the similar levels of cytosine methylation between induced pluripotent and embryonic stem cells, consistent with their respective patterns of transcription. Second, the mechanisms of de-differentiation (and by extension, differentiation) are very complex and cannot be easily duplicated, as seen by the significant number of differentially methylated regions between ES and iPS cell lines. Now that these two points have been established, we can examine some of the epigenetic mechanisms that are thought to regulate cellular differentiation.

Mechanisms of Epigenetic Regulation

Pioneering factors (Oct4, Sox2, *Nanog)*

Three transcription factors, OCT4, SOX2, and NANOG – the first two of which are used in iPSC reprogramming – are highly expressed in undifferentiated embryonic stem cells and are necessary for the maintenance of their pluripotency. It is thought that they achieve this through alterations in chromatin structure, such as histone modification and DNA methylation, to restrict or permit the transcription of target genes.

Polycomb Repressive Complex (PRC2)

In the realm of gene silencing, Polycomb repressive complex 2, one of two classes of the Polycomb group (PcG) family of proteins, catalyzes the di- and tri-methylation of histone H3 lysine 27 (H3K27me2/me3). By binding to the H3K27me2/3-tagged nucleosome, PRC1 (also a complex of PcG family proteins) catalyzes the mono-ubiquitinylation of histone H2A at lysine 119 (H2AK119Ub1), blocking RNA polymerase II activity and resulting in transcriptional suppression. PcG knockout ES cells do not differentiate efficiently into the three germ layers, and deletion of the PRC1 and PRC2 genes leads to increased expression of lineage-affiliated genes and unscheduled differentiation. Presumably, PcG complexes are responsible for transcriptionally repressing differentiation and development-promoting genes.

Trithorax Group Proteins (TrxG)

Alternately, upon receiving differentiation signals, PcG proteins are recruited to promoters of pluripotency transcription factors. PcG-deficient ES cells can begin differentiation but cannot maintain the differentiated phenotype. Simultaneously, differentiation and development-promoting genes are activated by Trithorax group (TrxG) chromatin regulators and lose their repression. TrxG proteins are recruited at regions of high transcriptional activity, where they catalyze the trimethylation of histone H3 lysine 4 (H3K4me3) and promote gene activation through histone acetylation. PcG and TrxG complexes engage in direct competition and are thought to be functionally antagonistic, creating at differentiation and development-promoting loci what is termed a "bivalent domain" and rendering these genes sensitive to rapid induction or repression.

DNA Methylation

Regulation of gene expression is further achieved through DNA methylation, in which the DNA methyltransferase-mediated methylation of cytosine residues in CpG dinucleotides maintains heritable repression by controlling DNA accessibility. The majority of CpG sites in embryonic stem cells are unmethylated and appear to be associated with H3K4me3-carrying nucleosomes. Upon differentiation, a small number of genes, including OCT4 and NANOG, are methylated and their promoters repressed to prevent their further expression. Consistently, DNA methylation-deficient embryonic stem cells rapidly enter apoptosis upon in vitro differentiation.

Nucleosome Positioning

While the DNA sequence of most cells of an organism is the same, the binding patterns of transcription factors and the corresponding gene expression patterns are different. To a large extent, differences in transcription factor binding are determined by the chromatin accessibility of their binding sites through histone modification and/or pioneer factors. In particular, it is important to know whether a nucleosome is covering a given genomic binding site or not. Recent studies have elucidated the role of nucleosome positioning during stem cell development.

Role of Signaling in Epigenetic Control

A final question to ask concerns the role of cell signaling in influencing the epigenetic processes governing differentiation. Such a role should exist, as it would be reasonable to think that extrinsic signaling can lead to epigenetic remodeling, just as it can lead to changes in gene expression through the activation or repression of different transcription factors. Interestingly, little direct data is available concerning the specific signals that influence the epigenome, and the majority of current knowledge consist of speculations on plausible candidate regulators of epigenetic remodeling. We will first discuss several major candidates thought to be involved in the induction and maintenance of both embryonic stem cells and their differentiated progeny, and then turn to one example of specific signaling pathways in which more direct evidence exists for its role in epigenetic change.

The first major candidate is Wnt signaling pathway. The Wnt pathway is involved in all stages of differentiation, and the ligand Wnt3a can substitute for the overexpression of c-Myc in the generation of induced pluripotent stem cells. On the other hand, disruption of ß-catenin, a component of the Wnt signaling pathway, leads to decreased proliferation of neural progenitors.

Growth factors comprise the second major set of candidates of epigenetic regulators of cellular differentiation. These morphogens are crucial for development, and include bone morphogenetic proteins, transforming growth factors (TGFs), and fibroblast growth factors (FGFs). TGFs and FGFs have been shown to sustain expression of OCT4, SOX2, and NANOG by downstream signaling to Smad proteins. Depletion of growth factors promotes the differentiation of ESCs, while genes with bivalent chromatin can become either more restrictive or permissive in their transcription.

Several other signaling pathways are also considered to be primary candidates. Cytokine leukemia inhibitory factors are associated with the maintenance of mouse ESCs in an undifferentiated state. This is achieved through its activation of the Jak-STAT3 pathway, which has been shown to be necessary and sufficient towards maintaining mouse ESC pluripotency. Retinoic acid can induce differentiation of human and mouse ESCs, and Notch signaling is involved in the proliferation and self-renewal of stem cells. Finally, Sonic hedgehog, in addition to its role as a morphogen, promotes embryonic stem cell differentiation and the self-renewal of somatic stem cells.

The problem, of course, is that the candidacy of these signaling pathways was inferred primarily on the basis of their role in development and cellular differentiation. While epigenetic regulation is necessary for driving cellular differentiation, they are certainly not sufficient for this process. Direct modulation of gene expression through modification of transcription factors plays a key role that must be distinguished from heritable epigenetic changes that can persist even in the absence of the original environmental signals. Only a few examples of signaling pathways leading to epigenetic changes that alter cell fate currently exist, and we will focus on one of them.

Expression of Shh (Sonic hedgehog) upregulates the production of BMI1, a component of the PcG complex that recognizes H3K27me3. This occurs in a Gli-dependent manner, as Gli1 and Gli2 are downstream effectors of the Hedgehog signaling pathway. In culture, Bmi1 mediates the Hedgehog pathway's ability to promote human mammary stem cell self-renewal. In both humans and mice, researchers showed Bmi1 to be highly expressed in proliferating immature cerebellar granule cell precursors. When Bmi1 was knocked out in mice, impaired cerebellar development resulted, leading to significant reductions in postnatal brain mass along with abnormalities in motor control and behavior. A separate study showed a significant decrease in neural stem cell proliferation along with increased astrocyte proliferation in Bmi null mice.

In summary, the role of signaling in the epigenetic control of cell fate in mammals is largely unknown, but distinct examples exist that indicate the likely existence of further such mechanisms.

Effect of Matrix Elasticity on Differentiation

In order to fulfill the purpose of regenerating a variety of tissues, adult stems are known to migrate from their niches, adhere to new extracellular matrices (ECM) and differentiate. The ductility of these microenvironments are unique to different tissue types. The ECM surrounding brain, muscle and bone tissues range from soft to stiff. The transduction of the stem cells into these cells types is not directed solely by chemokine cues and cell to cell signaling. The elasticity of the microenvironment can also affect the differentiation of mesenchymal stem cells (MSCs which originate in bone marrow.) When MSCs are placed on substrates of the same stiffness as brain, muscle and bone ECM, the MSCs take on properties of those respective cell types. Matrix sensing requires the cell to pull against the matrix at focal adhesions, which triggers a cellular mechano-transducer to generate a signal to be informed what force is needed to deform the matrix. To determine the key players in matrix-elasticity-driven lineage specification in MSCs, different matrix microenvironments were mimicked. From these experiments, it was concluded that focal adhesions of the MSCs were the cellular mechano-transducer sensing the differences of the matrix elasticity. The non-muscle myosin IIa-c isoforms generates the forces in the cell that lead to signaling of early commitment markers. Nonmuslce myosin IIa generates the least force increasing to non-muscle myosin IIc. There are also factors in the cell that inhibit non-muscle myosin II, such as blebbistatin. This makes the cell effectively blind to the surrounding matrix. Researchers have obtained some success in inducing stem cell-like properties in HEK 239 cells by providing a soft matrix without the use of diffusing factors. The stem-cell properties appear to be linked to tension in the cells' actin network. One identified mechanism for matrix-induced differentiation is tension-induced proteins, which remodel chromatin in response to mechanical stretch. The RhoA pathway is also implicated in this process.

EPIGENETICS IN STEM CELL DIFFERENTIATION

Embryonic stem cells are capable of self-renewing and differentiating to the desired fate depending on its position within the body. Stem cell homeostasis is maintained through epigenetic mechanisms that are highly dynamic in regulating the chromatin structure as well as specific gene transcription programs. Epigenetics has been used to refer to changes in gene expression, which are heritable through modifications not affecting the DNA sequence.

The mammalian epigenome undergoes global remodeling during early stem cell development that requires commitment of cells to be restricted to the desired lineage. There has been multiple evidence suggesting that the maintenance of the lineage commitment of stem cells are controlled by epigenetic mechanisms such as DNA methylation, histone modifications and regulation of ATP-dependent remolding of chromatin structure.Based on the *histone code* hypothesis, distinct covalent histone modifications can lead to functionally distinct chromatin structures that influence the fate of the cell.

This regulation of chromatin through epigenetic modifications is a molecular mechanism that will determine whether the cell will continue to differentiate into the desired fate. A research study performed by *Lee et al.* examined the effects of epigenetic modifications on the chromatin structure and the modulation of these epigenetic markers during stem cell differentiation through in vitro differentiation of murine embryonic stem (ES) cells.

Experimental Background

Embryonic stem cells exhibit dramatic and complex alterations to both global and site-specific chromatin structures. *Lee et al.* performed an experiment to determine the importance of deacetylation and acetylation for stem cell differentiation by looking at global acetylation and methylation levels at certain site-specific modification in histone sites *H3K9* and *H3K4*. Gene expression at these histones regulated by epigenetic modifications is critical in restricting the embryonic stem cell to desired cell lineages and developing cellular memory.

For mammalian cells, the maintenance of cytosine methylation is catalyzed by DNA methyltransferases and any disruption to these methyltransferases will cause a lethal phenotype to the embryo. Cytosine methylation is examined at *H3K9*, which is associated with inactive heterochromatin and occurs mainly at CpG dinucleotides while global acetylation is examined at *H3K4*, which is associated with active euchromatin. The mammalian zygotic genome undergoes active and passive global cytosine demethylation following fertilization that reaches a minimal point of 20% CpG methylation at the blastocyst stage to which is then followed by a wave of methylation that reprograms the chromatin structure in order to restore global levels of CpG methylation to 60%. Embryonic stem cells containing reduced or elevation levels of methylation are viable but unable to differentiate and therefore require critical regulation of cytosine methylation for mammalian development.

Effects of Global Histone Modifications During Embryonic Stem Cell Differentiation

Histones modifications in chromatin were analyzed at various time intervals (along a 6 day period) following the initiation of in vitro embryonic stem cell differentiation. Differentiation was triggered by the removal of Leukemia inhibitory factor (LIF) which inhibits differentiation. Representative data of the histone modifications at the specific sites were assessed using Western blotting. The data confirms that strong deacetylation at the *H3K4* and *H3K9* positions of histone H3 one day after LIF removal, followed by a small increase in acetylation by day two.

The histone *H3K4* methylation also decreased after one day of LIF removal but showed a rebound between days 2-4 of differentiation, finally ending with a decrease in methylation on day five. These results indicate a decrease in the level of active euchromatin epigenetic marks upon initiation of embryonic stem cell differentiation which is then followed immediately by reprogramming of the epigenome.

Histone modifications of *H3K9* position show a decrease in di- and tri-methylation of undifferentiated embryonic stem cells and had a gradual increase in methylation during the six-day time course of in vitro differentiation, which indicated that there is a global increase of inactive heterochromatin levels at this histone mark.

As the embryonic stem cell undergoes differentiation the markers for active euchromatin (histone acetylation and *H3K4* methylation) are decreased after the removal of LIF showing that the cell is indeed becoming more differentiated. The slight rebound in each of these marks allows for further differentiation to occur by allowing another opportunity to decrease the markers once again, bringing the cell closer to its desired fate. Since there is also an increase throughout the six-day period in *H3K9me*, a marker for active heterochromatin, once differentiation occurs it is concluded that the formation of heterochromatin occurs as the cell is differentiated into its desired fate making the cell inactive to prevent further differentiation.

DNA Methylation in differentiated Versus Undifferentiated Cells

Global levels of cytosine methylation were compared between undifferentiated and differentiated embryonic stem cells. Global 5-methylcytosine levels have been measured prior to differentiation and after in vitro differentiation. The global cytosine methylation pattern appears to be established prior to the reprogramming of the histone code that occurs upon in vitro differentiation of embryonic stem cells.

As the embryonic stem cell undergoes differentiation the level of DNA methylation increases. This is in agreement with findings that show that there is an increase in inactive heterochromatin during differentiation.

Supplemental effects of Methylation with DNMTs

In mammals, DNA methylation plays a role in regulating a key component of multipotency—the ability to rapidly self-renew. Khavari et al. discussed the fundamental mechanisms of DNA methylation and the interaction with several pathways regulating differentiation. New approaches studying the genomic status of DNA methylation in various states of differentiation have shown that methylation at CpG sites associated with putative enhancers are important in this process. DNA methylation can modulate the binding affinities of transcription factors by recruiting repressors such as *MeCP2* which display binding specificity for sequences containing methylated CpG dinucleotides. DNA methylation is controlled by certain methyltransferases, *DMNTs*, which perform different functions depending on each one. *DNMT3A* and *DNMT3B* have both been linked to a role in the establishment of DNA methylation pattern in the early development of the stem cell, whereas *DNMT1* is required to methylate a newly synthesized strand of DNA after the cell has undergone replication in order to sustain the epigenetic regulatory state. Numerous proteins can physically interact with *DNMTs* themselves, which help target *DNMT1* to hemi-methylated DNA.

Several new studies point to the central role of DNA methylation interacting with the regulation of cell cycles and DNA repair pathways in order to maintain the undifferentiated state. In embryonic stem cells, *DNMT1* depletion within the undifferentiated progenitor cell compartment led to cell cycle arrest, premature differentiation and a failure of tissue self-renewal. The loss of *DNMT1* occurred from profound effects associated with activation of differentiation genes and loss of genes promoting cell cycle progression, thus indicating that *DNMT1* and other *DNMTs* do not continuously suppress differentiation and thus maintain the pluripotent state.

These studies point to the important of the interaction of DNMTs in order to maintain stem cell states allowing for further differentiation and formation of heterochromatin to occur.

Epigenetic Modifications of Regulated Genes During ESC Differentiation

Okamoto et al. previously documented the expression level of the *Oct4* gene decreasing with embryonic stem cell differentiation. *Lee et al* performed a ChIP analysis of the Oct4 promoter, associated with undifferentiated cells, region to examine the epigenetic modifications of regulated genes undergoing development during embryonic stem cell differentiation. This promoter region decreased at *H3K4* methylation and *H3K9* acetylation sites and increased at the *H3K9* methylation site during differentiation. Analysis of a CpG motif of the *Oct4* gene promoter revealed a progressive increase of DNA methylation and was completely methylated at day 10 of differentiation as previously reported in Gidekel and Bergman.

These results indicate that there was a shift from the active eurchromatin to the inactive heterochromatin due to the decrease of acetylation of H3K4 and an increase of H3K9me. This means that the cell is becoming differentiated at the Oct4 gene, which is coincident with the silence of Oct4 gene expression.

Another site specific gene tested for histone modification was a *Brachyury* gene, a marker of mesoderm differentiation and is only slightly expressed in undifferentiated embryonic stem cells. "Brachyury was induced at day five of differentiation and completely silencing by day 10, corresponding to the last day of differentiation. The ChIP analysis of the Brachyury gene promoter revealed increase of expression in mono- and di-methylation of *H3K4* at day 0 and 5 of embryonic stem cell differentiation with a loss of gene expression at day 10. *H3K4* trimethylation coincides with the time of highest Brachyury gene expression since it only had gene expression on day 5. *H3K4* methylations in all forms are absent at day 10 of differentiation, which correlates with the silencing of Brachyury gene expression. Mono-methylation of both histones produced expression at day 0 indication a marker that is not useful for chromatin structure. Acetlyation of H3K9 does not correlate to *Brachyury* gene expression since it was down regulated at the induction of differentiation. Upon examining of DNA methylation expression, there was no formation of intermediate sized bad in the Southern analysis suggesting that CpG motifs upstream of the promoter region are not methylated in the absence of cytosine methylation at this site.

It is demonstrated from these studies that both *H3K9* di-and tri-methylation correlate with the DNA methylation and gene expression while *H3K4* tri-methylation is associated the highest gene expression stage of the *Brachyury* gene. A previous report from Santos-Rosa is in agreement with these data showing that active genes are associated with *H3K4* tri-methylation in yeast.

This data indicated the same results as for the *Oct4* gene, in that heterochromatin is forming as differentiation occurs again coinciding with the silence of *Brachyury* gene expression.

Effect of TSA on Stem Cell Differentiation

Leukemia inhibitory factor (LIF) was removed from all the cell lines. LIF inhibits cell differentiation, and its removal allows the cell lines to go through cell differentiation. The cell lines were treated with Trichostatin A (TSA) - a histone deacetylase inhibitor for 6 days. One group of cell lines was treated with 10nM of TSA. The western analysis showed the lack of initial deacetylation on Day-1 which, was observed in the control for the embryonic stem cell differentiation. The lack of histone deacetylase activity allowed the acetylation of H3K9 and histone H4. Embryonic stem cells were also analyzed morphologically to observe the formation of embryoid body formation as one of the measures of cell differentiation. The 10nM TSA treated cells failed to form the embryoid body by Day-6 as observed in the control cell line. This implies that the ES cells treated with TSA lacked the deacetylation on Day-1 and failed to differentiate after the removal of LIF. Second group,'-TSA Day4' was treated with TSA for 3days. As soon as the TSA treatment was stopped, on day 4 the deacetylation was observed and the acetylation recovered on Day-5. The morphological examination showed the formation of embryoid body formation by Day-6. In addition, Interestingly the embryoid body formation was faster than the control cell line. This suggests that the '-TSA Day4' lines were responding to the removal of LIF but, were unable to acquire any differentiation phenotype. They were able to acquire the differentiation phenotype after the cessation of TSA treatment and at rapid rate. The morphological examination of the third group,' 5 nM TSA' showed the intermediate effect between the control and 10nM-TSA group. The lower dose of TSA allowed the formation of some embryoid body formation. This experiment shows that TSA inhibits histone deacetylase and the activity of histone deacetylase is required for the embryonic stem cell differentiation. Without the initial deacetylation on Day-1, the ES cells cannot go through the differentiation.

Alkaline Phosphatase Activity

In normal stem cells, the activity of alkaline phosphatase activity is lowered upon differentiation. Trichostatin A causes the cells to maintain the activity of alkaline phosphatase. Significant increase in alkaline phosphatase extinction was observed when Trichostatin A was withdrawn after three days. Alkaline phosphatase activity correlates with the morphology changes. Initial deacetylation of histone is required for embryonic stem cell differentiation.

HDAC1, but not HDAC2 Controls Differentiation

Dovery et al. (2010) used HDAC knockout mice to demonstrate whether HDAC1 or HDAC2 was important for the embryonic stem cell differentiation. Examination of global histone acetylation in the absence of HDAC 1 showed an increase in acetylation. Global histone acetylation levels were unchanged by the loss of HDAC2. In order to analyze the process of HDAC knockout mouse in detail, the knockout mice embryonic stem cells were used to generate embryoid bodies. It showed that just before or during gastrulation, embryonic stem cells lacking HDAC1 acquired visible developmental defects. The continued culture of HDAC1 knockout embryonic stem cells showed that the embryoid bodies formed became irregular and reduced in size rather than uniformly spherical as in normal mice. Embryonic stem cell proliferation was unaffected by the loss of either HDAC1 or HDAC2 but the differentiation of embryonic stem cells were affected with that lack of HDAC 1. This shows that HDAC1 is required for cell fate determination during differentiation.

The Future

Any disturbance of a stable epigenetic regulation of gene expression mediated by DNA methylation is associated with a number of human disorders, including cancer as well as congenital diseases such as pseudohypoparathyroidism type IA, Beckwith-Wiedemann, Prader-Willi and Angelman syndromes, which are each caused by altered methylation-based imprinting at specific loci.

Perturbations of both global and gene-specific patterns of cytosine methylation are commonly observed in cancer while histone deacetylation is an important feature of nuclear reprogramming in oocytes during meiosis.

Recent studies have revealed that there is an array of different pathways that cooperates with one another in order to bestow proper epigenetic regulation by DNA methylation. Future studies will be needed to further clarify the certain mechanism pathways such as DNA binding proteins, DNA repair and noncoding RNAs serve in order to regulate DNA methylation to suppress differentiation and sustain self-renewal in somatic stem cells in the epidermis and other tissues. Addressing these questions will help extend insight into these recent findings for a central role in epigenetic regulators of DNA methylation in controlling embryonic stem cell differentiation.

REFERENCES

1. Villarreal, Luis P. (August 8, 2008) Are Viruses Alive? Scientific American

2. History of the Microscope. History-of-the-microscope.org, United Kingdom. Retrieved 24 April 2014.

3. Gest, H (2004). The discovery of microorganisms by Robert Hooke and Antoni Van Leeuwenhoek, fellows of the Royal Society. Notes and Records of the Royal Society of London. 58 *(2): 187–201. doi:10.1098/rsnr.2004.0055. PMID 15209075.*

4. Mazzarello, P. (1999). A unifying concept: the history of cell theory. Nature Cell Biology. 1 *(1): E13–5.* doi:10.1038/8964. PMID 10559875.

5. Inwood, Stephen (2003). The man who knew too much: the strange and inventive life of Robert Hooke, 1635–1703. London: Pan. p. 72. ISBN 0-330-48829-5.

6. Becker, Wayne M.; Kleinsmith, Lewis J.; Hardin, Jeff (2003). The World of the Cell. Benjamin/Cummings Publishing Company. p. 1. ISBN 978-0-8053-4854-5.

7. Robinson, Richard. History of Biology: Cell Theory and Cell Structure. Advameg, Inc. Retrieved 17 March 2014.

8. Dutrochet, Henri (1824) Recherches anatomiques et physiologiques sur la structure intime des animaux et des vegetaux, et sur leur motilite, par M.H. Dutrochet, avec deux planches

9. Kalenderblatt Dezember 2013 – Mathematisch-Naturwissenschaftliche Fakultät – Universität Rostock. Mathnat.uni-rostock.de (2013-11-28). Retrieved on 2015-10-15.

10. Sharp, L. W. (1921). Introduction To Cytology. New York: McGraw Hill Book Company Inc.

11. Schleiden, M. J. (1839). Beiträge zur Phytogenesis. Archiv für Anatomie, Physiologie und wissenschaftliche Medicin: 137–176.

12. Schwann, T. (1839). Mikroskopische Untersuchungen *über* die Uebereinstimmung in der Struktur und dem Wachsthum der Thiere und Pflanzen. Berlin: Sander.

13. Silver, GA (1987). Virchow, the heroic model in medicine: health policy by accolade. American Journal of Public Health. 77 *(1): 82–8. doi:10.2105/AJPH.77.1.82. PMC 1646803* . PMID 3538915.

14. The modern version of the Cell Theory. Retrieved 12 February 2015.

15. Ling, Gilbert N. (1984). In search of the physical basis of life. New York: Plenum Press. ISBN 0306414090.

16. Michaelis, L. (1925). Contribution to the Theory of Permeability of Membranes for Electrolytes. The Journal of General Physiology. 8 *(2): 33–59. doi:10.1085/jgp.8.2.33. PMC 2140746* . PMID 19872189.

17. Margulis, L. & Chapman, M.J. (2009). Kingdoms and Domains: An Illustrated Guide to the Phyla of Life on Earth ([4th ed.]. ed.). Amsterdam: Academic Press/Elsevier. p. 116.

18. Kryven, I.; Röblitz, S.; Schütte, Ch. (2015). “Solution of the chemical master equation by radial basis functions approximation with interface tracking”. BMC systems biology (9.1).

19. Slack, J.M.W. (2013) Essential Developmental Biology. Wiley-Blackwell, Oxford.

20. Slack, J.M.W. (2007). Metaplasia and transdifferentiation: from pure biology to the clinic. Nature Reviews Molecular Cell Biology. 8: 369–378. doi:10.1038/nrm2146.

21. NCI Dictionary of Cancer Terms. National Cancer Institute. Retrieved 1 November 2013.

22. Lodish, Harvey (2000). Molecular Cell Biology (4th ed.). New York: W. H. Freeman.

Section 14.2. ISBN 0-7167-3136-3.

23. Kumar, Rani (2008). Textbook of Human Embryology. I.K. International Publishing House. p. 22. ISBN 9788190675710.

24. D. Binder, Marc; Hirokawa, Nobutaka; Windhorst, Uwe (2009). Encyclopedia of Neuroscience. Springer. ISBN 3540237356.

25. Rakic, P (October 2009). Evolution of the neocortex: a perspective from developmental biology.. *Nature reviews. Neuroscience.* 10 *(10): 724–35.* doi:10.1038/nrn2719. PMC 2913577 . PMID 19763105.

26. Lui, JH; Hansen, DV; Kriegstein, AR (8 July 2011). Development and evolution of the human neocortex.. *Cell.* 146 *(1): 18–36. doi:10.1016/j.cell.2011.06.030. PMC 3610574* . PMID 21729779.

27. Rash, BG; Ackman, JB; Rakic, P (February 2016). Bidirectional radial Ca(2+) activity regulates neurogenesis and migration during early cortical column formation.. Science advances. 2 *(2): e1501733. doi:10.1126/sciadv.1501733. PMC 4771444* . PMID 26933693.

28. Stocum DL (2004). Amphibian regeneration and stem cells. Curr. Top. Microbiol. Immunol. Current Topics in Microbiology and Immunology. 280: 1–70. doi:10.1007/978-3-642-18846-6_1. ISBN 978-3-540-02238-1. PMID 14594207.

29. Casimir CM, Gates PB, Patient RK, Brockes JP (1988-12-01). Evidence for dedifferentiation and metaplasia in amphibian limb regeneration from inheritance of DNA methylation. Development. 104 *(4): 657–668. PMID 3268408.*

30. Giles KL. Dedifferentiation and Regeneration in Bryophytes: A Selective Review. New Zealand Journal of Botany. 9: 689–94. doi:10.1080/0028825x.1971.10430231.

31. Schnabel M, Marlovits S, Eckhoff G, et al. (January 2002). Dedifferentiation-associated changes in morphology and gene expression in primary human articular chondrocytes in cell culture. Osteoarthr. Cartil. 10 *(1): 62–70. doi:10.1053/joca.2001.0482. PMID 11795984.*

32. Sell S (December 1993). Cellular origin of cancer: dedifferentiation or stem cell maturation arrest?. *Environ. Health* Perspect. 101 *(*Suppl 5): 15–26. doi:10.2307/3431838. JSTOR 3431838. PMC 1519468 . PMID 7516873.

33. Tsonis PA (April 2004). Stem cells from differentiated cells. Mol. Interv. 4 *(2): 81–3. doi:10.1124/mi.4.2.4. PMID 15087480.*

34. Ben-Tabou de-Leon S, Davidson EH (2007). Gene regulation: gene control network in development.. Annu Rev Biophys Biomol Struct. 36 *(191): 191–212. doi:10.1146/annurev.biophys.35.040405.102002. PMID 17291181.*

35. Knisely, Karen; Gilbert, Scott F. (2009). Developmental Biology (8th ed.). Sunderland, Mass: Sinauer Associates. p. 147. ISBN 0-87893-371-9.

36. Rudel and Sommer; The evolution of developmental mechanisms. Developmental Biology 264, 15-37, 2003 Rudel, D.; Sommer, R. J. (2003). The evolution of developmental mechanisms. Developmental Biology. 264 *(1): 15–37.* doi:10.1016/S0012-1606(03)00353-1. PMID 14623229.

37. Yamamoto Y and WR Jeffery; Central role for the lens in cave fish eye degeneration. Science 289 (5479), 631-633, 2000 Yamamoto, Y.; Jeffery, W. R. (2000).

Central Role for the Lens in Cave Fish Eye Degeneration. Science. 289 *(5479): 631–633. Bibcode*:2000Sci...289..631Y. doi:10.1126/science.289.5479.631. PMID 10915628.

38. Kirk MM, A Ransick, SE Mcrae, DL Kirk; The relationship between cell size and cell fate in Volvox carteri. Journal of Cell Biology 123, 191-208, 1993 Kirk, M. M.; Ransick, A.; McRae, S. E.; Kirk, D. L. (1993). The relationship between cell size and cell fate in Volvox carteri. Journal of Cell Biology. 123 *(1): 191–208. doi:10.1083/jcb.123.1.191. PMC 2119814* . PMID 8408198.

39. Rabajante JF, Babierra AL (January 30, 2015). Branching and oscillations in the epigenetic landscape of cell-fate determination. Progress in Biophysics and Molecular Biology. doi:10.1016/j.pbiomolbio.2015.01.006. PMID 25641423.

40. Lister R; et al. (2011). Hotspots of aberrant epigenomic reprogramming in human induced pluripotent stem cells. Nature. **471** *(7336): 68–73. Bibcode*:2011Natur.471...68L. doi:10.1038/nature09798. PMC 3100360 . PMID 21289626.

41. Christophersen NS, Helin K (2010). Epigenetic control of embryonic stem cell fate. J Exp Med. **207** *(11): 2287–95. doi:10.1084/jem.20101438. PMC 2964577* . PMID 20975044.

42. Zhu, J.; et al. (2013). Genome-wide chromatin state transitions associated with developmental and environmental cues. Cell. **152** *(3): 642–654. doi:10.1016/j.cell.2012.12.033. PMID 23333102.*

43. Guenther MG, Young RA (2010). Repressive Transcription. Science. **329** *(5988): 150–1. Bibcode*:2010Sci...329..150G. doi:10.1126/science.1193995. PMC 3006433 . PMID 20616255.

44. Meissner A (2010). "Epigenetic modifications in pluripotent and differentiated cells". Nat Biotechnol. **28** *(10): 1079–88. doi:10.1038/nbt.1684. PMID 20944600.*

45. Teif VB, Vainshtein Y, Caudron-Herger M, Mallm JP, Marth C, Höfer T, Rippe K (2012). Genome-wide nucleosome positioning during embryonic stem cell development. Nat Struct Mol Biol. **19** *(11): 1185–92. doi:10.1038/nsmb.2419. PMID 23085715.*

46. Mohammad HP, Baylin SB (2010). Linking cell signaling and the epigenetic machinery. Nat Biotechnol. **28** *(10): 1033–8.* doi:10.1038/nbt1010-1033. PMID 20944593.

47. Niwa H, Burdon T, Chambers I, Smith A (1998). Self-renewal of pluripotent embryonic stem cells is mediated via activation of STAT3. Genes Dev. **12** *(13): 2048–60. doi:10.1101/gad.12.13.2048. PMC 316954* . PMID 9649508.

48. Liu S; et al. (2006). Hedgehog Signaling and Bmi-1 Regulate Self-renewal of Normal and Malignant Human Mammary Stem Cells. Cancer Res. **66** *(12): 6063–71. doi:10.1158/0008-5472.CAN-06-0054. PMID 16778178.*

49. Leung C; et al. (2004). Bmi1 is essential for cerebellar development and is overexpressed in human medulloblastomas. Nature. **428** *(6980): 337–41. Bibcode*:2004Natur.428..337L. doi:10.1038/nature02385. PMID 15029199.

50. Zencak D; et al. (2005). Bmi1 loss produces an increase in astroglial cells and a decrease in neural stem cell population and proliferation. J Neurosci. **25** *(24): 5774–83. doi:10.1523/JNEUROSCI.3452-04.2005. PMID 15958744.*

51. Engler, AJ; Sen, S; Sweeney, HL; Discher, DE (August 2006). Matrix Elasticity Directs Stem Cell Lineage Specification. Cell. **126**: 677–689. doi:10.1016/j.cell.2006.06.044. PMID 16923388. Retrieved 2016-04-22.

52. Guo, Jun; Wang, Yuexiu; Sachs, Frederick; Meng, Fanjie (2014-12-09). Actin stress in cell reprogramming. Proceedings of the National Academy of Sciences. **111** *(49): E5252–E5261. doi:10.1073/pnas.1411683111. ISSN 0027-8424. PMC 4267376* . PMID 25422450.

53. Guilak, Farshid; Cohen, Daniel M.; Estes, Bradley T.; Gimble, Jeffrey M.; Liedtke, Wolfgang; Chen, Christopher S. (2009-07-02). Control of Stem Cell Fate by Physical Interactions with the Extracellular Matrix. Cell Stem Cell. **5** *(1): 17–26. doi:10.1016/j.stem.2009.06.016. PMC 2768283* . PMID 19570510.

CHAPTER 3

Cytoskeleton

Chapter Contents

- **Introduction**
- **History**
- **Eukaryotic Cytoskeleton**
- **Microfilaments (Actin Filaments)**
- **Microtubule**
- **Spectrin**
- **In Saccharomyces Cerevisiae**
- **Prokaryotic Cytoskeleton**

The **cytoskeleton** of a cell is made up of microtubules, actin filaments, and intermediate filaments. These structures give the cell its shape and help organize the cell's parts. In addition, they provide a basis for movement and cell division.

INTRODUCTION

A **cytoskeleton** is present in all cells of all domains of life (archaea, bacteria, eukaryotes). It is a complex network of interlinking filaments and tubules that extend throughout the cytoplasm, from the nucleus to the plasma membrane The cytoskeletal systems of different organisms are composed of similar proteins. In eukaryotes, the cytoskeletal matrix is a dynamic structure composed of three main proteins, which are capable of rapid growth or disassembly dependent on the cell›s requirements at a certain period of time.

The structure, function and dynamic behaviour of the cytoskeleton can be very different, depending on organism and cell type. Even within one cell the cytoskeleton can change through association with other proteins and the previous history of the network.

There is a multitude of functions that the cytoskeleton can perform. Primarily, it gives the cell shape and mechanical resistance to deformation, and through association with extracellular connective tissue and other cells it stabilizes entire tissues. The cytoskeleton can also actively contract, thereby deforming the cell and the cell's environment and allowing cells to migrate. Moreover, it is involved in many cell signaling pathways, in the uptake of extracellular material (endocytosis), segregates chromosomes during cellular division, is involved in cytokinesis (the division of a mother cell into two daughter cells), provides a scaffold to organize the contents of the cell in space and for intracellular transport (for example, the movement of vesicles and organelles within the cell); and can be a template for the construction of a cell wall. Furthermore, it forms specialized structures, such as flagella, cilia, lamellipodia and podosomes.

A large-scale example of an action performed by the cytoskeleton is muscle contraction. During contraction of a muscle, within each muscle cell, myosin molecular motors collectively exert forces on parallel actin filaments. This action contracts the muscle cell, and through the synchronous process in many muscle cells, the entire muscle.

HISTORY

In 1903, Nikolai K. Koltsov proposed that the shape of cells was determined by a network of tubules that he termed the cytoskeleton. The concept of a protein mosaic that dynamically coordinated cytoplasmic biochemistry was proposed by Rudolph Peters in 1929 while the term (*cytosquelette*, in French) was first introduced by French embryologist Paul Wintrebert in 1931.

EUKARYOTIC CYTOSKELETON

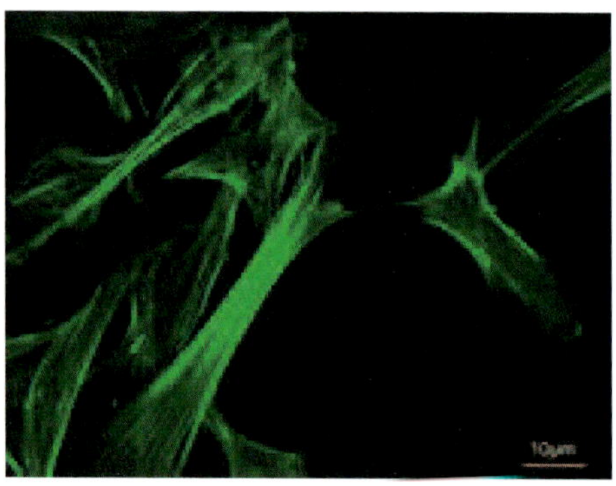

Actin cytoskeleton of mouseembryo fibroblasts, stained with phalloidin

Eukaryotic cells contain three main kinds of cytoskeletal filaments: microfilaments, microtubules, and intermediate filaments. Each cytoskeletal filament has a shape and intracellular distribution. Additionally, the filaments are formed by polymerization of different types of sub-units. The microfilament consist of the polymers of the protein actin which has a diameter of 7 nm. The microtubules are made up of the protein called tubulin which has a diameter of 25 nm. Intermediate filaments are made up of various proteins which varies depending on the cell type. These type of filament normally have diameters ranging from 8-12 nm.The cytoskeleton provides the cell with structure and shape, and by excluding macromolecules from some of the cytosol, it adds to the level of macromolecular crowding in this compartment. Cytoskeletal elements interact extensively and intimately with cellular membranes. A number of small molecule cytoskeletal drugs have been discovered that interact with actin and microtubules. These compounds have proven useful in studying the cytoskeleton and several have clinical applications.

All filaments interact with accessory proteins that regulate and link the filaments to other cell compounds and each other. The accessory proteins are essential for controlled assembly of cytoskeletal filaments in particular locations, and they include motor proteins.

MICROFILAMENTS (ACTIN FILAMENTS)

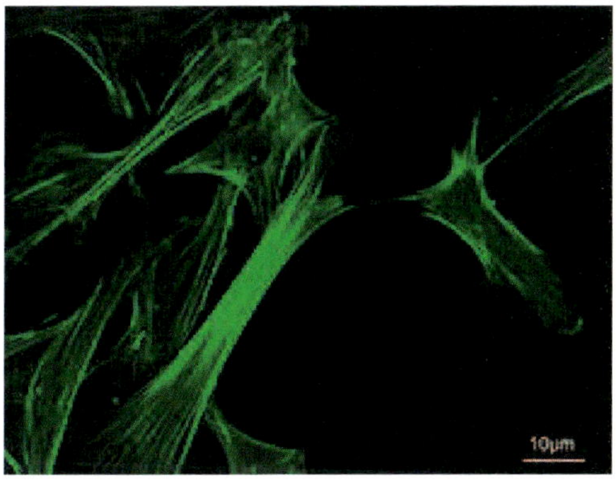

Actin cytoskeleton of mouse embryofibroblasts, stained with Fluorescein isothiocyanate-phalloidin

Microfilaments, also called **actin filaments**, are filamentous structures in the cytoplasm of eukaryoticcells and form part of the cytoskeleton. They are primarily composed of polymers of actin, but in cells are modified by and interact with numerous other proteins. Microfilaments are usually about 7 nm in diameter and composed of two strands of actin. Microfilament functions include cytokinesis, amoeboid movement and cell motility in general, changes in cell shape, endocytosis and exocytosis, cell contractility and mechanical stability. Microfilaments are flexible and relatively strong, resisting buckling by multi-piconewton compressive forces and filament fracture by nanonewton tensile forces. In inducing cell motility, one end of the actin filament elongates while the other end contracts, presumably by myosin II molecular motors. Additionally, they function as part of actomyosin-driven contractile molecular motors, wherein the thin filaments serve as tensile platforms for myosin›s ATP-dependent pulling action in muscle contraction and pseudopod advancement. Microfilaments have a tough, flexible framework which helps the cell in movement.

Actin and microfilament-mediated processes were subject of research since long time ago. Engelmann (1879) suggested that many kinds of movement observed in plants and protozoa like cytoplasmic streaming and amoeboid movement were in fact a primitive version of the movements of muscle contraction.

In the 1930s, Szent-Györgyi and collaborators, violating one of the canons of biochemistry, started to study the residue instead of the extract», that is, structural proteins and not enzymes, leading to the many discoveries related to microfilaments.

Organization

Actin filaments are assembled in two general types of structures: bundles and networks. Bundles can be composed of polar filament arrays, in which all barbed ends point to the same end of the bundle, or non-polar arrays, where the barbed ends point towards both ends. A class of actin-binding proteins, called cross-linking proteins, dictate the formation of these structures. Cross-linking proteins determine filament orientation and spacing in the bundles and networks. These structures are regulated by many other classes of actin-binding proteins, including motor proteins, branching proteins, severing proteins, polymerization promoters, and capping proteins.

In vitro self-assembly

Measuring approximately 6 nm in diameter, microfilaments are the thinnest fibers of the cytoskeleton. They are polymers of actin subunits (globular actin, or G-actin), which as part of the fiber are referred to as filamentous actin, or F-actin. Each microfilament is made up of two helical, interlaced strands of subunits. Much like microtubules, actin filaments are polarized. Electron micrographs have provided evidence of their fast-growing barbed-ends and their slow-growing pointed-end. This polarity has been determined by the pattern created by the binding of myosin S1 fragments: they themselves are subunits of the larger myosin II protein complex. The pointed end is commonly referred to as the minus (−) end and the barbed end is referred to as the plus (+) end.

In vitro actin polymerization, or nucleation, starts with the self-association of three G-actin monomers to form a trimer. ATP-bound actin then itself binds the barbed end, and the ATP is subsequently hydrolyzed. ATP hydrolysis occurs with a half time of about 2 seconds, while the half time for the dissociation of the inorganic phosphate is about 6 minutes. This autocatalyzed event reduces the binding strength between neighboring subunits, and thus generally destabilizes the filament. *In vivo* actin polymerization is catalyzed by a class of filament end-tracking molecular motors known as actoclampins. Recent evidence suggests that the rate of ATP hydrolysis and the rate of monomer incorporation are strongly coupled.

Subsequently, ADP-actin dissociates slowly from the pointed end, a process significantly accelerated by the actin-binding protein, cofilin. ADP bound cofilin severs ADP-rich regions nearest the (−)-ends. Upon release, the free actin monomer slowly dissociates from ADP, which in turn rapidly binds to the free ATP diffusing in the cytosol, thereby forming the ATP-actin monomeric units needed for further barbed-end filament elongation. This rapid turnover is important for the cell›s movement. End-capping proteins such as CapZ prevent the addition or loss of monomers at the filament end where actin turnover is unfavorable, such as in the muscle apparatus.

Actin polymerization together with capping proteins were recently used to control the 3-dimensional growth of protein filament so as to perform 3D topologies useful in technology and the making of electrical interconnect. Electrical conductivity is obtained by metallisation of the protein 3D structure.

Mechanism of Force Generation

As a result of ATP hydrolysis, filaments elongate approximately 10 times faster at their barbed ends than their pointed ends. At steady-state, the polymerization rate at the barbed end matches the depolymerization rate at the pointed end, and microfilaments are said to be treadmilling. Treadmilling results in elongation in the barbed end and shortening in the pointed-end, so that the filament in total moves. Since both processes are energetically favorable, this means force is generated, the energy ultimately coming from ATP.

Actin in Cells

Intracellular actin cytoskeletal assembly and disassembly are tightly regulated by cell signaling mechanisms. Many signal transduction systems use the actin cytoskeleton as a scaffold, holding them at or near the inner face of the peripheral membrane. This subcellular location allows immediate responsiveness to transmembrane receptor action and the resulting cascade of signal-processing enzymes.

Because actin monomers must be recycled to sustain high rates of actin-based motility during chemotaxis, cell signalling is believed to activate cofilin, the actin-filament depolymerizing protein which binds to ADP-rich actin subunits nearest the filament›s pointed-end and promotes filament fragmentation, with concomitant depolymerization in order to liberate actin monomers. In most animal cells, monomeric actin is bound to profilin and thymosin beta-4, both of which preferentially bind with one-to-one stoichiometry to ATP-containing monomers. Although thymosin beta-4 is strictly a monomer-sequestering protein, the behavior of profilin is far more complex. Profilin enhances the ability of monomers to assemble by stimulating the exchange of actin-bound ADP for solution-phase ATP to yield actin-ATP and ADP. Profilin is transferred to the leading edge by virtue of its PIP_2 binding site, and it employs its poly-L-proline binding site to dock onto end-tracking proteins. Once bound, profilin-actin-ATP is loaded into the monomer-insertion site of actoclampin motors.

Another important component in filament formation is the Arp2/3 complex, which binds to the side of an already existing filament (or «mother filament»), where it nucleates the formation of a new daughter filament at a 70 degree angle relative to the mother filament, effecting a fan-like branched filament network.

Associated Proteins

In non-muscle cells, actin filaments are formed proximal to membrane surfaces. Their formation and turnover are regulated by many proteins, including:

- Filament end-tracking protein (e.g., formins, VASP, N-WASP)
- Filament-nucleator known as the Actin-Related Protein-2/3 (or Arp2/3) complex
- Filament cross-linkers (e.g., α-actinin, fascin, and fimbrin)
- Actin monomer-binding proteins profilin and thymosin β4
- Filament barbed-end cappers such as Capping Protein and CapG, *etc.*
- Filament-severing proteins like gelsolin.
- Actin depolymerizing proteins such as ADF/cofilin.

The actin filament network in non-muscle cells is highly dynamic. The actin filament network is arranged with the barbed-end of each filament attached to the cell's peripheral membrane by means of clamped-filament elongation motors, the above-mentioned "actoclampins", formed from a filament barbed-end and a clamping protein (formins, VASP, Mena, WASP, and N-WASP). The primary substrate for these elongation motors is profilin-actin-ATP complex which is directly transferred to elongating filament ends. The pointed-end of each filament is oriented toward the cell's interior. In the case of lamellipodial growth, the Arp2/3 complex generates a branched network, and in filopodia a parallel array of filaments is formed.

Actin Acts as a Track for Myosin Motor Motility

Myosin motors are intracellular ATP-dependent enzymes that bind to and move along actin filaments. Various classes of myosin motors have very different behaviors, including exerting tension in the cell and transporting cargo vesicles.

A proposed model – actoclampins track filament ends

One proposed model suggests the existence of actin filament barbed-end-tracking molecular motors termed «actoclampin». The proposed actoclampins generate the propulsive forces needed for actin-based motility of lamellipodia, filopodia, invadipodia, dendritic spines, intracellularvesicles, and motile processes in endocytosis, exocytosis, podosome formation, and phagocytosis. Actoclampin motors also propel such intracellular pathogens as *Listeria monocytogenes*, *Shigella flexneri*, *Vaccinia* and *Rickettsia*. When assembled under suitable conditions, these end-tracking molecular motors can also propel biomimetic particles.

The term actoclampin is derived from *acto-* to indicate the involvement of an actin filament, as in actomyosin, and *clamp* to indicate a clasping device used for strengthening flexible/moving objects and for securely fastening two or more components, followed by the suffix *-in* to indicate its protein origin. An actin filament end-tracking protein may thus be termed a clampin.

Dickinson and Purich recognized that prompt ATP hydrolysis could explain the forces achieved during actin-based motility. They proposed a simple mechanoenzymatic sequence known as the Lock, Load & Fire Model, in which an end-tracking protein remains tightly bound ("locked" or clamped) onto the end of one sub-filament of the double-stranded actin filament. After binding to Glycyl-Prolyl-Prolyl-Prolyl-Prolyl-Prolyl-registers on tracker proteins, Profilin-ATP-actin is delivered («loaded») to the unclamped end of the other sub-filament, whereupon ATP within the already clamped terminal subunit of the other subfragment is hydrolyzed («fired»), providing the energy needed to release that arm of the end-tracker, which then can bind another Profilin-ATP-actin to begin a new monomer-addition round.

Steps Involved

The following steps describe one force-generating cycle of an actoclampin molecular motor:

- The polymerization cofactor profilin and the ATP·actin combine to form a profilin-ATP-actin complex that then binds to the end-tracking unit

- The cofactor and monomer are transferred to the barbed-end of an actin already clamped filament
- The tracking unit and cofactor dissociate from the adjacent protofilament, in a step that can be facilitated by ATP hydrolysis energy to modulate the affinity of the cofactor and/or the tracking unit for the filament; and this mechanoenzymatic cycle is then repeated, starting this time on the other sub-filament growth site.

When operating with the benefit of ATP hydrolysis, AC motors generate per-filament forces of 8–9 pN, which is far greater than the per-filament limit of 1–2 pN for motors operating without ATP hydrolysis. The term actoclampin is generic and applies to all actin filament end-tracking molecular motors, irrespective of whether they are driven actively by an ATP-activated mechanism or passively.

Some actoclampins (e.g., those involving Ena/VASP proteins, WASP, and N-WASP) apparently require Arp2/3-mediated filament initiation to form the actin polymerization nucleus that is then "loaded" onto the end-tracker before processive motility can commence. To generate a new filament, Arp2/3 requires a «mother» filament, monomeric ATP-actin, and an activating domain from Listeria ActA or the VCA region of N-WASP. The Arp2/3 complex binds to the side of the mother filament, forming a Y-shaped branch having a 70 degree angle with respect to the longitudinal axis of the mother filament. Then upon activation by ActA or VCA, the Arp complex is believed to undergo a major conformational change, bringing its two actin-related protein subunits near enough to each other to generate a new filament gate. Whether ATP hydrolysis may be required for nucleation and/or Y-branch release is a matter under active investigation.

Intermediate filaments

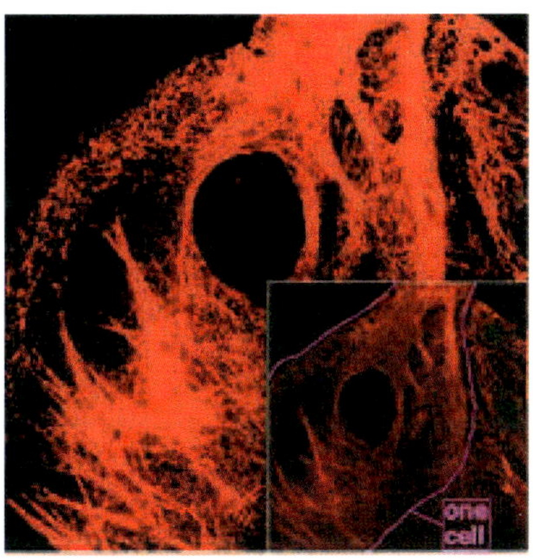

Microscopy of keratin filaments inside cells

MICROTUBULE

Microtubules (*micro-* + *tube* + *-ule*) are a component of the cytoskeleton, found throughout the cytoplasm. These tubular polymers of tubulin can grow as long as 50 micrometres and are highly dynamic. The outer diameter of a microtubule is about 24 nm while the inner diameter is about 12 nm. They are found in eukaryotic cells, as well as some bacteria, and are formed by the polymerization of a dimer of two globular proteins, alpha and beta tubulin.

Microtubules are very important in a number of cellular processes. They are involved in maintaining the structure of the cell and, together with microfilaments and intermediate filaments, they form the cytoskeleton. They also make up the internal structure of cilia and flagella.They provide platforms for intracellular transport and are involved in a variety of cellular processes, including the movement of secretory vesicles, organelles, and intracellular macromolecular assemblies (see entries for dynein and kinesin). They are also involved in chromosome separation (mitosis and meiosis), and are the major constituents of mitotic spindles, which are used to pull apart eukaryotic chromosomes.

Microtubules are nucleated and organized by microtubule organizing centers (MTOCs), such as the centrosome found in the center of many animal cells or the basal bodies found in cilia and flagella, or the spindle pole bodies found in most fungi.

There are many proteins that bind to microtubules, including the motor proteins kinesin and dynein, severing proteins like katanin, and other proteins important for regulating microtubule dynamics. Recently an actin-like protein has been found in a gram-positive bacterium *Bacillus thuringiensis*, which forms a microtubule-like structure and is involved in plasmid segregation.

Tubulin and microtubule-mediated processes, like cell locomotion, were seen by early microscopists, like Leeuwenhoek (1677). However, the fibrous nature of flagella and other structures were discovered two centuries later, with improved light microscopes, and confirmed in the 20th with electron microscope and biochemical studies.

Structure

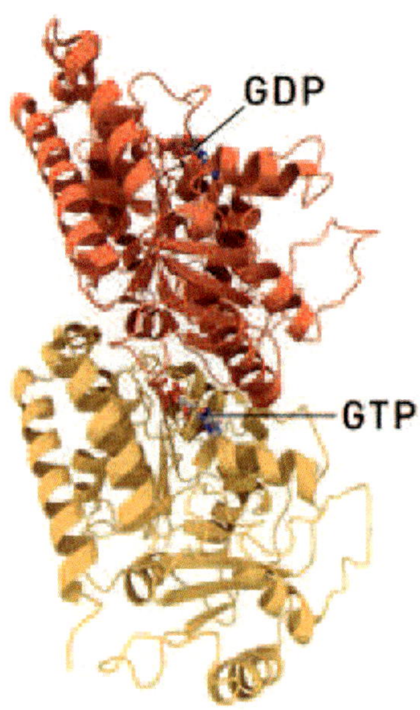

Cartoon representation of the structure of α(yellow)/β(red)-tubulin heterodimer, GTP and GDP.

In eukaryotes, microtubules are long, hollow cylinders made up of polymerised α- and β-tubulin dimers.The inner space of the hollow microtubule cylinders is referred to as the lumen. The α and β-tubulin subunits are approximately 50% identical at the amino acid level, and each have a molecular weight of approximately 50 kDa.

These α/β-tubulin dimers polymerize end-to-end into linear protofilaments that associate laterally to form a single microtubule, which can then be extended by the addition of more α/β-tubulin dimers. Typically, microtubules are formed by the parallel association of thirteen protofilaments, although microtubules composed of fewer or more protofilaments have been observed *in vitro*.

Microtubules have a distinct polarity that is critical for their biological function. Tubulin polymerizes end to end, with the β-subunits of one tubulin dimer contacting the α-subunits of the next dimer. Therefore, in a protofilament, one end will have the α-subunits exposed while the other end will have the β-subunits exposed. These ends are designated the (−) and (+) ends, respectively. The protofilaments bundle parallel to one another with the same polarity, so, in a microtubule, there is one end, the (+) end, with only β-subunits exposed, while the other end, the (−) end, has only α-subunits exposed. While microtubule elongation can occur at both the (+) and (-) ends, it is significantly more rapid at the (+) end.

The lateral association of the protofilaments generates a pseudo-helical structure, with one turn of the helix containing 13 tubulin dimers, each from a different protofilament. In the most common "13-3" architecture, the 13th tubulin dimer interacts with the next tubulin dimer with a vertical offset of 3 tubulin monomers due to the helicity of the turn. There are other alternative architectures, such as 11-3, 12-3, 14-3, 15-4, or 16-4 , that have been detected at a much lower occurrence. Microtubules can also morph into other forms such as helical filaments, which are observed in protist organisms like foraminifera. There are two distinct types of interactions that can occur between the subunits of lateral protofilaments within the microtubule called the A-type and B-type lattices. In the A-type lattice, the lateral associations of protofilaments occur between adjacent α and β-tubulin subunits (i.e. an α-tubulin subunit from one protofilament interacts with a β-tubulin subunit from an adjacent protofilament). In the B-type lattice, the α and β-tubulin subunits from one protofilament interact with the α and β-tubulin subunits from an adjacent protofilament, respectively. Experimental studies have shown that the B-type lattice is the primary arrangement within microtubules. However, in most microtubules there is a seam in which tubulin subunits interact α-β.

Some species of *Prosthecobacter* also contain microtubules. The structure of these bacterial microtubules is similar to that of eukaryotic microtubules, consisting of a hollow tube of protofilaments assembled from heterodimers of bacterial tubulin A (BtubA) and bacterial tubulin B (BtubB). Both BtubA and BtubB share features of both α- and β-tubulin. Unlike eukaryotic microtubules, bacterial microtubules do not require chaperones to fold. In contrast to the 13 protofilaments of eukaryotic microtubules, bacterial microtubules comprise only five.

Intracellular Organization

Microtubules are part of a structural network (the cytoskeleton) within the cell's cytoplasm. Roles of the microtubule cytoskeleton include mechanical support, organization of the cytoplasm, transport, motility and chromosome segregation. A microtubule is capable of growing and shrinking in order to generate force, and there are motor proteins that allow organelles and other cellular components to be carried along a microtubule. This combination of roles makes microtubules important for organizing and moving intracellular constituents.

The organization of microtubules in the cell is cell-type specific. In epithelia, the minus-ends of the microtubule polymer are anchored near the site of cell-cell contact and organized along the apical-basal axis. After nucleation, the minus-ends are released and then re-anchored in the periphery by factors such as ninein and Nezha/PLEKHA7. In this manner, they can facilitate the transport of proteins, vesicles and organelles along the apical-basal axis of the cell. In fibroblasts and other mesenchymal cell-types, microtubules are anchored at the centrosome and radiate with their plus-ends outwards towards the cell periphery (as shown in the first figure). In these cells, the microtubules play important roles in cell migration. Moreover, the polarity of microtubules is acted upon by motor proteins, which organize many components of the cell, including the Endoplasmic Reticulum and the Golgi Apparatus.

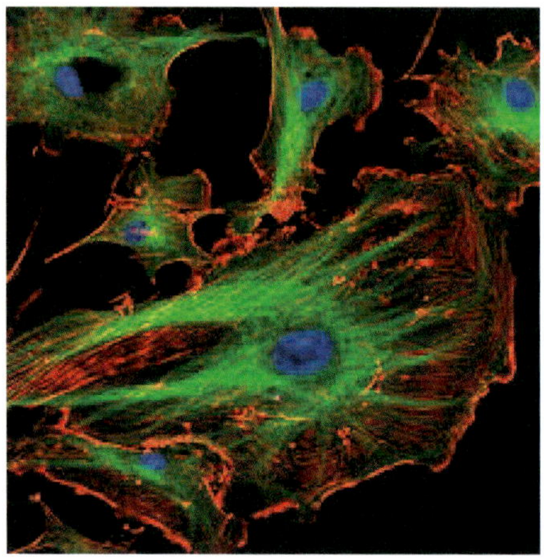

Components of the eukaryotic cytoskeleton. Actin filaments are shown in red, microtubules are in green, and the nuclei are in blue. The cystoskeleton provides the cell with an inner framework and enables it to move and change shape.

Microtubule Polymerization

Nucleation

Microtubules are typically nucleated and organized by dedicated organelles called microtubule-organizing centres (MTOCs). Contained within the MTOC is another type of tubulin, γ-tubulin, which is distinct from the α- and β-subunits of the microtubules themselves. The γ-tubulin combines with several other associated proteins to form a lock washer-like structure known as the «γ-tubulin ring complex» (γ-TuRC). This complex acts as a template for α/β-tubulin dimers to begin polymerization; it acts as a cap of the (−) end while microtubule growth continues away from the MTOC in the (+) direction.

The centrosome is the primary MTOC of most cell types. However, microtubules can be nucleated from other sites as well. For example, cilia and flagella have MTOCs at their base termed basal bodies. In addition, work from the Kaverina group at Vanderbilt, as well as others, suggests that the Golgi apparatus can serve as an important platform for the nucleation of microtubules. Because nucleation from the centrosome is inherently symmetrical, Golgi-associated microtubule nucleation may allow the cell to establish asymmetry in the microtubule network. In recent studies, the Vale group at UCSF identified the protein complex augmin as a critical factor for centrosome-dependent, spindle-based microtubule generation. It that has been shown to interact with γ-TuRC and increase microtubule density around the mitotic spindle origin.

Some cell types, such as plant cells, do not contain well defined MTOCs. In these cells, microtubules are nucleated from discrete sites in the cytoplasm. Other cell types, such as trypanosomatid parasites, have a MTOC but it is permanently found at the base of a flagellum. Here, nucleation of microtubules for structural roles and for generation of the mitotic spindle is not from a canonical centriole-like MTOC.

Polymerization

Following the initial nucleation event, tubulin monomers must be added to the growing polymer. The process of adding or removing monomers depends on the concentration of αβ-tubulin dimers in solution in relation to the critical concentration, which is the steady state concentration of dimers at which there is no longer any net assembly or disassembly at the end of the microtubule. If the dimer concentration is greater than the critical concentration, the microtubule will polymerize and grow. If the concentration is less than the critical concentration, the length of the microtubule will decrease.

Microtubule Dynamics

Dynamic Instability

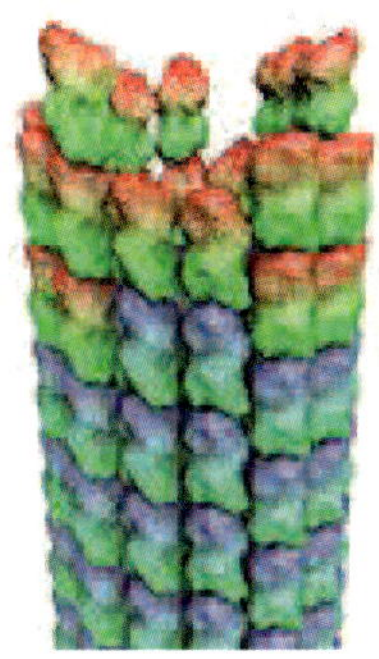

Animation of the microtubule dynamic instability. Tubulin dimers bound to GTP (red) bind to the growing end of a microtubule and subsequently hydrolyze GTP into GDP (blue).

Dynamic instability refers to the coexistence of assembly and disassembly at the ends of a microtubule. The microtubule can dynamically switch between growing and shrinking phases in this region. Tubulin dimers can bind two molecules of GTP, one of which can be hydrolyzed subsequent to assembly. During polymerization, the tubulin dimers are in the GTP-bound state. The GTP bound to α-tubulin is stable and it plays a structural function in this bound state. However, the GTP bound to β-tubulin may be hydrolyzed to GDP shortly after assembly. The assembly properties of GDP-tubulin are different from those of GTP-tubulin, as GDP-tubulin is more prone to depolymerization. A GDP-bound tubulin subunit at the tip of a microtubule will tend to fall off, although a GDP-bound tubulin in the middle of a microtubule cannot spontaneously pop out of the polymer. Since tubulin adds onto the end of the microtubule in the GTP-bound state, a cap of GTP-bound tubulin is proposed to exist at the tip of the microtubule, protecting it from disassembly. When hydrolysis catches up to the tip of the microtubule, it begins a rapid depolymerization and shrinkage. This switch from growth to shrinking is called a catastrophe. GTP-bound tubulin can begin adding to the tip of the microtubule again, providing a new cap and protecting the microtubule from shrinking. This is referred to as «rescue».

"Search and Capture" Model

In 1986, Marc Kirschner and Tim Mitchison proposed that microtubules use their dynamic properties of growth and shrinkage at their plus ends to probe the three dimensional space of the cell. Plus ends that encounter kinetochores or sites of polarity become captured and no longer display growth or shrinkage. In contrast to normal dynamic microtubules, which have a half-life of 5–10 minutes, the captured microtubules can last for hours. This idea is commonly known as the "search and capture" model. Indeed, work since then has largely validated this idea. At the kinetochore, a variety of complexes have been shown to capture microtubule (+)-ends. Moreover, a (+)-end capping activity for interphase microtubules has also been described. This later activity is mediated by formins, the adenomatous polyposis coli protein, and EB1, a protein that tracks along the growing plus ends of microtubules.

Regulation of Microtubule Dynamics

Post-translational Modifications

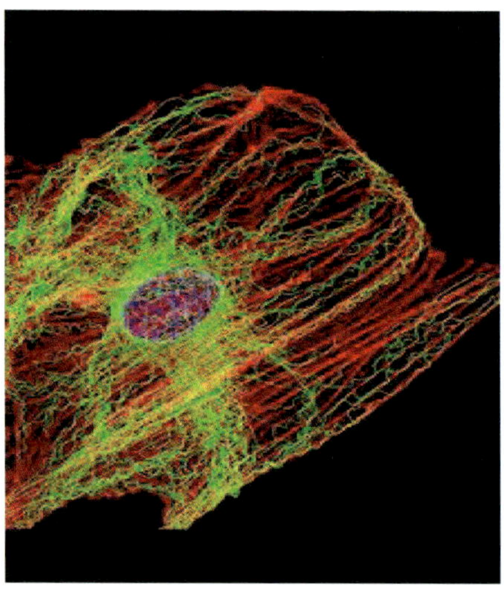

Image of a fibroblast cell containing fluorescently labeled actin (red) and microtubules (green).

Although most microtubules have a half-life of 5-10 min, certain microtubules can remain stable for hours. These stabilized microtubules accumulate post-translational modifications on their tubulin subunits by the action of microtubule-bound enzymes. However, once the microtubule depolymerizes, most of these modifications are rapidly reversed by soluble enzymes. Since most modification reactions are slow while their reverse reactions are rapid, modified tubulin is only detected on long-lived stable microtubules. Most of these modifications occur on the C-terminal region of alpha-tubulin. This region, which is rich in negatively charged glutamate, forms relativey unstructured tails that project out from the microtubule and form contacts with motors. Thus, it is believed that tubulin modifications regulate the interaction of motors with the microtubule. Since these stable modified microtubules are typically oriented towards the site of cell polarity in interphase cells, this subset of modified microtubules provide a specialized route that helps deliver vesicles to these polarized zones. These modifications include:

- Detyrosination: the removal of the C-terminal tyrosine from alpha-tubulin. This reaction exposes a glutamate at the new C-terminus. As a result, microtubules that accumulate this modification are often referred to as Glu-microtubules. Although the tubulin carboxypeptidase has yet to be identified, the tubulin—tyrosine ligase (TTL) is known.
- Delta2: the removal of the last two residues from the C-terminus of alpha-tubulin. Unlike detyrosination, this reaction is thought to be irreversible and has only been documented in neurons.
- Acetylation: the addition of an acetyl group to lysine 40 of alpha-tubulin. This modification occurs on a lysine that is accessible only from the inside of the microtubule, and it remains unclear how enzymes access the lysine residue. The nature of the tubulin acetyltransferase remains controversial, but it has been found that in mammals the major acetyltransferase is ATAT1. however, the reverse reaction is known to be catalyzed by HDAC6.
- Polyglutamylation: the addition of a glutamate polymer (typically 4-6 residues long) to the gamma-carboxyl group of any one of five glutamates found near the end of alpha-tubulin. Enzymes related to TTL add the initial branching glutamate (TTL4,5 and 7), while other enzymes that belong to the same family lengthen the polyglutamate chain (TTL6,11 and 13).
- Polyglycylation: the addition of a glycine polymer (2-10 residues long) to the gamma-carboxyl group of any one of five glutamates found near the end of beta-tubulin. TTL3 and 8 add the initial branching glycine, while TTL10 lengthens the polyglycine chain.

Tubulin is also known to be phosphorylated, ubiquitinated, sumoylated, and palmitoylated.

Tubulin-Binding Drugs and Chemical Effects

A wide variety of drugs are able to bind to tubulin and modify its assembly properties. These drugs can have an effect at intracellular concentrations much lower than that of tubulin. This interference with microtubule dynamics can have the effect of stopping a cell's cell cycle and can lead to programmed cell death or apoptosis. However, there are data to suggest that interference of microtubule dynamics is insufficient to block the cells undergoing mitosis. These studies have demonstrated that suppression of dynamics occurs at concentrations lower than those needed to block mitosis. Suppression of microtubule dynamics by tubulin mutations or by drug treatment have been shown to inhibit cell migration. Both microtubule stabilizers and destabilizers can suppress microtubule dynamics.

The drugs that can alter microtubule dynamics include:

- The cancer-fighting taxane class of drugs (paclitaxel (taxol) and docetaxel) block dynamic instability by stabilizing GDP-bound tubulin in the microtubule. Thus, even when hydrolysis of GTP reaches the tip of the microtubule, there is no depolymerization and the microtubule does not shrink back.
- The epothilones, e.g. Ixabepilone, work in a similar way to the taxanes.
- Nocodazole, vincristine, and colchicine have the opposite effect, blocking the polymerization of tubulin into microtubules.
- Eribulin binds to the (+) growing end of the microtubules. Eribulin exerts its anticancer effects by triggering apoptosis of cancer cells following prolonged and irreversible mitotic blockade.

Expression of β3-tubulin has been reported to alter cellular responses to drug-induced suppression of microtubule dynamics. In general the dynamics are normally suppressed by low, subtoxic concentrations of microtubule drugs that also inhibit cell migration. However, incorporating β3-tubulin into microtubules increases the concentration of drug that is needed to suppress dynamics and inhibit cell migration. Thus, tumors that express β3-tubulin are not only resistant to the cytotoxic effects of microtubule targeted drugs, but also to their ability to suppress tumor metastasis. Moreover, expression of β3-tubulin also counteracts the ability of these drugs to inhibit angiogenesis which is normally another important facet of their action.

Microtubule polymers are extremely sensitive to various environmental effects. Very low levels of free calcium can destabilize microtubules and this prevented early researchers from studying the polymer in vitro. Cold temperatures also cause rapid depolymerization of microtubules. In contrast, heavy water promotes microtubule polymer stability.

Proteins that Interact with Microtubules

Microtubule-associated Proteins (MAPs)

MAPs have been shown to play a crucial role in the regulation of microtubule dynamics *in-vivo*. The rates of microtubule polymerization, depolymerization, and catastrophe vary depending on which microtubule-associated proteins (MAPs) are present. The originally identified MAPs from brain tissue can be classified into two groups based on their molecular weight. This first class comprises MAPs with a molecular weight below 55-62 kDa, and are called τ (tau) proteins. *In-vitro*, tau proteins have been shown to directly bind microtubules, promote nucleation and prevent disassembly, and to induce the formation of parallel arrays. Additionally, tau proteins have also been shown to stabilize microtubules in axons and have been implicated in Alzheimer's disease. The second class is composed of MAPs with a molecular weight of 200-1000 kDa, of which there are four known types: MAP-1, MAP-2, MAP-3 and MAP-4. MAP-1 proteins consists of a set of three different proteins: A, B and C. The C protein plays an important role in the retrograde transport of vesicles and is also known as cytoplasmic dynein. MAP-2 proteins are located in the dendrites and in the body of neurons, where they bind with other cytoskeletal filaments. The MAP-4 proteins are found in the majority of cells and stabilize microtubules. In addition to MAPs that have a stabilizing effect on microtubule structure, other MAPs can have a destabilizing effect either by cleaving or by inducing depolymerization of microtubules. Three proteins called katanin, spastin, and fidgetin have been observed to regulate the number and length of microtubules via their destabilizing activities.

Plus-end Tracking Proteins (+TIPs)

Plus end tracking proteins are MAP proteins which bind to the tips of growing microtubules and play an important role in regulating microtubule dynamics. For example, +TIPs have been observed to participate in the interactions of microtubules with chromosomes during mitosis. The first MAP to be identified as a +TIP was CLIP170 (cytoplasmic linker protein), which has been shown to play a role in microtubule depolymerization rescue events. Additional examples of +TIPs include EB1, EB2, EB3, p150Glued, Dynamitin, Lis1, CLIP115, CLASP1, and CLASP2.

Motor proteins

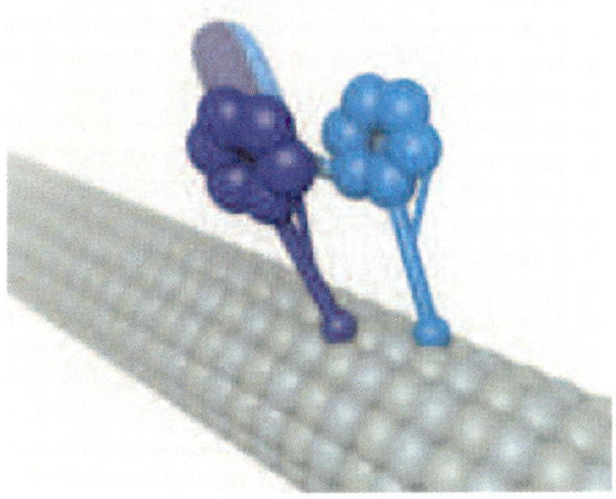

A cytoplasmic dynein motor bound to a microtubule.

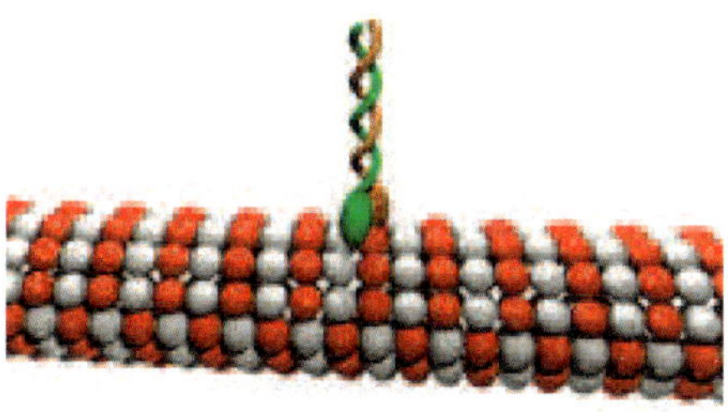

A kinesin molecule bound to a microtubule.

Microtubules can act as substrates for motor proteins that are involved in important cellular functions such as vesicle trafficking and cell division. Unlike other microtubule-associated proteins, motor proteins utilize the energy from ATP hydrolysis to generate mechanical work that moves the protein along the substrate. The major motor proteins that interact with microtubules are kinesin, which usually moves toward the (+) end of the microtubule, and dynein, which moves toward the (−) end.

- Dynein is composed of two identical heavy chains, which make up two large globular head domains, and a variable number of intermediate and light chains. Dynein-mediated transport takes place from the (+) end towards the (-) end of the microtubule. ATP hydrolysis occurs in the globular head domains, which share similarities with the AAA+ (ATPase associated with various cellular activities) protein family. ATP hydolysis in these domains is coupled to movement along the microtubule via the microtubule-binding domains. Dynein transports vesicles and organelles throughout the cytoplasm. In order to do this, dynein molecules bind organelle membranes via a protein complex that contains a number of elements including dynactin.
- Kinesin has a similar structure to dynein. Kinesin is involved in the transport of a variety of intracellular cargoes, including vesicles, organelles, protein complexes, and mRNAs toward the microtubule›s (+) end.

Some viruses (including retroviruses, herpesviruses, parvoviruses, and adenoviruses) that require access to the nucleus to replicate their genomes attach to motor proteins.

Functions

Cell Migration

Microtubule plus ends are often localized to particular structures. In polarized interphase cells, microtubules are disproportionately oriented from the MTOC toward the site of polarity, such as the leading edge of migrating fibroblasts. This configuration is thought to help deliver microtubule-bound vesicles from the Golgi to the site of polarity.

Dynamic instability of microtubules is also required for the migration of most mammalian cells that crawl. Dynamic microtubules regulate the levels of key G-proteins such as RhoA and Rac1, which regulate cell contractility and cell spreading. Dynamic microtubules are also required to trigger focal adhesion disassembly, which is necessary for migration.

Mitosis

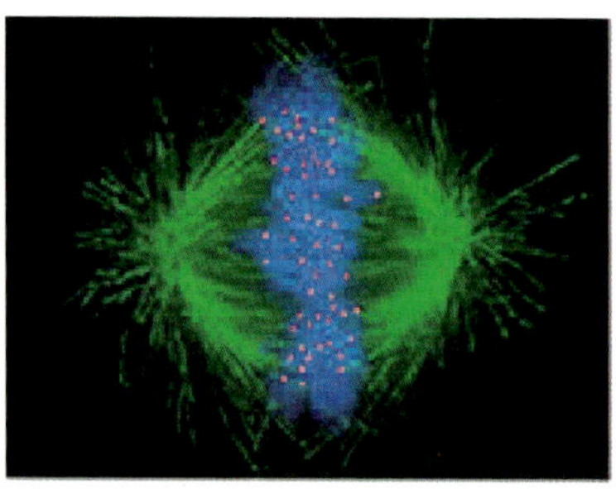

Image of the mitotic spindle in a human cell showing microtubules in green, (chromosomes (DNA) in blue, and kinetochores in red).

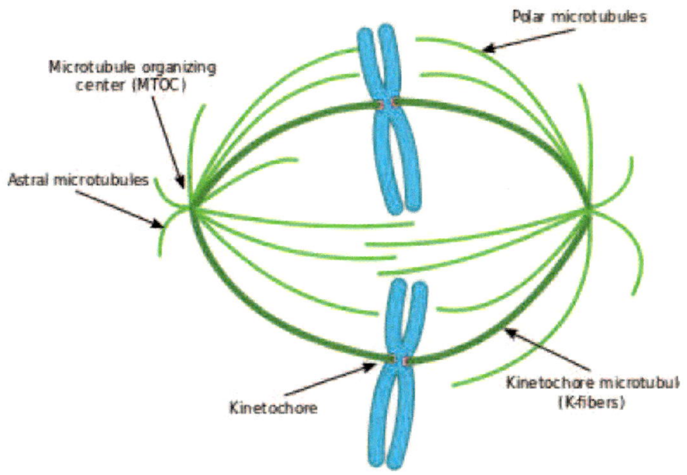

This diagram depicts the organization of a typical mitotic spindle found in animal cells. Chromosomes are attached to kinetochore microtubules via a multiprotein complex called the kinetochore. Polar microtubules interdigitate at the spindle midzone and push the spindle poles apart via motor proteins. Astral microtubules anchor the spindle poles to the cell membrane. Microtubule polymerization is nucleated at the microtubule organizing center.

A notable structure composed largely of microtubules is the mitotic spindle, used by eukaryotic cells to segregate their chromosomes during cell division. The mitotic spindle includes the spindle microtubules, microtubule-associated proteins (MAPs), and the MTOC. The microtubules originate in the MTOC and fan out into the cell; each cell has two MTOCs, as shown in the diagram.

The process of mitosis is facilitated by three main subgroups of microtubules, known as astral, polar, and kinetochore microtubules. An astral microtubule is a microtubule originating from the MTOC that does not connect to a chromosome. Astral microtubules instead interact with the cytoskeleton near the cell membrane and function in concert with specialized dynein motors. Dynein motors pull the MTOC toward the cell membrane, thus assisting in correct positioning and orientation of the entire apparatus.

Kinetochore microtubules directly connect to the chromosomes, at the kinetochores. To clarify the terminology, each chromosome has two chromatids, and each chromatid has a kinetochore. The two kinetochores associated with a region of the chromosome called the centromere. The polar microtubules from one MTOC intertwine with the microtubules from the other MTOC; motor proteins make them push against each other and assist in the separation of the chromosomes to the two daughter cells.

Cell division in a typical eukaryote finishes with the generation of a final cytoplasmic bridge between the two daughter cells termed the midbody. This structure is built of microtubules that originally made up part of the mitotic spindle.

Cilia and Flagella

Microtubules have a major structural role in eukaryotic cilia and flagella. Cilia and flagella always extend directly from a MTOC, in this case termed the basal body. The action of the dynein motor proteins on the various microtubule strands that run along a cilium or flagellum allows the organelle to bend and generate force for swimming, moving extracellular material, and other roles. Prokaryotes possess tubulin-like proteins including FtsZ. However, prokaryotic flagella are entirely different in structure from eukaryotic flagella and do not contain microtubule-based structures.

Development

The cytoskeleton formed by microtubules is essential to the morphogenetic process of an organism's development. For example, a network of polarized microtubules is required within the oocyte of *Drosophila melanogaster* during its embryogenesis in order to establish the axis of the egg. Signals sent between the follicular cells and the oocyte (such as factors similar to epidermal growth factor) cause the reorganization of the microtubules so that their (-) ends are located in the lower part of the oocyte, polarizing the structure and leading to the appearance of an anterior-posterior axis. This involvement in the body's architecture is also seen in mammals.

Another area where microtubules are essential is the formation of the nervous system in higher vertebrates, where tubulin's dynamics and those of the associated proteins (such as the MAPs) is finely controlled during the development of the nervous system.

Gene Regulation

The cellular cytoskeleton is a dynamic system that functions on many different levels: In addition to giving the cell a particular form and supporting the transport of vesicles and organelles, it can also influence gene expression. However, the signal transduction mechanisms involved in this communication are little understood. Notwithstanding this, the relationship between the drug-mediated depolymerization of microtubules and the specific expression of transcription factors has been described, which has provided information on the differential expression of the genes depending on the presence of these factors. This communication between the cytoskeleton and the regulation of the cellular response is also related to the action of growth factors: for example, this relation exists for connective tissue growth factor.

Comparison

Cytoskeleton type	Diameter (nm)	Structure	Subunit examples
Microfilaments	6	double helix	actin
Intermediate filaments	10	two anti-parallel helices/ dimers, forming tetramers	• vimentin (mesenchyme) • glial fibrillary acidic protein (glial cells) • neurofilament proteins (neuronal processes) • keratins (epithelial cells) • nuclear lamins

Microtubules	23	protofilaments, in turn consisting of tubulin subunits in complex with stathmin	α- and β-tubulin

Septins

Spectrin

From Wikipedia, the free encyclopedia

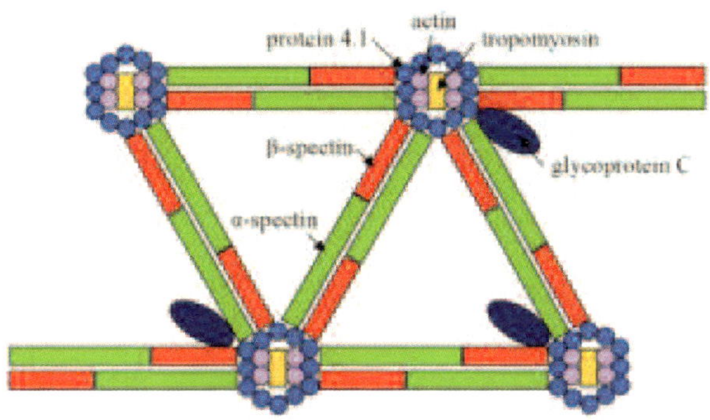

A schematic diagram of spectrin and other cytoskeletal molecules

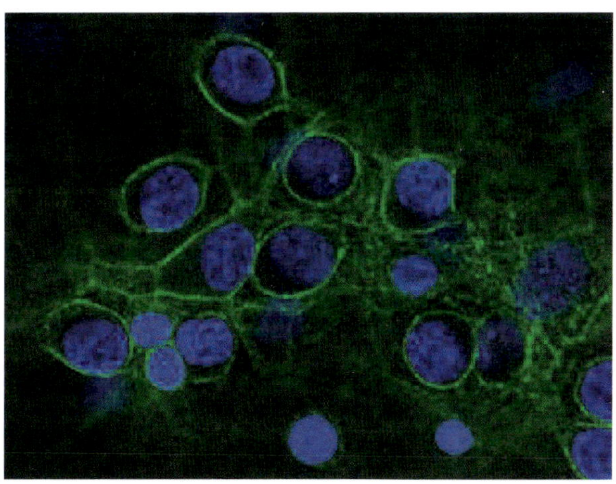

Localization of alpha-II spectrin in green under the plasma membrane of rat neurons in tissue culture as shown with confocal microscopy and immunofluorescence. The nuclei of the cells is revealed in blue by the DNA dye DAPI.

Spectrin is a cytoskeletal protein that lines the intracellular side of the plasma membrane in eukaryotic cells. Spectrin forms pentagonal or hexagonal arrangements, forming a scaffolding and playing an important role in maintenance of plasma membrane integrity and cytoskeletal structure. The hexagonal arrangements are formed by tetramers of spectrin subunits associating with short actin filaments at either end of the tetramer. These short actin filaments act as junctional complexes allowing the formation of the hexagonal mesh. The protein is named spectrin since it was first isolated as a major protein component of human red blood cells which had been treated with mild detergents; the detergents lysed the cells and the hemoglobin and other cytoplasmic components were washed out. In the light microscope the basic shape of the red blood cell could still be seen as the spectrin containing submembranous cytoskeleton preserved the shape of the cell in outline. This became known as a red blood cell «ghost» (spectre), and so the major protein of the ghost was named spectrin.

In certain types of brain injury such as diffuse axonal injury, spectrin is irreversibly cleaved by the proteolytic enzyme calpain, destroying the cytoskeleton. Spectrin cleavage causes the membrane to form blebs and ultimately to be degraded, usually leading to the death of the cell. Spectrin subunits may also be cleaved by caspase family enzymes, and calpain and caspase produce different spectrin breakdown products which can be detected by western blotting with appropriate antibodies. Calpain cleavage may indicate activation of necrosis, while caspase cleavage may indicate apoptosis.

Spectrin in Erythrocytes

The convenience of using erythrocytes compared to other cell types means they have become the standard model for the investigation of the spectrin cytoskeleton. Dimeric spectrin is formed by the lateral association of αI and βI monomers to form a dimer. Dimers then associate in a head-to-head formation to produce the tetramer. End-to-end association of these tetramers with short actin filaments produces the hexagonal complexes observed.

In humans, association with the intracellular face of the plasma membrane is by indirect interaction, through direct interactions with protein 4.1 and ankyrin, with the transmembrane ion transporter band 3 Protein 4.2 binds the spectrin tail region to the transmembrane protein glycophorin A. In animals, spectrin forms the meshwork that provides red blood cells their shape.

The erythrocyte model demonstrates the importance of the spectrin cytoskeleton in that mutations in spectrin commonly cause hereditary defects of the erythrocyte, including hereditary elliptocytosis and rarely hereditary spherocytosis.

Spectrin in Invertebrates

There are three spectrins in invertebrates, α,β and β_H. Mutations in β_H spectrin in *C. elegans* cause defects in morphogenesis resulting in a significantly shorter, but otherwise mostly normal, animal that moves and reproduces. These animals are called "sma» for their small phenotype and carry mutations in the *C. elegans* sma-1 gene. A mutation in β spectrin in *C. elegans* results in an uncoordinated phenotype in which the worms are paralysed and much shorter than wild-type. In addition to the morphological effects, the Unc-70 mutation also produce defective neurons. Neuron numbers are normal but neuronal outgrowth was defective.

Similarly, spectrin plays a role in *Drosophila* neurons. Knock-out of α or β spectrin in *D. melanogaster* results in neurons that are morphologically normal but have reduced neurotransmission at the neuromuscular junction. In animals, spectrin forms the meshwork that provides red blood cells their shape.

Spectrin in Vertebrates

Vertebrate Spectrin Genes

The spectrin gene family has undergone expansion during evolution. Rather than the one α and two β genes in invertebrates, there are two α spectrins (αI and αII) and five β spectrins (βI to V), named in the order of discovery.

In humans, the genes are:

- Alpha: SPTA1, SPTAN1
- Beta: SPTB, SPTBN1, SPTBN2, SPTBN4, SPTBN5

The production of spectrin is promoted by the transcription factor GATA1.

Role of Spectrin in Muscle Tissue

Some evidence for the role of spectrins in muscle tissues exist. In myocardial cells, aII spectrin distribution is coincident with Z-discs and the plasma membrane of myofibrils. Additionally, mice with an ankyrin (ankB) knock-out have disrupted calcium homeostasis in the myocardia. Affected mice have disrupted z-band and sarcomere morphology. In this experimental model ryanodine and IP_3 receptors have abnormal distribution in cultured myocytes. The calcium signaling of the cultured cells is disrupted. In humans, a mutation within the AnkB gene results in the long QT syndrome and sudden death, strengthening the evidence for a role for the spectrin cytoskeleton in excitable tissue.

SPECTRIN

Septins are a group of GTP-binding proteins found primarily in eukaryotic cells of fungi and animals, but also in some green algae. Different septins form protein complexes with each other. These complexes can further assemble into filaments, rings and gauzes. Assembled as such, septins function in cells by localizing other proteins, either by providing a scaffold to which proteins can attach, or by preventing diffusion of molecules from one compartment of the cell to another.

Septins have been implicated in the localization of cellular processes at the site of cell division, at the plasma membrane, at sites where specialized structures like cilia or flagella are attached to the cell body. In yeast cells, they compartmentalize parts of the cell and build scaffolding to provide structural support during cell division at the septum, from which they derive their name. Recent research in human cells suggests that septins build cages around bacterial pathogens, immobilizing the harmful microbes and preventing them from invading other cells.

As filament forming proteins, septins can be considered part of the cytoskeleton. Apart from forming non-polar filaments, septins associate with cell membranes, actin filaments and microtubules. Although present in most eukaryotes, septins have not been observed in plants.

Structure

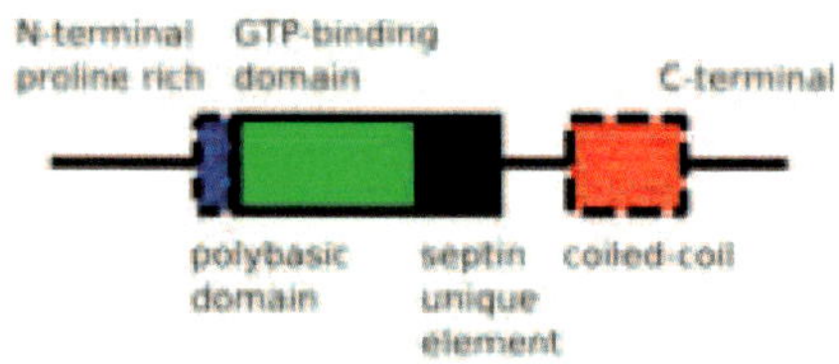

Schematic Domain Structure of Septin Polypeptide Chain

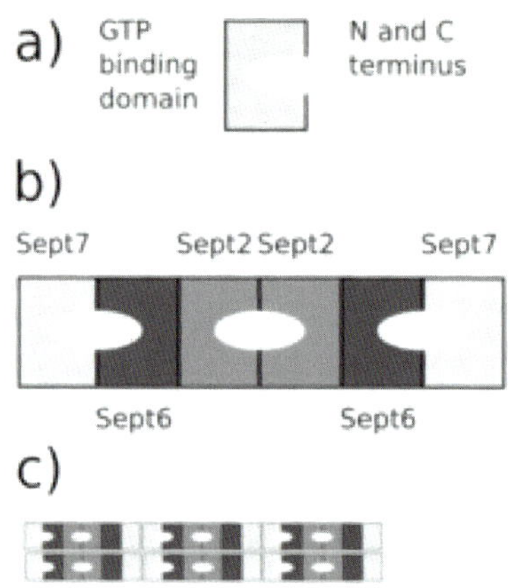

- schematic of septin molecule with GTP binding domain to one side and the N and C termini of the polypeptide chain to the other
- schematic of septin heterohexameric complex (of human septins), where different septins bind to each other via their GTP binding domains or via the N and C termini. Note the symmetry of the complex
- schematic how septin complexes could align to form septin filaments

Septins are P-Loop-NTPase proteins that range in weight from 30-65 kDa. Septins are highly conserved between different eukaryotic species. They are composed of a variable-length proline rich N-terminus with a basic phosphoinositide binding motif important for membrane association, a GTP-binding domain, a highly conserved Septin Unique Element domain, and a C-terminal extension including a coiled coil domain of varying length.

Septins interact either via their respective GTP-binding domains, or via both their N- and C-termini. Different organisms express a different number of septins, and from those symmetric oligomeres are formed. For example, in humans Sept7-Sept6-Sept2-Sept2-Sept6-Sept7 form one complex, and in yeast Cdc11-Cdc12-Cdc3-Cdc10-Cdc10-Cdc3-Cdc12-Cdc11 form another one. These complexes then associate to form non-polar filaments, filament bundles, cages or ring structures in cells.

Occurrence

Septins are found in fungi, animals and some green algae, but not in higher plants.

IN SACCHAROMYCES CEREVISIAE

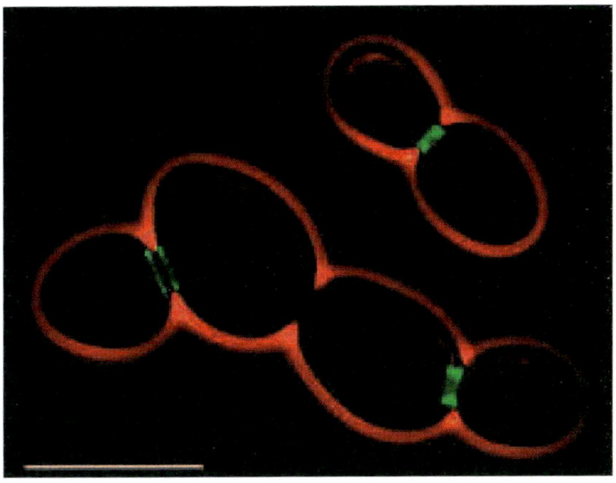

Septins in Saccharomyces cerevisiae (fluorescent micrograph)

- Green: septins (*AgSEP7-GFP*)
- Red: cell outline (phase contrast)
- Scale bar: 10 μm

There are seven different septins in Saccharomyces cerevisiae. Five of those are involved in mitosis, while two (Spr3 and Spr28) are specific to sporulation. Mitotic septins (Cdc3, Cdc10, Cdc11, Cdc12, Shs1) form a ring structure at the bud neck during cell division. They are involved in the selection of the bud-site, the positioning of the mitotic spindle, polarized growth, and cytokinesis. The sporulating septins (Spr3, Spr28) localize together with Cdc3 and Cdc11 to the edges of prospore membranes.

Organization

The septin cortex undergoes several changes throughout the cell cycle: The first visible septin structure is a distinct ring which appears ~15 min before bud emergence. After bud emergence, the ring broadens to assume the shape of an hourglass around the mother-bud neck. During cytokinesis, the septin cortex splits into a double ring which eventually disappears. How can the septin cortex undergo such dramatic changes, although some of its functions may require it to be a stable structure? FRAP analysis has revealed that the turnover of septins at the neck undergoes multiple changes during the cell cycle. The predominant, functional conformation is characterized by a low turnover rate (frozen state), during which the septins are phosphorylated. Structural changes require a destabilization of the septin cortex (fluid state) induced by dephosphorylation prior to bud emergence, ring splitting and cell separation.

The composition of the septin cortex does not only vary throughout the cell cycle but also along the mother-bud axis. This polarity of the septin network allows concentration of some proteins primarily to the mother side of the neck, some to the center and others to the bud site.

Functions

Scaffold

The septins act as a scaffold, recruiting many proteins. These protein complexes are involved in cytokinesis, chitin deposition, cell polarity, spore formation, in the morphogenesis checkpoint, spindle alignment checkpoint and bud site selection.

Cytokinesis

Budding yeast cytokinesis is driven through two septin dependent, redundant processes: recruitment and contraction of the actomyosin ring and formation of the septum by vesicle fusion with the plasma membrane. In contrast to septin mutants, disruption of one single pathway only leads to a delay in cytokinesis, not complete failure of cell division. Hence, the septins are predicted to act at the most upstream level of cytokinesis.

Cell polarity

After the isotropic-apical switch in budding yeast, cortical components, supposedly of the exocyst and polarisome, are delocalized from the apical pole to the entire plasma membrane of the bud, but not the mother cell. The septin ring at the neck serves as a cortical barrier that prevents membrane diffusion of these factors between the two compartments. This asymmetric distribution is abolished in septin mutants.

Some conditional septin mutants do not form buds at their normal axial location. Moreover, the typical localization of some bud-site-selection factors in a double ring at the neck is lost or disturbed in these mutants. This indicates that the septins may serve as anchoring site for such factors in axially budding cells.

In Filamentous Fungi

Since their discovery in *S. cerevisiae,* septin homologues have been found in other eukaryotic species, including filamentous fungi. Septins in filamentous fungi display a variety of different shapes within single cells, where they control aspects of filamentous morphology.

Candida albicans

The genome of *C. albicans* encodes homologues to all *S. cerevisiae* septins. Without Cdc3 and Cdc12 genes *Candida albicans* cannot proliferate, other septins affect morphology and chitin deposition, but are not essential. *Candida albicans* can display different morphologies of vegetative growth, which determines the appearance of septin structures. Newly forming hyphae form a septin ring at the base, Double rings form at sites of hyphal septation, and a septin cap forms at hyphal tips. Elongated septin-filaments encircle the spherical chlamydospores.

Double rings of septins at the septation site also bear growth polarity, with the growing tip ring disassembling, while the basal ring remaining intact.

Aspergillus nidulans

Five septins are found in *A. nidulans* (AnAspAp, AnAspBp, AnAspCp, AnAspDp, AnAspEp). AnAspBp forms single rings at septation sites that eventually split into double rings. Additionally, AnAspBp forms a ring at sites of branch emergence which broadens into a band as the branch grows. Like in *C. albicans,* double rings reflect polarity of the hypha. In the case of *Aspergillus nidulans* polarity is conveyed by disassembly of the more basal ring (the ring further away from the hyphal growth tip), leaving the apical ring intact, potentially as a growth guidance cue.

Ashbya Gossypii

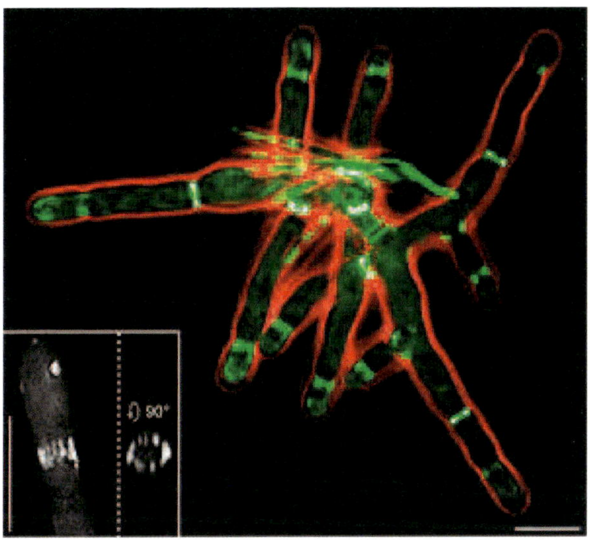

Septins in Ashbya gossypii (fluorescent micrograph) • Green: septins (AgSEP7-GFP)

- Red: cell outline (phase contrast)
- Inlay: 3D reconstruction of a discontinuous septin ring
- Scale bars: 10 µm

The *ascomycete A. gossypii* possesses homologues to all *S. cerevisiae* septins, with one being duplicated (*AgCDC3, AgCDC10, AgCDC11A, AgCDC11B, AgCDC12, AgSEP7*). *In vivo* studies of AgSep7p-GFP have revealed that septins assemble into discontinuous hyphal rings close to growing tips and sites of branch formation, and into asymmetric structures at the base of branching points. Rings are made of filaments which are long and diffuse close to growing tips and short and compact further away from the tip. During septum formation, the septin ring splits into two to form a double ring. *Agcdc3* , *Agcdc10* and *Agcdc12* deletion mutants display aberrant morphology and are defective for actin-ring formation, chitin-ring formation, and sporulation. Due to the lack of septa, septin deletion mutants are highly sensitive, and damage of a single hypha can result in complete lysis of a young mycelium.

In metazoa

In contrast to septins in yeast, and in contrast to other cytoskeletal components of metazoa, septins do not form a continuous network in metazoan cells, but several dispersed ones in the cortical cytoplasm. These are integrated with actin bundles and microtubules. For example, the actin bundling protein anillin is required for correct spatial control of septin organization.In the sperm cells of mammals, septins form a stable ring called annulus in the tail. In mice (and potentially in humans, too), defective annulus formation leads to male infertility.

Human

In humans, septins are involved in cytokinesis, cilium formation and neurogenesis through the capability to recruit other proteins or serve as a diffusion barrier. There are 13 different human genes coding for septins. The septin proteins produced by these genes are grouped into four subfamilies each named after its founding member:

- SEPT2 (SEPT1, SEPT4, SEPT5),
- SEPT3 (SEPT9, SEPT12),
- SEPT6 (SEPT8, SEPT10, SEPT11, SEPT14), and
- SEPT7. Septin protein complexes are assembled to form either hetero-hexamers (incorporating monomers selected from three different groups and the monomer from each group is present in two copies; 3 x 2 = 6) or hetero-octamers (monomers from four different groups, each monomer present in two copies; 4 x 2 = 8). These hetero-oligomers in turn form higher-order structures such as filaments and rings.

Septins form cage-like structures around bacterial pathogens, immobilizing harmful microbes and preventing them from invading healthy cells. This cellular defence system could potentially be exploited to create therapies for dysentery and other illnesses. For example, *Shigella* is a bacterium that causes lethal diarrhoea in humans. To propagate from cell to cell, *Shigella* bacteria develop actin-polymer 'tails', which propel the microbes and allow them to gain entry into neighbouring host cells. As part of the immune response, human cells produce a cell-signalling protein called TNF-α which trigger thick bundles of septin filaments to encircle the microbes within the infected host cell. Microbes that become trapped in these septin cages are broken down by autophagy. Disruptions in septins and mutations in the genes that code for them could be involved in causing leukaemia, colon cancer and neurodegenerative conditions such as Parkinson›s disease and Alzheimer›s disease. Potential therapies for these, as well as for bacterial conditions such as dysentery caused by *Shigella*, might bolster the body's immune system with drugs that mimic the behaviour of TNF-α and allow the septin cages to proliferate.

Caenorhabditis elegans

In the nematode worm *Caenorhabditis elegans* there are two genes coding for septins, and septin complexes contain the two different septins in a tetrameric UNC59-UNC61-UNC61-UNC59 complex. Septins in *C.elegans* concentrate at the cleavage furrow and the spindle midbody during cell division. Septins are also involved in cell migration and axon guidance in *C.elegans*.

In Mitochondria

The septin localized in the mitochondria is called mitochondrial septin (M-septin). It was identified as a CRMP/CRAM-interacting protein in developing mouse brain.

Yeast cytoskeleton

In budding yeast (an important model organism), actin forms cortical patches, actin cables, and a cytokinetic ring and the cap. Cortical patches are discrete actin bodies on the membrane and are important for endocytosis, especially the recycling of glucan synthase which is important for cell wall synthesis. Actin cables are bundles of actin filaments and are involved in the transport of vesicles towards the cap (which contains a number of different proteins to polarize cell growth) and in the positioning of mitochondria. The cytokinetic ring forms and constricts around the site of cell division.

PROKARYOTIC CYTOSKELETON

The cytoskeleton was once thought to be a feature only of eukaryotic cells, but homologues to all the major proteins of the eukaryotic cytoskeleton have been found in prokaryotes. Although the evolutionary relationships are so distant that they are not obvious from protein sequence comparisons alone, the similarity of their three-dimensional structures and similar functions in maintaining cell shape and polarity provides strong evidence that the eukaryotic and prokaryotic cytoskeletons are truly homologous. However, some structures in the bacterial cytoskeleton may not have been identified as of yet.

FtsZ

FtsZ was the first protein of the prokaryotic cytoskeleton to be identified. Like tubulin, FtsZ forms filaments in the presence of guanosine triphosphate (GTP), but these filaments do not group into tubules. During cell division, FtsZ is the first protein to move to the division site, and is essential for recruiting other proteins that synthesize the new cell wall between the dividing cells.

MreB and ParM

Prokaryotic actin-like proteins, such as MreB, are involved in the maintenance of cell shape. All non-spherical bacteria have genes encoding actin-like proteins, and these proteins form a helical network beneath the cell membrane that guides the proteins involved in cell wall biosynthesis.

Some plasmids encode a partitioning system that involves an actin-like protein ParM. Filaments of ParM exhibit dynamic instability, and may partition plasmid DNA into the dividing daughter cells by a mechanism analogous to that used by microtubules during eukaryotic mitosis.

Crescentin

The bacterium *Caulobacter crescentus* contains a third protein, crescentin, that is related to the intermediate filaments of eukaryotic cells. Crescentin is also involved in maintaining cell shape, such as helical and vibrioid forms of bacteria, but the mechanism by which it does this is currently unclear.

Common features and differences between prokaryotes and eukaryotes

By definition, the cytoskeleton is composed of proteins that can form longitudinal arrays (fibres) in all organisms. These filament forming proteins have been classified into 4 classes. Tubulin-like, actin-like, Walker A cytoskeletal ATPases (WACA-proteins), and intermediate filaments.

Tubulin-like proteins are tubulin in eukaryotes and FtsZ, TubZ, RepX in prokaryotes. Actin-like proteins are actin in eukaryotes and MreB, FtsA in prokaryotes. An example of a WACA-proteins, which are mostly found in prokaryotes, is MinD. Examples for intermediate filaments, which have almost exclusively been found in animals (i.e. eukaryotes) are the lamins, keratins, vimentin, neurofilaments, desmin.

Although tubulin-like proteins share some amino acid sequence similarity, their similarity in protein-fold and the similarity in the GTP binding site is more striking. The same holds true for the actin-like proteins and their structure and ATP binding domain.

Cytoskeletal proteins are usually associated with cell shape, DNA segregation and cell division in prokaryotes and eukaryotes. Which proteins fulfill which task is very different. For example, DNA segregation in all eukaryotes happens through use of tubulin, but in prokaryotes either WACA proteins, actin-like or tubulin-like proteins can be used. Cell division is mediated in eukaryotes by actin, but in prokaryotes usually by tubulin-like (often FtsZ-ring) proteins and sometimes (Crenarchaeota) ESCRT-III, which in eukaryotes still has a role in the last step of division.

REFERENCES

1. Hardin, Jeff; Bertoni, Gregory; Kleinsmith, Lewis J. (2015). Becker›s World of the Cell (8th ed.). New York: Pearson. pp. 422–446. ISBN 978013399939-6.
2. McKinley, Michael; Dean O›Loughlin, Valerie; Pennefather-O›Brien, Elizabeth; Harris, Ronald (2015). Human Anatomy (4th ed.). New York: McGraw Hill Education. p. 29. ISBN 0-07-352573-1.
3. Alberts, Bruce; et al. (2008). Molecular Biology of the Cell (5th ed.). New York: Garland Science. ISBN 978-0-8153-4105-5.
4. Wickstead B, Gull K (Aug 2011). «The evolution of the cytoskeleton». The Journal of Cell Biology. 194 *(4): 513–25. doi:10.1083/jcb.201102065. PMC 3160578* . PMID 21859859.
5. Fletcher DA, Mullins RD (Jan 2010). «Cell mechanics and the cytoskeleton». Nature. 463 *(7280): 485–92. Bibcode*:2010Natur.463..485F. doi:10.1038/nature08908. PMC 2851742 . PMID 20110992.

6. Herrmann H, Bär H, Kreplak L, Strelkov SV, Aebi U (Jul 2007). «Intermediate filaments: from cell architecture to nanomechanics». Nature Reviews Molecular Cell Biology. 8 *(7): 562–73.* doi:10.1038/nrm2197. PMID 17551517.

7. Geli MI, Riezman H (Apr 1998). «Endocytic internalization in yeast and animal cells: similar and different». Journal of Cell Science. 111 *(8): 1031–7. PMID 9512499.*

8. Peters RA. «The Harben Lectures, 1929. Reprinted in: Peters, R. A. (1963) Biochemical lesions and lethal synthesis, p. 216. Pergamon Press, Oxford.».

9. Frixione E (Jun 2000). «Recurring views on the structure and function of the cytoskeleton: a 300-year epic». Cell Motility and the Cytoskeleton. 46 *(2): 73–94.* doi:10.1002/1097-0169(200006)46:2<73::AID-CM1>3.0.CO;2-0. PMID 10891854.

10. Minton AP (Oct 1992). «Confinement as a determinant of macromolecular structure and reactivity». Biophysical Journal. 63 *(4): 1090–100. Bibcode*:1992BpJ....63.1090M. doi:10.1016/S0006-3495(92)81663-6. PMC 1262248 . PMID 1420928.

11. Doherty GJ, McMahon HT (2008). «Mediation, modulation, and consequences of membrane-cytoskeleton interactions». Annual Review of Biophysics. 37: 65–95. doi:10.1146/annurev.biophys.37.032807.125912. PMID 18573073.

12. Alberts, Bruce (2015). Molecular Biology of the Cell. Garland Science. p. 889. ISBN 978-0-8153-4464-3.

13. Gunning PW, Ghoshdastider U, Whitaker S, Popp D, Robinson RC (Jun 2015). «The evolution of compositionally and functionally distinct actin filaments». Journal of Cell Science. 128 *(11): 2009–2019. doi:10.1242/jcs.165563. PMID 25788699.*

14. Paulin D, Li Z (Nov 2004). «Desmin: a major intermediate filament protein essential for the structural integrity and function of muscle». Experimental Cell Research. 301 *(1): 1–7. doi:10.1016/j.yexcr.2004.08.004. PMID 15501438.*

15. Unless else specified in boxes, then ref is:Walter F., PhD. Boron (2003). Medical Physiology: A Cellular And Molecular Approaoch. Elsevier/Saunders. p. 1300. ISBN 1-4160-2328-3. Page 25

16. Fuchs E, Cleveland DW (Jan 1998). «A structural scaffolding of intermediate filaments in health and disease». Science. 279 *(5350): 514–9. Bibcode*:1998Sci...279..514F. doi:10.1126/science.279.5350.514. PMID 9438837.

17. Steinmetz MO (May 2007). «Structure and thermodynamics of the tubulin-stathmin interaction». Journal of Structural Biology. 158 *(2): 137–47. doi:10.1016/j.jsb.2006.07.018. PMID 17029844.*

18. Mostowy S, Cossart P (Mar 2012). «Septins: the fourth component of the cytoskeleton». Nature Reviews Molecular Cell Biology. 13 *(3): 183–94.* doi:10.1038/nrm3284. PMID 22314400.

19. Mascarelli A (December 2011). «Septin proteins take bacterial prisoners: A cellular defence against microbial pathogens holds therapeutic potential». Nature. doi:10.1038/nature.2011.9540.

20. Calpain proteolysis of alpha-II-spectrin in the normal adult human brain». Neurosci. Lett. 316 *(1): 41–4. December 2001.* doi:10.1016/S0304-3940(01)02371-0. PMID 11720774.

21. Pruyne D, Bretscher A (Feb 2000). «Polarization of cell growth in yeast». Journal of Cell Science. 113 *(4): 571–85. PMID 10652251.*

22. Shih YL, Rothfield L (Sep 2006). «The bacterial cytoskeleton». Microbiology and Molecular Biology Reviews. 70 *(3): 729–54. doi:10.1128/MMBR.00017-06. PMC 1594594* . PMID 16959967.

23. Michie KA, Löwe J (2006). «Dynamic filaments of the bacterial cytoskeleton» (PDF). Annual Review of Biochemistry. 75: 467–92. doi:10.1146/annurev.biochem.75.103004.142452. PMID 16756499.

24. Briegel A, Dias DP, Li Z, Jensen RB, Frangakis AS, Jensen GJ (Oct 2006). «Multiple large filament bundles observed in Caulobacter crescentus by electron cryotomography». Molecular Microbiology. 62 *(1): 5–14. doi:10.1111/j.1365-2958.2006.05355.x. PMID 16987173.*

25. Popp D, Narita A, Maeda K, Fujisawa T, Ghoshdastider U, Iwasa M, Maéda Y, Robinson RC (2010). «Filament structure, organization, and dynamics in MreB sheets». The Journal of Biological Chemistry. 285 *(21): 15858–65. doi:10.1074/jbc.M109.095901. PMC 2871453* . PMID 20223832.

26. Popp D, Narita A, Lee LJ, Ghoshdastider U, Xue B, Srinivasan R, Balasubramanian MK, Tanaka T, Robinson RC (2012). «Novel actin-like filament structure from Clostridium tetani». The Journal of Biological Chemistry. 287 *(25): 21121–9. doi:10.1074/jbc. M112.341016. PMC 3375535* . PMID 22514279.

27. Ausmees N, Kuhn JR, Jacobs-Wagner C (Dec 2003). «The bacterial cytoskeleton: an intermediate filament-like function in cell shape». Cell. 115 *(6): 705–13.* doi:10.1016/ S0092-8674(03)00935-8. PMID 14675535.

CHAPTER 4

Endomembrane System

Chapter Contents

The endomembrane system is a series of compartments that work together to package, label, and ship proteins and molecules. In your cells, the endomembrane system is made up of both the endoplasmic reticulum and the Golgi apparatus. These compartments are folds of membranes that form tubes and sacs in your cells.

INTRODUCTION

The **endomembrane system** is composed of the different membranes that are suspended in the cytoplasm within a eukaryotic cell. These membranes divide the cell into functional and structural compartments, or organelles. In eukaryotes the organelles of the endomembrane system include: the nuclear membrane, the endoplasmic reticulum, the Golgi apparatus, lysosomes, vesicles, endosomes and the cell membrane. The system is defined more accurately as the set of membranes that form a single functional and developmental unit, either being connected directly, or exchanging material through vesicle transport. Importantly, the endomembrane system does not include the membranes of chloroplasts or mitochondria, but might have evolved from the latter (see below: Evolution).

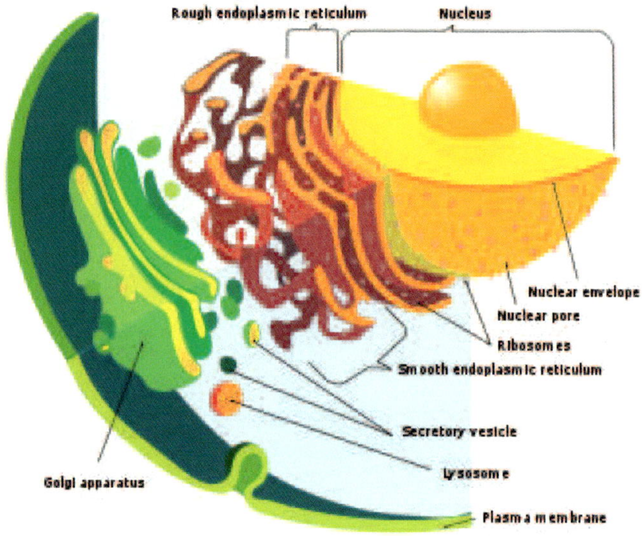

Detail of the endomembrane system and its components

The nuclear membrane contains two lipid bilayers that encompass the contents of the nucleus. The endoplasmic reticulum (ER) is a synthesis and transport organelle that branches into the cytoplasm in plant and animal cells. The Golgi apparatus is a series of multiple compartments where molecules are packaged for delivery to other cell components or for secretion from the cell. Vacuoles, which are found in both plant and animal cells (though much bigger in plant cells), are responsible for maintaining the shape and structure of the cell as well as storing waste products. A vesicle is a relatively small, membrane-enclosed sac that stores or transports substances. The cell membrane is a protective barrier that regulates what enters and leaves the cell. There is also an organelle known as the Spitzenkörper that is only found in fungi, and is connected with hyphal tip growth.

In prokaryotes endomembranes are rare, although in many photosynthetic bacteria the plasma membrane is highly folded and most of the cell cytoplasm is filled with layers of light-gathering membrane. These light-gathering membranes may even form enclosed structures called chlorosomes in green sulfur bacteria.

The organelles of the endomembrane system are related through direct contact or by the transfer of membrane segments as vesicles. Despite these relationships, the various membranes are not identical in structure and function. The thickness, molecular composition, and metabolic behavior of a membrane are not fixed, they may be modified several times during the membrane›s life. One unifying characteristic the membranes share is a lipid bilayer, with proteins attached to either side or traversing them.

HISTORY OF THE CONCEPT

Most lipids are synthesized in yeast either in the endoplasmic reticulum, lipid particles, or the mitochondrion, with little or no lipid synthesis occurring in the plasma membrane or nuclear membrane. Sphingolipid biosynthesis begins in the endoplasmic reticulum, but is completed in the Golgi apparatus. The situation is similar on mammals, with the exception of the first few steps in ether lipid biosynthesis, which occur in peroxisomes. The various membranes that enclose the other subcellular organelles must therefore be constructed by transfer of lipids from these sites of synthesis. However, although it is clear that lipid transport is a central process in organelle biogenesis, the mechanisms by which lipids are transported through cells remain poorly understood.

The first proposal that the membranes within cells form a single system that exchanges material between its components was by Morré and Mollenhauer in 1974. This proposal was made as a way of explaining how the various lipid membranes are assembled in the cell, with these membranes being assembled through *lipid flow* from the sites of lipid synthesis. The idea of lipid flow through a continuous system of membranes and vesicles was an alternative to the various membranes being independent entities that are formed from transport of free lipid components, such as fatty acids and sterols, through the cytosol. Importantly, the transport of lipids through the cytosol and lipid flow through a continuous endomembrane system are not mutually exclusive processes and both may occur in cells.

COMPONENTS OF THE SYSTEM

Nuclear Envelope

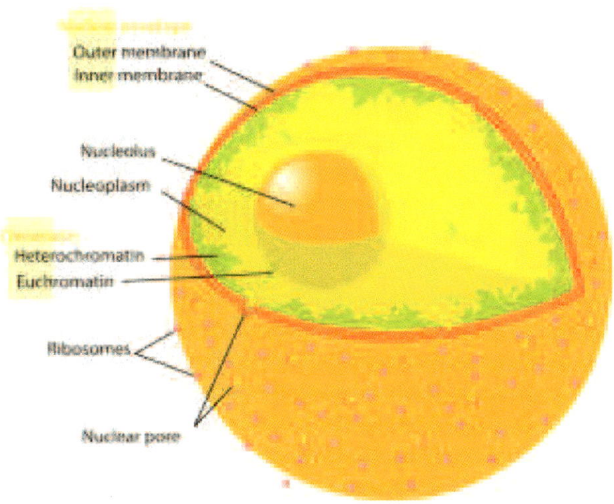

Diagram of the nucleus with the nuclear envelope shown as the orange portion.

The nuclear envelope surrounds the nucleus, separating its contents from the cytoplasm. It has two membranes, each a lipid bilayer with associated proteins. The outer nuclear membrane is continuous with the rough endoplasmic reticulum membrane, and like that structure, features ribosomes attached to the surface. The outer membrane is also continuous with the inner nuclear membrane since the two layers are fused together at numerous tiny holes called nuclear pores that perforate the nuclear envelope. These pores are about 120 nm in diameter and regulate the passage of molecules between the nucleus and cytoplasm, permitting some to pass through the membrane, but not others. Since the nuclear pores are located in an area of high traffic, they play an important role in the physiology of cells. The space between the outer and inner membranes is called the perinuclear space and is joined with the lumen of the rough ER.

The nuclear envelope›s structure is determined by a network of intermediate filaments (protein filaments). This network is organized into lining similar to mesh called the nuclear lamina, which binds to chromatin, integral membrane proteins, and other nuclear components along the inner surface of the nucleus. The nuclear lamina is thought to help materials inside the nucleus reach the nuclear pores and in the disintegration of the nuclear envelope during mitosis and its reassembly at the end of the process.

The nuclear pores are highly efficient at selectively allowing the passage of materials to and from the nucleus, because the nuclear envelope has a considerable amount of traffic. RNA and ribosomal subunits must be continually transferred from the nucleus to the cytoplasm. Histones, gene regulatory proteins, DNA and RNA polymerases, and other substances essential for nuclear activities must be imported from the cytoplasm. The nuclear envelope of a typical mammalian cell contains 3000–4000 pore complexes. If the cell is synthesizing DNA each pore complex needs to transport about 100 histone molecules per minute. If the cell is growing rapidly, each complex also needs to transport about 6 newly assembled large and small ribosomal subunits per minute from the nucleus to the cytosol, where they are used to synthesize proteins.

Endoplasmic Reticulum

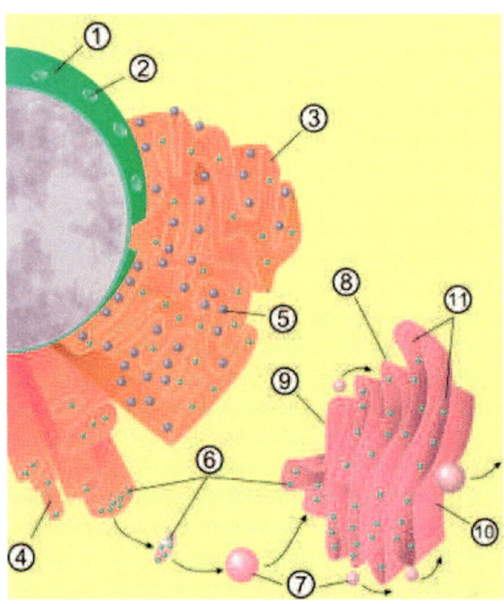

1 Nucleus **2** Nuclear pore **3** Rough endoplasmic reticulum (RER) **4** Smooth endoplasmic reticulum (SER) **5** Ribosome on the rough ER **6** Proteins that are transported **7** Transport vesicle **8** Golgi apparatus **9** Cis face of the Golgi apparatus **10** Trans face of the Golgi apparatus **11** Cisternae of the Golgi apparatus.

The endoplasmic reticulum (ER) is a membranous synthesis and transport organelle that is an extension of the nuclear envelope. More than half the total membrane in eukaryotic cells is accounted for by the ER. The ER is made up of flattened sacs and branching tubules that are thought to interconnect, so that the ER membrane forms a continuous sheet enclosing a single internal space. This highly convoluted space is called the ER lumen and is also referred to as the ER cisternal space. The lumen takes up about ten percent of the entire cell volume. The endoplasmic reticulum membrane allows molecules to be selectively transferred between the lumen and the cytoplasm, and since it is connected to the nuclear envelope, it provides a channel between the nucleus and the cytoplasm.

The ER has a central role in producing, processing, and transporting biochemical compounds for use inside and outside of the cell. Its membrane is the site of production of all the transmembrane proteins and lipids for most of the cell's organelles, including the ER itself, the Golgi apparatus, lysosomes, endosomes, mitochondria, peroxisomes, secretory vesicles, and the plasma membrane. Furthermore, almost all of the proteins that will exit the cell, plus those destined for the lumen of the ER, Golgi apparatus, or lysosomes, are originally delivered to the ER lumen. Consequently, many of the proteins found in the cisternal space of the endoplasmic reticulum lumen are there only temporarily as they pass on their way to other locations. Other proteins, however, constantly remain in the lumen and are known as endoplasmic reticulum resident proteins. These special proteins contain a specialized retention signal made up of a specific sequence of amino acids that enables them to be retained by the organelle. An example of an important endoplasmic reticulum resident protein is the chaperon protein known as BiP which identifies other proteins that have been improperly built or processed and keeps them from being sent to their final destinations.

The ER is involved in cotranslational sorting of proteins. A polypeptide which contains an ER signal sequence is recognised by a signal recognition protein which halts the production of the protein. the SRP transports the polypeptide to the ER membrane where its released in through a membrane pore and translation resumes.

There are two distinct, though connected, regions of ER that differ in structure and function: smooth ER and rough ER. The rough endoplasmic reticulum is so named because the cytoplasmic surface is covered with ribosomes, giving it a bumpy appearance when viewed through an electron microscope. The smooth ER appears smooth since its cytoplasmic surface lacks ribosomes.

Functions of the Smooth ER

In the great majority of cells, smooth ER regions are scarce and are often partly smooth and partly rough. They are sometimes called transitional ER because they contain ER exit sites from which transport vesicles carrying newly synthesized proteins and lipids bud off for transport to the Golgi apparatus. In certain specialized cells, however, the smooth ER is abundant and has additional functions. The smooth ER of these specialized cells functions in diverse metabolic processes, including synthesis of lipids, metabolism of carbohydrates, and detoxification of drugs and poisons.

Enzymes of the smooth ER are vital to the synthesis of lipids, including oils, phospholipids, and steroids. Sex hormones of vertebrates and the steroid hormones secreted by the adrenal glands are among the steroids produced by the smooth ER in animal cells. The cells that synthesize these hormones are rich in smooth ER.

Liver cells are another example of specialized cells that contain an abundance of smooth ER. These cells provide an example of the role of smooth ER in carbohydrate metabolism. Liver cells store carbohydrates in the form of glycogen. The breakdown of glycogen eventually leads to the release of glucose from the liver cells, which is important in the regulation of sugar concentration in the blood. However, the primary product of glycogen breakdown is glucose-1-phosphate. This is converted to glucose-6-phosphate and then an enzyme of the liver cell's smooth ER removes the phosphate from the glucose, so that it can then leave the cell.

Enzymes of the smooth ER can also help detoxify drugs and poisons. Detoxification usually involves the addition of a hydroxyl group to a drug, making the drug more soluble and thus easier to purge from the body. One extensively studied detoxification reaction is carried out by the cytochrome P450 family of enzymes, which catalyze water-insoluble drugs or metabolites that would otherwise accumulate to toxic levels in cell membrane.

Muscle cells have another specialized function of smooth ER. The ER membrane pumps calcium ions from the cytosol into the cisternal space. When a muscle cell becomes stimulated by a nerve impulse, calcium goes back across the ER membrane into the cytosol and generates the contraction of the muscle cell.

Functions of the Rough ER

Many types of cells export proteins produced by ribosomes attached to the rough ER. The ribosomes assemble amino acids into protein units, which are carried into the rough ER for further adjustments. These proteins may be either transmembrane proteins, which become embedded in the membrane of the endoplasmic reticulum, or water-soluble proteins, which are able to pass through the membrane into the lumen. Those that reach the inside of the endoplasmic reticulum are folded into the correct three-dimensional conformation. Chemicals, such as carbohydrates or sugars, are added, then the endoplasmic reticulum either transports the completed proteins, called secretory proteins, to areas of the cell where they are needed, or they are sent to the Golgi apparatus for further processing and modification.

Once secretory proteins are formed, the ER membrane separates them from the proteins that will remain in the cytosol. Secretory proteins depart from the ER enfolded in the membranes of vesicles that bud like bubbles from the transitional ER. These vesicles in transit to another part of the cell are called transport vesicles. An alternative mechanism for transport of lipids and proteins out of the ER are through lipid transfer proteins at regions called membrane contact sites where the ER becomes closely and stably associated with the membranes of other organelles, such as the plasma membrane, Golgi or lysosomes.

In addition to making secretory proteins, the rough ER makes membranes that grows in place from the addition of proteins and phospholipids. As polypeptides intended to be membrane proteins grow from the ribosomes, they are inserted into the ER membrane itself and are kept there by their hydrophobic portions. The rough ER also produces its own membrane phospholipids; enzymes built into the ER membrane assemble phospholipids. The ER membrane expands and can be transferred by transport vesicles to other components of the endomembrane system.

Golgi Apparatus

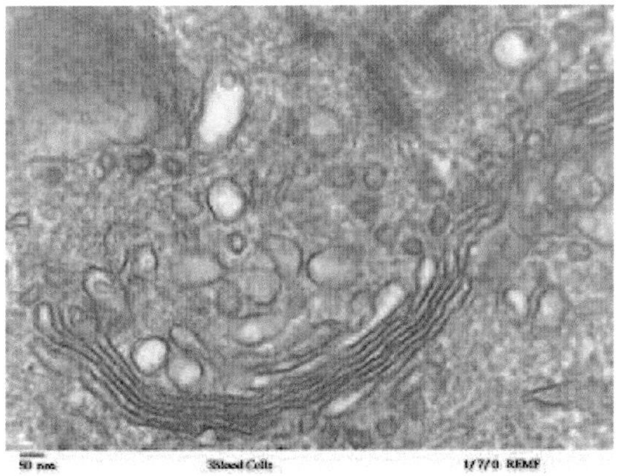

Micrograph of Golgi apparatus, visible as a stack of semicircular black rings near the bottom. Numerous circular vesicles can be seen in proximity to the organelle

The Golgi apparatus (also known as the Golgi body and the Golgi complex) is composed of interconnected sacs called cisternae. Its shape is similar to a stack of pancakes. The number of these stacks varies with the specific function of the cell. The Golgi apparatus is used by the cell for further protein modification. The section of the Golgi apparatus that receives the vesicles from the ER is known as the cis face, and is usually near the ER. The opposite end of the Golgi apparatus is called the trans face, this is where the modified compounds leave. The trans face is usually facing the plasma membrane, which is where most of the substances the Golgi apparatus modifies are sent.

Vesicles sent off by the ER containing proteins are further altered at the Golgi apparatus and then prepared for secretion from the cell or transport to other parts of the cell. Various things can happen to the proteins on their journey through the enzyme covered space of the Golgi apparatus. The modification and synthesis of the carbohydrate portions of glycoproteins is common in protein processing. The Golgi apparatus removes and substitutes sugar monomers, producing a large variety of oligosaccharides. In addition to modifying proteins, the Golgi also manufactures macromolecules itself. In plant cells, the Golgi produces pectins and other polysaccharides needed by the plant structure.

Once the modification process is completed, the Golgi apparatus sorts the products of its processing and sends them to various parts of the cell. Molecular identification labels or tags are added by the Golgi enzymes to help with this. After everything is organized, the Golgi apparatus sends off its products by budding vesicles from its trans face.

Vacuoles

Vacuoles, like vesicles, are membrane-bound sacs within the cell. They are larger than vesicles and their specific function varies. The operations of vacuoles are different for plant and animal vacuoles.

In plant cells, vacuoles cover anywhere from 30% to 90% of the total cell volume. Most mature plant cells contain one large central vacuole encompassed by a membrane called the tonoplast. Vacuoles of plant cells act as storage compartments for the nutrients and waste of a cell. The solution that these molecules are stored in is called the cell sap. Pigments that color the cell are sometime located in the cell sap. Vacuoles can also increase the size of the cell, which elongates as water is added, and they control the turgor pressure (the osmotic pressure that keeps the cell wall from caving in). Like lysosomes of animal cells, vacuoles have an acidic pH and contain hydrolytic enzymes. The pH of vacuoles enables them to perform homeostatic procedures in the cell. For example, when the pH in the cells environment drops, the H^+ ions surging into the cytosol can be transferred to a vacuole in order to keep the cytosol's pH constant.

In animals, vacuoles serve in exocytosis and endocytosis processes. Endocytosis refers to when substances are taken into the cell, whereas for exocytosis substances are moved from the cell into the extracellular space. Material to be taken-in is surrounded by the plasma membrane, and then transferred to a vacuole. There are two types of endocytosis, phagocytosis (cell eating) and pinocytosis (cell drinking). In phagocytosis, cells engulf large particles such as bacteria. Pinocytosis is the same process, except the substances being ingested are in the fluid form.

Vesicles

Vesicles are small membrane-enclosed transport units that can transfer molecules between different compartments. Most vesicles transfer the membranes assembled in the endoplasmic reticulum to the Golgi apparatus, and then from the Golgi apparatus to various locations.

There are various types of vesicles each with a different protein configuration. Most are formed from specific regions of membranes. When a vesicle buds off from a membrane it contains specific proteins on its cytosolic surface. Each membrane a vesicle travels to contains a marker on its cytosolic surface. This marker corresponds with the proteins on the vesicle traveling to the membrane. Once the vesicle finds the membrane, they fuse.

There are three well known types of vesicles. They are clathrin-coated, COPI-coated, and COPII-coated vesicles. Each performs different functions in the cell. For example, clathrin-coated vesicles transport substances between the Golgi apparatus and the plasma membrane. COPI- and COPII-coated vesicles are frequently used for transportation between the ER and the Golgi apparatus.

LYSOSOMES

A **lysosome** is a membrane-bound organelle found in nearly all animal cells. They are spherical vesicles which contain hydrolytic enzymes that can break down virtually all kinds of biomolecules. Simply stated, a lysosome is a type of vesicle with specific composition, of both its membrane proteins, and proteins of its lumen. The lumen›s pH (4.5 - 5.0) is optimal for the enzymes involved in hydrolysis, analogous to the activity of the stomach. Besides degradation of polymers, the lysosome is involved in various cell processes, including secretion, plasma membrane repair, cell signalling, and energy metabolism.

The lysosomes also act as the waste disposal system of the cell by digesting unwanted materials in the cytoplasm, both from outside of the cell and obsolete components inside the cell. Material from the outside of the cell is taken-up through endocytosis, while material from the inside of the cell is digested through autophagy. Their sizes can be very different—the biggest ones can be more than 10 times bigger than the smallest ones. They were discovered and named by Belgian biologist Christian de Duve, who eventually received the Nobel Prize in Physiology or Medicine in 1974.

Lysosomes are known to contain more than 60 different enzymes. Enzymes of the lysosomes are synthesised in the rough endoplasmic reticulum. The enzymes are imported from the Golgi apparatus in small vesicles, which fuse with larger acidic vesicles. Enzymes destined for a lysosome are specifically tagged with the molecule mannose 6-phosphate, so that they are properly sorted into acidified vesicles.

Synthesis of lysosomal enzymes is controlled by nuclear genes. Mutations in the genes for these enzymes are responsible for more than 30 different human genetic diseases, which are collectively known as lysosomal storage diseases. These diseases result from an accumulation of specific substrates, due to the inability to break them down. These genetic defects are related to several neurodegenerative disorders, cancer, cardiovascular diseases, and ageing-related diseases.

Discovery

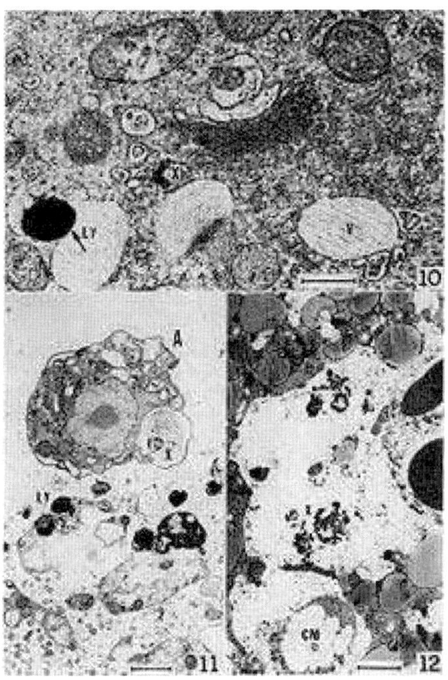

TEM views of various vesicular compartments. Lysosomes are denoted by "Ly". They are dyed dark due to their acidity; in the center of the top image, a Golgi Apparatus can be seen, distal from the cell membrane relative to the lysosomes.

Christian de Duve, then chairman of the Laboratory of Physiological Chemistry at the Catholic University of Louvain in Belgium, had been studying the mechanism of action of a pancreatic hormone insulin in liver cells. By 1949, he and his team had focused on the enzyme called glucose 6-phosphatase, which is the first crucial enzyme in sugar metabolism and the target of insulin. They already suspected that this enzyme played a key role in regulating blood sugar levels. However, even after a series of experiments, they failed to purify and isolate the enzyme from the cellular extracts. Therefore, they tried a more arduous procedure of cell fractionation, by which cellular components are separated based on their sizes using centrifugation.

They succeeded in detecting the enzyme activity from the microsomal fraction. This was the crucial step in the serendipitous discovery of lysosomes. To estimate this enzyme activity, they used that of standardised enzyme acid phosphatase, and found that the activity was only 10% of the expected value. One day, the enzyme activity of purified cell fractions which had been refrigerated for five days was measured. Surprisingly, the enzyme activity was increased to normal of that of the fresh sample. The result was the same no matter how many times they repeated the estimation, and led to the conclusion that a membrane-like barrier limited the accessibility of the enzyme to its substrate, and that the enzymes were able to diffuse after a few days (and react with their substrate). They described this membrane-like barrier as a saclike structure surrounded by a membrane and containing acid phosphatase.

It became clear that this enzyme from the cell fraction came from a membranous fractions, which were definitely cell organelles, and in 1955 De Duve named them lysosomes to reflect their digestive properties. The same year, Alex B. Novikoff from the University of Vermont visited de Duve′s laboratory, and successfully obtained the first electron micrographs of the new organelle. Using a staining method for acid phosphatase, de Duve and Novikoff confirmed the location of the hydrolytic enzymes of lysosomes using light and electron microscopic studies. de Duve won the Nobel Prize in Physiology or Medicine in 1974 for this discovery.

Originally, De Duve had termed the organelles the suicide bags or suicide sacs of the cells, for their hypothesized role in apoptosis. However, it has since been concluded that they only play a minor role in cell death.

Function and Structure

Lysosomes contain a variety of enzymes, enabling the cell to break down various biomolecules it engulfs, including peptides, nucleic acids, carbohydrates, and lipids. The enzymes responsible for this hydrolysis require an acidic environment for optimal activity.

In addition to being able to break down polymers, lysosomes are capable of fusing with other organelles & digesting large structures or cellular debris; through cooperation with phagosomes, they are able to conduct autophagy, clearing out damaged structures. Similarly, they are able to break-down virus particles or bacteria in phagocytosis of macrophages.

The size of lysosomes varies from 0.1 μm to 1.2 μm. With a pH ranging from 4.5 to 5.0, the interior of the lysosomes is acidic compared to the slightly basic cytosol (pH 7.2). The lysosomal membrane protects the cytosol, and therefore the rest of the cell, from the degradative enzymes within the lysosome. The cell is additionally protected from any lysosomal acid hydrolases that drain into the cytosol, as these enzymes are pH-sensitive and do not function well or at all in the alkaline environment of the cytosol. This ensures that cytosolic molecules and organelles are not destroyed in case there is leakage of the hydrolytic enzymes from the lysosome.

The lysosome maintains its pH differential by pumping in protons (H^+ ions) from the cytosol across the membrane via proton pumps and chloride ion channels. Vacuolar H+-ATPases are responsible for transport of protons, while the counter transport of chloride ions is performed by ClC-7 Cl^-/H^+ antiporter. In this way a steady acidic environment is maintained.

It sources its versatile capacity for degradation by import of enzymes with specificity for different substrates; cathepsins are the major class of hydrolytic enzymes, while lysosomal alpha-glucosidase (GAA) is responsible for carbohydrates, and ACP_2 is necessary to release phosphate groups of phospholipids.

Formation

The Lysosome is shown in purple, as an endpoint in Endocytotic sorting. AP2 is necessary for vesicle formation, whereas the Mannose-6-receptor is necessary for sorting Hydrolase into the Lysosome's lumen.

Many components of animal cells are recycled by transferring them inside or embedded in sections of membrane. For instance, in endocytosis (more specifically, macropinocytosis), a portion of the cell's plasma membrane pinches off to form a vesicle that will eventually fuse with an organelle within the cell. Without active replenishment, the plasma membrane would continuously decrease in size. It is thought that lysosomes participate in this dynamic membrane exchange system and are formed by a gradual maturation process from endosomes.

The production of lysosomal proteins suggests one method of lysosome sustainment. Lysosomal protein genes are transcribed in the nucleus. mRNA transcripts exit the nucleus into the cytosol, where they are translated by ribosomes. The nascent peptide chains are translocated into the rough endoplasmic reticulum, where they are modified. Upon exiting the endoplasmic reticulum and entering the Golgi apparatus via vesicular transport, a specific lysosomal tag, mannose 6-phosphate, is added to the peptides. The presence of these tags allow for binding to mannose 6-phosphate receptors in the Golgi apparatus, a phenomenon that is crucial for proper packaging into vesicles destined for the lysosomal system.

Upon leaving the Golgi apparatus, the lysosomal enzyme-filled vesicle fuses with a late endosome, a relatively acidic organelle with an approximate pH of 5.5. This acidic environment causes dissociation of the lysosomal enzymes from the mannose 6-phosphate receptors. The enzymes are packed into vesicles for further transport to established lysosomes. The late endosome itself can eventually grow into a mature lysosome, as evidenced by the transport of endosomal membrane components from the lysosomes back to the endosomes.

Pathogen Entry

Cholera gaining entry into a cell via endocytosis.

As the endpoint of endocytosis, the lysosome also acts as a safeguard in preventing pathogens from being able to reach the cytoplasm before being degraded. Pathogens often hijack endocytotic pathways such as pinocytosis in order to gain entry into the cell. The lysosome prevents easy entry into the cell by hydrolyzing the biomolecules of pathogens necessary for their replication strategies; reduced Lysosomal activity results in an increase in viral infectivity, including HIV. In addition, AB_5 toxins such as cholera hijack the endosomal pathway while evading lysosomal degradation.

Disease

Lysosomes are responsible for a group of genetically inherited disorders called lysosomal storage diseases (LSD). They are a type of inborn errors of metabolism caused by malfunction of one of the enzymes. The rate of incidence is estimated to be 1 in 5,000 live births, and the true figure expected to be higher as many cases are likely to be undiagnosed or misdiagnosed. The primary cause is deficiency of an acidic hydrolase (a hydrolase which functions best in acidic environments). Other conditions are due to defects in lysosomal membrane proteins that fail to transport the enzyme, non-enzymatic soluble lysosomal proteins. The initial effect of such disorders is accumulation of specific macromolecules or monomeric compounds inside the endosomal–autophagic–lysosomal system. This results in abnormal signaling pathways, calcium homeostasis, lipid biosynthesis and degradation and intracellular trafficking, ultimately leading to pathogenetic disorders. The organs most affected are brain, viscera, bone and cartilage.

There is no direct medical treatment to cure LSDs. The most common LSD is Gaucher›s disease, which is due to deficiency of the enzyme glucocerebrosidase. Consequently, the enzyme substrate, the fatty acid glucosylceramide accumulates, particularly in white blood cells, which in turn affects spleen, liver, kidneys, lungs, brain and bone marrow. The disease is characterized by bruises, fatigue, anaemia, low blood platelets, osteoporosis, and enlargement of the liver and spleen.

Metachromatic leukodystrophy is another lysosomal storage disease that also affects sphingolipid metabolism.

Lysosomotropism

Weak bases with lipophilic properties accumulate in acidic intracellular compartments like lysosomes. While the plasma and lysosomal membranes are permeable for neutral and uncharged species of weak bases, the charged protonated species of weak bases do not permeate biomembranes and accumulate within lysosomes. The concentration within lysosomes may reach levels 100 to 1000 fold higher than extracellular concentrations. This phenomenon is called "lysosomotropism" or "acid trapping". The amount of accumulation of lysosomotropic compounds may be estimated using a cell-based mathematical model.

A significant part of the clinically approved drugs are lipophilic weak bases with lysosomotropic properties. This explains a number of pharmacological properties of these drugs, such as high tissue-to-blood concentration gradients or long tissue elimination half-lifes; these properties have been found for drugs such as haloperidol, levomepromazine, and amantadine. However, high tissue concentrations and long elimination half-lives are explained also by lipophilicity and absorption of drugs to fatty tissue structures. Important lysosomal enzymes, such as acid sphingomyelinase, may be inhibited by lysosomally accumulated drugs. Such compounds are termed FIASMAs (functional inhibitor of acid sphingomyelinase) and include for example fluoxetine, sertraline, or amitriptyline.

Ambroxol is a lysosomotropic drug of clinical use to treat conditions of productive cough for its mucolytic action. Ambroxol triggers the exocytosis of lysosomes via neutralization of lysosomal pH and calcium release from acidic calcium stores. Presumably for this reason, Ambroxol was also found to improve cellular function in some disease of lysosomal origin such as Parkinson›s or lysosomal storage disease.

Controversy in Botany

By scientific convention, the term lysosome is applied to those vesicular organelles only in animals, and vacuoles to plants, fungi and algae. Discoveries in plant cells since the 1970s started to challenge this definition. Plant vacuoles are found to be much more diverse in structure and function than previously thought. Some vacuoles contain their own hydrolytic enzymes and perform the classic lysosomal activity, which is autophagy. These vacuoles are therefore seen as fulfilling the role of the animal lysosome. Based on de Duve's description that "only when considered as part of a system involved directly or indirectly in intracellular digestion does the term lysosome describe a physiological unit", some botanists strongly argued that these vacuoles are lysosomes. However, this is not universally accepted as the vacuoles are strictly not similar to lysosomes, such as in their specific enzymes and lack of phagocytic functions. Vacuoles do not have catabolic activity and do not undergo exocytosis as lysosomes do.

Etymology and Pronunciation

The word *lysosome* is New Latin that uses the combining forms *lyso-* (referring to lysis and derived from the Latin *lysis*, meaning "to loosen", via Ancient Greek λύσις [lúsis]), and *-some*, from *soma*, "body", yielding "body that lyses" or "lytic body". The adjectival form is *lysosomal*. The forms **lyosome* and **lyosomal* are much rarer; they use the *lyo-* form of the prefix but are often treated by readers and editors as mere unthinking replications of typos, which has no doubt been true as often as not.

Spitzenkörper

The spitzenkörper is a component of the endomembrane system found only in fungi, and is associated with hyphal tip growth. It is a phase-dark body that is composed of an aggregation of membrane-bound vesicles containing cell wall components, serving as a point of assemblage and release of such components intermediate between the Golgi and the cell membrane. The spitzenkörper is motile and generates new hyphal tip growth as it moves forward.

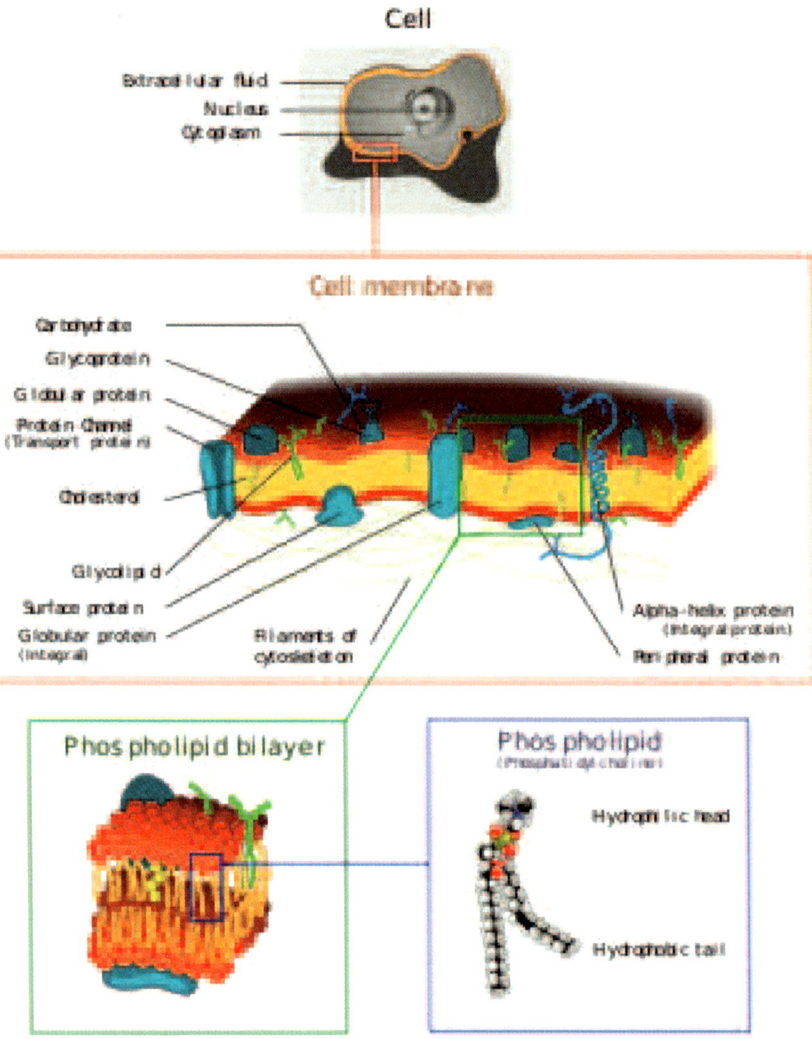

Detailed illustration of the plasma membrane. Including the structure of a phospholipid.

Plasma Membrane

The plasma membrane is a phospholipid bilayer membrane that separates the cell from its environment and regulates the transport of molecules and signals into and out of the cell. Embedded in the membrane are proteins that perform the functions of the plasma membrane. The plasma membrane is not a fixed or rigid structure, the molecules that compose the membrane are capable of lateral movement. This movement and the multiple components of the membrane are why it is referred to as a fluid mosaic. Smaller molecules such as carbon dioxide, water, and oxygen can pass through the plasma membrane freely by diffusion or osmosis. Larger molecules needed by the cell are assisted by proteins through active transport.

The plasma membrane of a cell has multiple functions. These include transporting nutrients into the cell, allowing waste to leave, preventing materials from entering the cell, averting needed materials from leaving the cell, maintaining the pH of the cytosol, and preserving the osmotic pressure of the cytosol. Transport proteins which allow some materials to pass through but not others are used for these functions. These proteins use ATP hydrolysis to pump materials against their concentration gradients.

In addition to these universal functions, the plasma membrane has a more specific role in multicellular organisms. Glycoproteins on the membrane assist the cell in recognizing other cells, in order to exchange metabolites and form tissues. Other proteins on the plasma membrane allow attachment to the cytoskeleton and extracellular matrix; a function that maintains cell shape and fixes the location of membrane proteins. Enzymes that catalyze reactions are also found on the plasma membrane. Receptor proteins on the membrane have a shape that matches with a chemical messenger, resulting in various cellular responses.

Evolution

The origin of the endomembrane system is linked to the origin of eukaryotes themselves and the origin of eukaryoties to the endosymbiotic origin of mitochondria. Many models have been put forward to explain the origin of the endomembrane system (reviewed in). The most recent concept suggests that the endomembrane system evolved from outer membrane vesicles the endosymbiotic mitochondrion secreted. This OMV-based model for the origin of the endomembrane system is currently the one that requires the least amount of novel inventions at eukaryote origin and explains the many connections of mitochondria with other compartments of the cell.

PEROXISOME

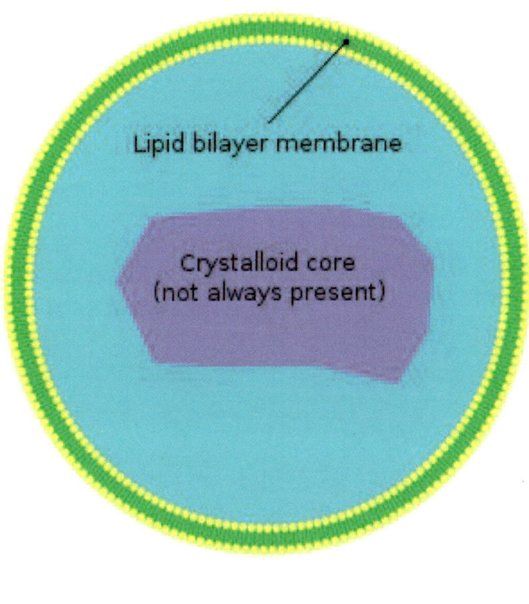

Basic structure of a peroxisome

Distribution of peroxisomes (white) in HEK 293 cells during mitosis.

Peroxisome in rat neonatal cardiomyocyte

Peroxisomes IPA: (also called **microbodies**) are organelles found in virtually all eukaryotic cells. They are involved in catabolism of very long chain fatty acids, branched chain fatty acids, D-amino acids, and polyamines, reduction of reactive oxygen species – specifically hydrogen peroxide – and biosynthesis of plasmalogens, i.e. ether phospholipids critical for the normal function of mammalian brains and lungs. They also contain approximately 10% of the total activity of two enzymes in the pentose phosphate pathway, which is important for energy metabolism. It is vigorously debated if peroxisomes are involved in isoprenoid and cholesterol synthesis in animals. Other known peroxisomal functions include the glyoxylate cycle in germinating seeds ("glyoxysomes), photorespiration in leaves, glycolysis in trypanosomes ("glycosomes), and methanol and/or amine oxidation and assimilation in some yeasts.

Peroxisomes were identified as organelles by the Belgian cytologist Christian de Duve in 1967 after they had been first described by a Swedish doctoral student, J. Rhodin in 1954.

Metabolic Functions

A major function of the peroxisome is the breakdown of very long chain fatty acids through beta-oxidation. In animal cells, the long fatty acids are converted to medium chain fatty acids, which are subsequently shuttled to mitochondria where they are eventually broken down to carbon dioxide and water. In yeast and plant cells, this process is carried out exclusively in peroxisomes.

The first reactions in the formation of plasmalogen in animal cells also occur in peroxisomes. Plasmalogen is the most abundant phospholipid in myelin. Deficiency of plasmalogens causes profound abnormalities in the myelination of nerve cells, which is one reason why many peroxisomal disorders affect the nervous system. Peroxisomes also play a role in the production of bile acids important for the absorption of fats and fat-soluble vitamins, such as vitamins A and K. Skin disorders are features of genetic disorders affecting peroxisome function as a result.

Peroxisomes contain oxidative enzymes, such as catalase, D-amino acid oxidase, and uric acid oxidase.However the last enzyme is absent in humans, explaining the disease known as gout, caused by the accumulation of uric acid. Certain enzymes within the peroxisome, by using molecular oxygen, remove hydrogen atoms from specific organic substrates (labeled as R), in an oxidative reaction, producing hydrogen peroxide (H_2O_2, itself toxic):

$$RH_2 + O_2 \rightarrow R + H_2O_2$$

Catalase, another peroxisomal enzyme, uses this H_2O_2 to oxidize other substrates, including phenols, formic acid, formaldehyde, and alcohol, by means of the peroxidation reaction:

$H_2O_2 + R'H_2 \rightarrow R' + 2H_2O$, thus eliminating the poisonous hydrogen peroxide in the process.

This reaction is important in liver and kidney cells, where the peroxisomes detoxify various toxic substances that enter the blood. About 25% of the ethanol alcohol humans drink is oxidized to acetaldehyde in this way. In addition, when excess H_2O_2 accumulates in the cell, catalase converts it to H_2O through this reaction:

$$2H_2O_2 \rightarrow 2H_2O + O_2$$

In higher plants, peroxisomes contain also a complex battery of antioxidative enzymes such as superoxide . dismutase, the components of the ascorbate-glutathione cycle, and the NADP-dehydrogenases of the pentose-phosphate pathway. It has been demonstrated that peroxisomes generate superoxide ($O_2^{\bullet-}$) and nitric oxide ($^{\bullet}NO$) radicals.

The peroxisome of plant cells is polarised when fighting fungal penetration. Infection causes a glucosinolate molecule to play an antifungal role to be made and delivered to the outside of the cell through the action of the peroxisomal proteins (PEN2 and PEN3).

Peroxisome Assembly

Peroxisomes can be derived from the endoplasmic reticulum and replicate by fission. Peroxisome matrix proteins are translated in the cytoplasm prior to import. Specific amino acid sequences (PTS or peroxisomal targeting signal) at the *C-terminus* (PTS1) or *N-terminus* (PTS2) of peroxisomal matrix proteins signals them to be imported into the organelle. There are at least 32 known peroxisomal proteins, called peroxins,which participate in the process of peroxisome assembly. Proteins do not have to unfold to be imported into the peroxisome. The protein receptors, the peroxins *PEX5* and *PEX7*, accompany their cargoes (containing a PTS1 or a PTS2 amino acid sequence, respectively) all the way into the peroxisome where they release the cargo and then return to the cytosol – a step named *recycling*. A model describing the import cycle is referred to as the *extended shuttle mechanism*. There is now evidence that ATP hydrolysis is required for the recycling of receptors to the cytosol. Also, ubiquitination appears to be crucial for the export of PEX5 from the peroxisome, to the cytosol.

Associated Medical Conditions

Peroxisomal disorders are a class of medical conditions that typically affect the human nervous system as well as many other organ systems. Two common examples are X-linked adrenoleukodystrophy and peroxisome biogenesis disorders.

Genes

PEX genes encode the protein machinery ("peroxins") required for proper peroxisome assembly, as described above. Membrane assembly and maintenance requires three of these (peroxins 3, 16, and 19) and may occur without the import of the matrix (lumen) enzymes. Proliferation of the organelle is regulated by Pex11p.

Genes that encode peroxin proteins include: PEX1, PEX2 – PXMP3, PEX3, PEX5, PEX6, PEX7, PEX10, PEX11A, PEX11B, PEX11G, PEX12, PEX13, PEX14, PEX16, PEX19, PEX26, PEX28, PEX30, and PEX31.

Evolutionary Origins

The protein content of peroxisomes varies across species, but the presence of proteins common to many species has been used to suggest an endosymbiotic origin; that is, peroxisomes evolved from bacteria that invaded larger cells as parasites, and very gradually evolved a symbiotic relationship. However, this view has been challenged by recent discoveries. For example, peroxisome-less mutants can restore peroxisomes upon introduction of the wild-type gene.

Two independent evolutionary analyses of the peroxisomal proteome found homologies between the peroxisomal import machinery and the ERAD pathway in the endoplasmic reticulum, along with a number of metabolic enzymes that were likely recruited from the mitochondria. Recently, it has been suggested that the peroxisome may have had an actinobacterial origin, however, this is controversial.

Other related organelles

Other organelles of the microbody family related to peroxisomes include glyoxysomes of plants and filamentous fungi, glycosomes of kinetoplastids and Woronin bodies of filamentous fungi.

MITOCHONDRION

The **mitochondrion** (plural **mitochondria**) is a double membrane-bound organelle found in all eukaryotic organisms, although some cells in some organisms may lack them (e.g. red blood cells). A number of organisms have reduced or transformed their mitochondria into other structures. To date, only one eukaryote, *Monocercomonoides*, is known to have completely lost its mitochondria. The word mitochondrion comes from the Greek μίτος, *mitos*, i.e. "thread", and *chondrion*, i.e. "granule" or "grain-like". Mitochondria generate most of the cell's supply of adenosine triphosphate (ATP), used as a source of chemical energy.

Mitochondria are commonly between 0.75 and 3μm in diameter but vary considerably in size and structure. Unless specifically stained, they are not visible. In addition to supplying cellular energy, mitochondria are involved in other tasks, such as signaling, cellular differentiation, and cell death, as well as maintaining control of the cell cycle and cell growth. Mitochondrial biogenesis is in turn temporally coordinated with these cellular processes. Mitochondria have been implicated in several human diseases, including mitochondrial disorders, cardiac dysfunction, heart failure and autism.

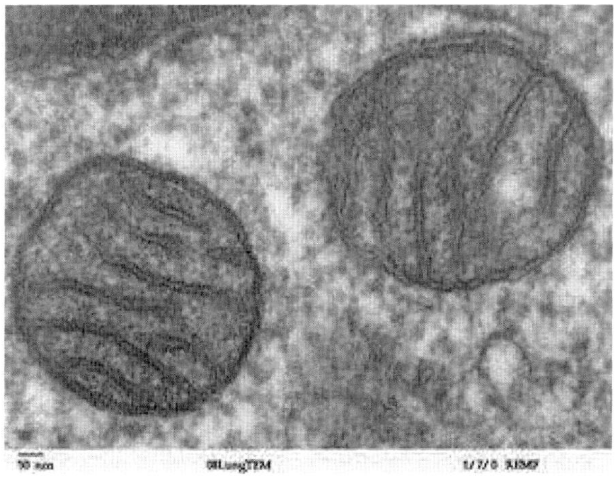

Two mitochondria from mammalian lung tissue displaying their matrix and membranes as shown by electron microscopy

The number of mitochondria in a cell can vary widely by organism, tissue, and cell type. For instance, red blood cells have no mitochondria, whereas liver cells can have more than 2000. The organelle is composed of compartments that carry out specialized functions. These compartments or regions include the outer membrane, the intermembrane space, the inner membrane, and the cristae and matrix. Mitochondrial proteins vary depending on the tissue and the species. In humans, 615 distinct types of protein have been identified from cardiac mitochondria, whereas in rats, 940 proteins have been reported. The mitochondrial proteome is thought to be dynamically regulated. Although most of a cell's DNA is contained in the cell nucleus, the mitochondrion has its own independent genome that shows substantial similarity to bacterial genomes.

The first observations of intracellular structures that probably represented mitochondria were published in the 1840s. Richard Altmann, in 1890, established them as cell organelles and called them bioblasts. The term mitochondria was coined by Carl Benda in 1898.Leonor Michaelis discovered that Janus green can be used as a supravital stain for mitochondria in 1900. In 1904, Friedrich Meves, made the first recorded observation of mitochondria in plants in cells of the white waterlily, *Nymphaea alba* and in 1908, along with Claudius Regaud, suggested that they contain proteins and lipids. Benjamin F. Kingsbury, in 1912, first related them with cell respiration, but almost exclusively based on morphological observations. In 1913, particles from extracts of guinea-pig liver were linked to respiration by Otto Heinrich Warburg, which he called grana. Warburg and Heinrich Otto Wieland, who had also postulated a similar particle mechanism, disagreed on the chemical nature of the respiration. It was not until 1925, when David Keilin discovered cytochromes, that the respiratory chain was described.

In 1939, experiments using minced muscle cells demonstrated that cellular respiration using one oxygen atom can form two adenosine triphosphate (ATP) molecules, and, in 1941, the concept of the phosphate bonds of ATP being a form of energy in cellular metabolism was developed by Fritz Albert Lipmann. In the following years, the mechanism behind cellular respiration was further elaborated, although its link to the mitochondria was not known. The introduction of tissue fractionation by Albert Claude allowed mitochondria to be isolated from other cell fractions and biochemical analysis to be conducted on them alone. In 1946, he concluded that cytochrome oxidase and other enzymes responsible for the respiratory chain were isolated to the mitchondria. Eugene Kennedy and Albert Lehninger discovered in 1948 that mitochondria are the site of oxidative phosphorylation in eukaryotes. Over time, the fractionation method was further developed, improving the quality of the mitochondria isolated, and other elements of cell respiration were determined to occur in the mitochondria.

The first high-resolution electron micrographs appeared in 1952, replacing the Janus Green stains as the preferred way of visualising the mitochondria. This led to a more detailed analysis of the structure of the mitochondria, including confirmation that they were surrounded by a membrane. It also showed a second membrane inside the mitochondria that folded up in ridges dividing up the inner chamber and that the size and shape of the mitochondria varied from cell to cell.

The popular term powerhouse of the cell was coined by Philip Siekevitz in 1957.

In 1967, it was discovered that mitochondria contained ribosomes. In 1968, methods were developed for mapping the mitochondrial genes, with the genetic and physical map of yeast mitochondrial DNA being completed in 1976.

ORIGIN AND EVOLUTION

There are two hypotheses about the origin of mitochondria: endosymbiotic and autogenous. The endosymbiotic hypothesis suggests that mitochondria were originally prokaryotic cells, capable of implementing oxidative mechanisms that were not possible for eukaryotic cells; they became endosymbionts living inside the eukaryote. In the autogenous hypothesis, mitochondria were born by splitting off a portion of DNA from the nucleus of the eukaryotic cell at the time of divergence with the prokaryotes; this DNA portion would have been enclosed by membranes, which could not be crossed by proteins. Since mitochondria have many features in common with bacteria, the endosymbiotic hypothesis is more widely accepted.

A mitochondrion contains DNA, which is organized as several copies of a single, circular chromosome. This mitochondrial chromosome contains genes for redox proteins, such as those of the respiratory chain. The CoRR hypothesis proposes that this co-location is required for redox regulation. The mitochondrial genome codes for some RNAs of ribosomes, and the 22 tRNAs necessary for the translation of messenger RNAs into protein. The circular structure is also found in prokaryotes. The proto-mitochondrion was probably closely related to the *Rickettsia*. However, the exact relationship of the ancestor of mitochondria to the alphaproteobacteria and whether the mitochondrion was formed at the same time or after the nucleus, remains controversial.

A recent study by researchers of the University of Hawaii at Manoa and the Oregon State University indicates that the SAR11 clade of bacteria shares a relatively recent common ancestor with the mitochondria existing in most eukaryotic cells.

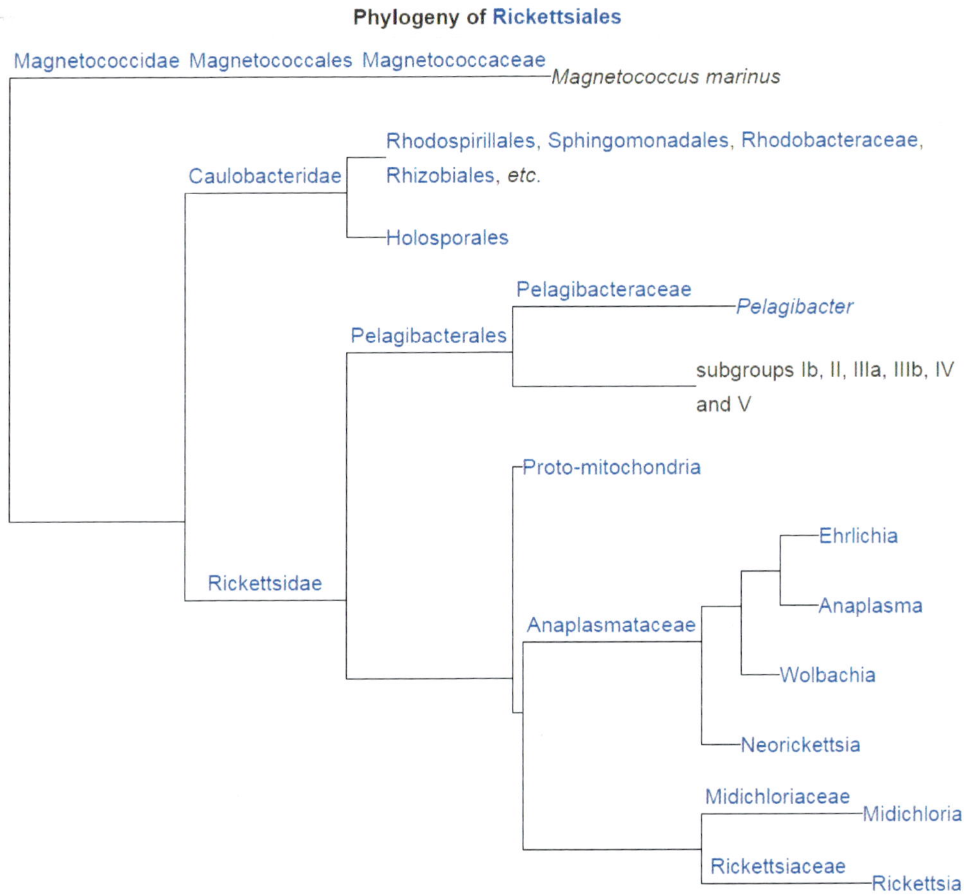

The ribosomes coded for by the mitochondrial DNA are similar to those from bacteria in size and structure. They closely resemble the bacterial 70S ribosome and not the 80S cytoplasmic ribosomes, which are coded for by nuclear DNA.

The endosymbiotic relationship of mitochondria with their host cells was popularized by Lynn Margulis. The endosymbiotic hypothesis suggests that mitochondria descended from bacteria that somehow survived endocytosis by another cell, and became incorporated into the cytoplasm. The ability of these bacteria to conduct respiration in host cells that had relied on glycolysis and fermentation would have provided a considerable evolutionary advantage. This symbiotic relationship probably developed 1.7 to 2 billion years ago.

A few groups of unicellular eukaryotes have only vestigial mitochondria or derived structures: the microsporidians, metamonads, and archamoebae. These groups appear as the most primitive eukaryotes on phylogenetic trees constructed using rRNA information, which once suggested that they appeared before the origin of mitochondria. However, this is now known to be an artifact of long-branch attraction—they are derived groups and retain genes or organelles derived from mitochondria (e.g., mitosomes and hydrogenosomes).

Monocercomonoides appear to have lost their mitochondria completely and at least some of the mitochondrial functions seem to be carried out by cytoplasmic proteins now.

Structure

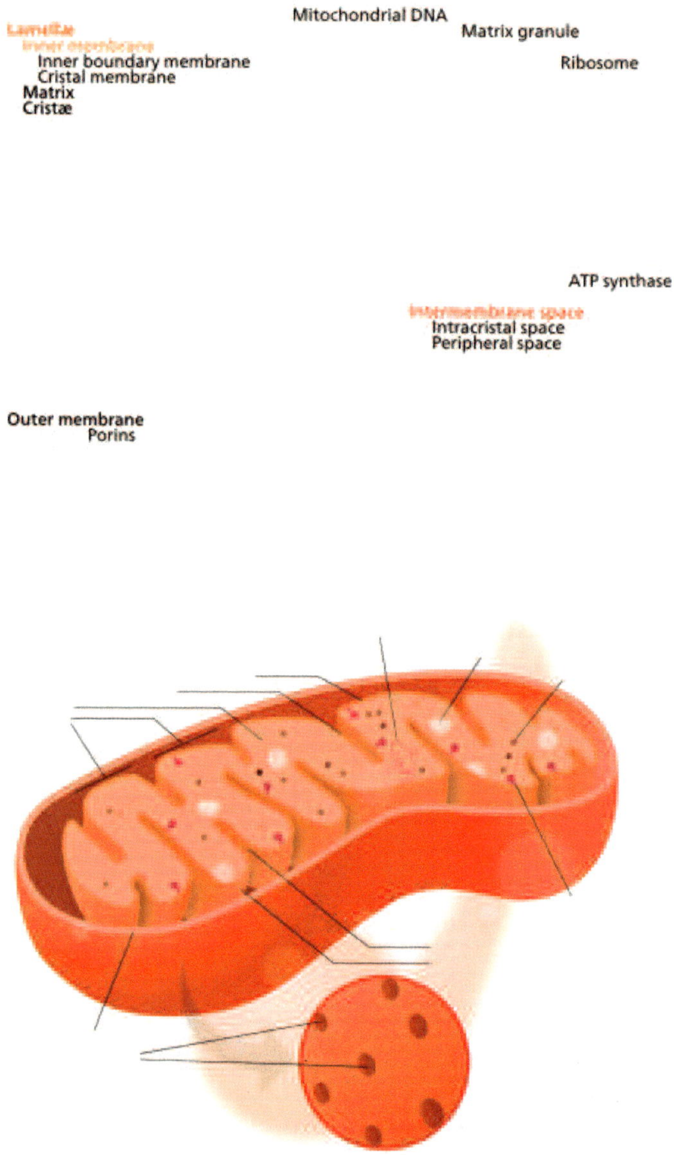

Mitochondrion ultrastructure *(interactive diagram)* A mitochondrion has a double membrane; the inner one contains its chemiosmotic apparatus and has deep grooves which increase its surface area. While commonly depicted as an "orange sausage with a blob inside of it" (like it is here), mitochondria can take many shapes and their intermembrane space is quite thin.

A mitochondrion contains outer and inner membranes composed of phospholipid bilayers and proteins. The two membranes have different properties. Because of this double-membraned organization, there are five distinct parts to a mitochondrion. They are:

- the outer mitochondrial membrane,
- the intermembrane space (the space between the outer and inner membranes),
- the inner mitochondrial membrane,
- the cristae space (formed by infoldings of the inner membrane), and
- the matrix (space within the inner membrane).

Mitochondria stripped of their outer membrane are called mitoplasts.

Outer membrane

The **outer mitochondrial membrane**, which encloses the entire organelle, is 60 to 75 angstroms (Å) thick. It has a protein-to-phospholipid ratio similar to that of the eukaryotic plasma membrane (about 1:1 by weight). It contains large numbers of integral membrane proteins called porins. These porins form channels that allow molecules of 5000 daltons or less in molecular weight to freely diffuse from one side of the membrane to the other. Larger proteins can enter the mitochondrion if a signaling sequence at their N-terminus binds to a large multisubunit protein called translocase of the outer membrane, which then actively moves them across the membrane.Mitochondrial pro-proteins are imported through specialised translocation complexes. The outer membrane also contains enzymes involved in such diverse activities as the elongation of fatty acids, oxidation of epinephrine, and the degradation of tryptophan. These enzymes include monoamine oxidase, rotenone-insensitive NADH-cytochrome c-reductase, kynurenine hydroxylase and fatty acid Co-A ligase. Disruption of the outer membrane permits proteins in the intermembrane space to leak into the cytosol, leading to certain cell death. The mitochondrial outer membrane can associate with the endoplasmic reticulum (ER) membrane, in a structure called MAM (mitochondria-associated ER-membrane). This is important in the ER-mitochondria calcium signaling and is involved in the transfer of lipids between the ER and mitochondria. Outside the outer membrane there are small (diameter: 60Å) particles named sub-units of Parson.

Intermembrane space

The intermembrane space is the space between the outer membrane and the inner membrane. It is also known as perimitochondrial space. Because the outer membrane is freely permeable to small molecules, the concentrations of small molecules, such as ions and sugars, in the intermembrane space is the same as in the cytosol. However, large proteins must have a specific signaling sequence to be transported across the outer membrane, so the protein composition of this space is different from the protein composition of the cytosol. One protein that is localized to the intermembrane space in this way is cytochrome c.

Inner membrane

The inner mitochondrial membrane contains proteins with five types of functions:

- Those that perform the redox reactions of oxidative phosphorylation
- ATP synthase, which generates ATP in the matrix
- Specific transport proteins that regulate metabolite passage into and out of the matrix
- Protein import machinery

- Mitochondrial fusion and fission protein

It contains more than 151 different polypeptides, and has a very high protein-to-phospholipid ratio (more than 3:1 by weight, which is about 1 protein for 15 phospholipids). The inner membrane is home to around 1/5 of the total protein in a mitochondrion. In addition, the inner membrane is rich in an unusual phospholipid, cardiolipin. This phospholipid was originally discovered in cow hearts in 1942, and is usually characteristic of mitochondrial and bacterial plasma membranes. Cardiolipin contains four fatty acids rather than two, and may help to make the inner membrane impermeable. Unlike the outer membrane, the inner membrane doesn't contain porins, and is highly impermeable to all molecules. Almost all ions and molecules require special membrane transporters to enter or exit the matrix. Proteins are ferried into the matrix via the translocase of the inner membrane (TIM) complex or via Oxa1. In addition, there is a membrane potential across the inner membrane, formed by the action of the enzymes of the electron transport chain.

Cristae

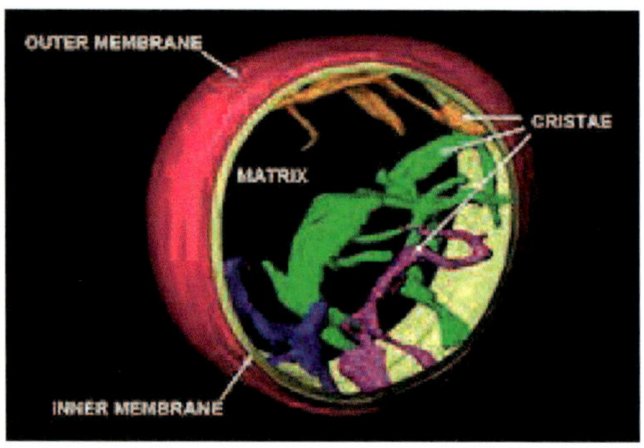

Cross-sectional image of cristae in rat liver mitochondrion to demonstrate the likely 3D structure and relationship to the inner membrane

The inner mitochondrial membrane is compartmentalized into numerous cristae, which expand the surface area of the inner mitochondrial membrane, enhancing its ability to produce ATP. For typical liver mitochondria, the area of the inner membrane is about five times as large as the outer membrane. This ratio is variable and mitochondria from cells that have a greater demand for ATP, such as muscle cells, contain even more cristae. These folds are studded with small round bodies known as F_1 particles or oxysomes. These are not simple random folds but rather invaginations of the inner membrane, which can affect overall chemiosmotic function.

One recent mathematical modeling study has suggested that the optical properties of the cristae in filamentous mitochondria may affect the generation and propagation of light within the tissue.

Matrix

The matrix is the space enclosed by the inner membrane. It contains about 2/3 of the total protein in a mitochondrion. The matrix is important in the production of ATP with the aid of the ATP synthase contained in the inner membrane. The matrix contains a highly concentrated mixture of hundreds of enzymes, special mitochondrial ribosomes, tRNA, and several copies of the mitochondrial DNA genome. Of the enzymes, the major functions include oxidation of pyruvate and fatty acids, and the citric acid cycle.

Mitochondria have their own genetic material, and the machinery to manufacture their own RNAs and proteins (*see: protein biosynthesis*). A published human mitochondrial DNA sequence revealed 16,569 base pairs encoding 37 genes: 22 tRNA, 2 rRNA, and 13 peptide genes. The 13 mitochondrial peptides in humans are integrated into the inner mitochondrial membrane, along with proteins encoded by genes that reside in the host cell's nucleus.

Mitochondria-Associated ER Membrane (MAM)

The mitochondria-associated ER membrane (MAM) is another structural element that is increasingly recognized for its critical role in cellular physiology and homeostasis. Once considered a technical snag in cell fractionation techniques, the alleged ER vesicle contaminants that invariably appeared in the mitochondrial fraction have been re-identified as membranous structures derived from the MAM—the interface between mitochondria and the ER. Physical coupling between these two organelles had previously been observed in electron micrographs and has more recently been probed with fluorescence microscopy. Such studies estimate that at the MAM, which may comprise up to 20% of the mitochondrial outer membrane, the ER and mitochondria are separated by a mere 10–25 nm and held together by protein tethering complexes.

Purified MAM from subcellular fractionation has been shown to be enriched in enzymes involved in phospholipid exchange, in addition to channels associated with Ca^{2+} signaling. These hints of a prominent role for the MAM in the regulation of cellular lipid stores and signal transduction have been borne out, with significant implications for mitochondrial-associated cellular phenomena, as discussed below. Not only has the MAM provided insight into the mechanistic basis underlying such physiological processes as intrinsic apoptosis and the propagation of calcium signaling, but it also favors a more refined view of the mitochondria. Though often seen as static, isolated 'powerhouses' hijacked for cellular metabolism through an ancient endosymbiotic event, the evolution of the MAM underscores the extent to which mitochondria have been integrated into overall cellular physiology, with intimate physical and functional coupling to the endomembrane system.

Phospholipid Transfer

The MAM is enriched in enzymes involved in lipid biosynthesis, such as phosphatidylserine synthase on the ER face and phosphatidylserine decarboxylase on the mitochondrial face. Because mitochondria are dynamic organelles constantly undergoing fission and fusion events, they require a constant and well-regulated supply of phospholipids for membrane integrity. But mitochondria are not only a destination for the phospholipids they finish synthesis of; rather, this organelle also plays a role in inter-organelle trafficking of the intermediates and products of phospholipid biosynthetic pathways, ceramide and cholesterol metabolism, and glycosphingolipid anabolism.

Such trafficking capacity depends on the MAM, which has been shown to facilitate transfer of lipid intermediates between organelles. In contrast to the standard vesicular mechanism of lipid transfer, evidence indicates that the physical proximity of the ER and mitochondrial membranes at the MAM allows for lipid flipping between opposed bilayers. Despite this unusual and seemingly energetically unfavorable mechanism, such transport does not require ATP. Instead, in yeast, it has been shown to be dependent on a multiprotein tethering structure termed the ER-mitochondria encounter structure, or ERMES, although it remains unclear whether this structure directly mediates lipid transfer or is required to keep the membranes in sufficiently close proximity to lower the energy barrier for lipid flipping.

The MAM may also be part of the secretory pathway, in addition to its role in intracellular lipid trafficking. In particular, the MAM appears to be an intermediate destination between the rough ER and the Golgi in the pathway that leads to very-low-density lipoprotein, or VLDL, assembly and secretion. The MAM thus serves as a critical metabolic and trafficking hub in lipid metabolism.

Calcium Signaling

A critical role for the ER in calcium signaling was acknowledged before such a role for the mitochondria was widely accepted, in part because the low affinity of Ca^{2+} channels localized to the outer mitochondrial membrane seemed to fly in the face of this organelle's purported responsiveness to changes in intracellular Ca^{2+} flux. But the presence of the MAM resolves this apparent contradiction: the close physical association between the two organelles results in Ca^{2+} microdomains at contact points that facilitate efficient Ca^{2+} transmission from the ER to the mitochondria. Transmission occurs in response to so-called "Ca^{2+} puffs" generated by spontaneous clustering and activation of IP3R, a canonical ER membrane Ca^{2+} channel.

The fate of these puffs—in particular, whether they remain restricted to isolated locales or integrated into Ca^{2+} waves for propagation throughout the cell—is determined in large part by MAM dynamics. Although reuptake of Ca^{2+} by the ER (concomitant with its release) modulates the intensity of the puffs, thus insulating mitochondria to a certain degree from high Ca^{2+} exposure, the MAM often serves as a firewall that essentially buffers Ca^{2+} puffs by acting as a sink into which free ions released into the cytosol can be funneled. This Ca^{2+} tunneling occurs through the low-affinity Ca^{2+} receptor VDAC1, which recently has been shown to be physically tethered to the IP3R clusters on the ER membrane and enriched at the MAM. The ability of mitochondria to serve as a Ca^{2+} sink is a result of the electrochemical gradient generated during oxidative phosphorylation, which makes tunneling of the cation an exergonic process. Normally, mild calcium influx from cytosol into the mitochondrial matrix causes transient depolarization that is corrected by pumping out protons.

But transmission of Ca^{2+} is not unidirectional; rather, it is a two-way street. The properties of the Ca^{2+} pump SERCA and the channel IP3R present on the ER membrane facilitate feedback regulation coordinated by MAM function. In particular, the clearance of Ca^{2+} by the MAM allows for spatio-temporal patterning of Ca^{2+} signaling because Ca^{2+} alters IP3R activity in a biphasic manner. SERCA is likewise affected by mitochondrial feedback: uptake of Ca^{2+} by the MAM stimulates ATP production, thus providing energy that enables SERCA to reload the ER with Ca^{2+} for continued Ca^{2+} efflux at the MAM.

Thus, the MAM is not a passive buffer for Ca^{2+} puffs; rather it helps modulate further Ca^{2+}signaling through feedback loops that affect ER dynamics.

Regulating ER release of Ca^{2+} at the MAM is especially critical because only a certain window of Ca^{2+} uptake sustains the mitochondria, and consequently the cell, at homeostasis. Sufficient intraorganelle Ca^{2+} signaling is required to stimulate metabolism by activating dehydrogenase enzymes critical to flux through the citric acid cycle. However, once Ca^{2+} signaling in the mitochondria passes a certain threshold, it stimulates the intrinsic pathway of apoptosis in part by collapsing the mitochondrial membrane potential required for metabolism. Studies examining the role of pro- and anti-apoptotic factors support this model; for example, the anti-apoptotic factor Bcl-2 has been shown to interact with IP3Rs to reduce Ca^{2+} filling of the ER, leading to reduced efflux at the MAM and preventing collapse of the mitochondrial membrane potential post-apoptotic stimuli. Given the need for such fine regulation of Ca^{2+} signaling, it is perhaps unsurprising that dysregulated mitochondrial Ca^{2+} has been implicated in several neurodegenerative diseases, while the catalogue of tumor suppressors includes a few that are enriched at the MAM.

Molecular basis for tethering

Recent advances in the identification of the tethers between the mitochondrial and ER membranes suggest that the scaffolding function of the molecular elements involved is secondary to other, non-structural functions. In yeast, ERMES, a multiprotein complex of interacting ER- and mitochondrial-resident membrane proteins, is required for lipid transfer at the MAM and exemplifies this principle. One of its components, for example, is also a constituent of the protein complex required for insertion of transmembrane beta-barrel proteins into the lipid bilayer. However, a homologue of the ERMES complex has not yet been identified in mammalian cells. Other proteins implicated in scaffolding likewise have functions independent of structural tethering at the MAM; for example, ER-resident and mitochondrial-resident mitofusins form heterocomplexes that regulate the number of inter-organelle contact sites, although mitofusins were first identified for their role in fission and fusion events between individual mitochondria. Glucose-related protein 75 (grp75) is another dual-function protein. In addition to the matrix pool of grp75, a portion serves as a chaperone that physically links the mitochondrial and ER Ca^{2+} channels VDAC and IP3R for efficient Ca^{2+} transmission at the MAM. Another potential tether is Sigma-1R, a non-opioid receptor whose stabilization of ER-resident IP3R may preserve communication at the MAM during the metabolic stress response.

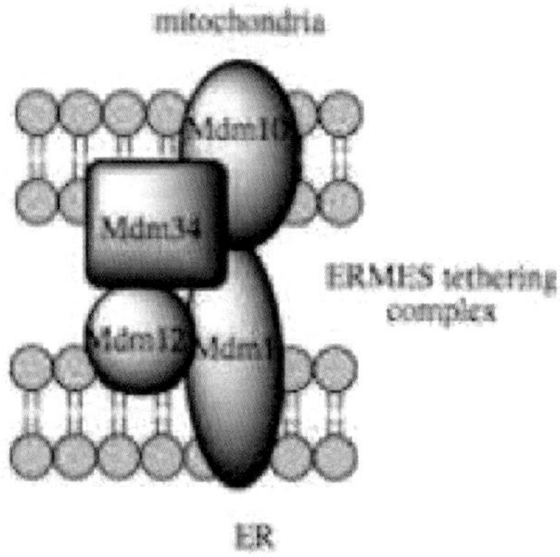

Model of the yeast multimeric tethering complex, ERMES

Perspective

The MAM is a critical signaling, metabolic, and trafficking hub in the cell that allows for the integration of ER and mitochondrial physiology. Coupling between these organelles is not simply structural but functional as well and critical for overall cellular physiology and homeostasis. The MAM thus offers a perspective on mitochondria that diverges from the traditional view of this organelle as a static, isolated unit appropriated for its metabolic capacity by the cell. Instead, this mitochondrial-ER interface emphasizes the integration of the mitochondria, the product of an endosymbiotic event, into diverse cellular processes.

ORGANIZATION AND DISTRIBUTION

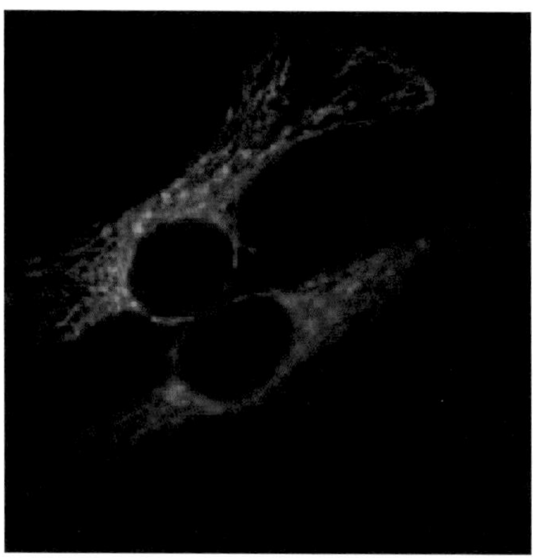

Typical mitochondrial network (green) in two human cells (HeLa cells)

Mitochondria (and related structures) are found in all eukaryotes (except one—the Oxymonad*Monocercomonoides* sp.). Although commonly depicted as bean-like structures they form a highly dynamic network in the majority of cells where they constantly undergo fission and fusion. Mitochondria vary in number and location according to cell type. A single mitochondrion is often found in unicellular organisms. Conversely, numerous mitochondria are found in human liver cells, with about 1000–2000 mitochondria per cell, making up 1/5 of the cell volume. The mitochondrial content of otherwise similar cells can vary substantially in size and membrane potential, with differences arising from sources including uneven partitioning at cell divisions, leading to extrinsic differences in ATP levels and downstream cellular processes. The mitochondria can be found nestled between myofibrils of muscle or wrapped around the sperm flagellum. Often, they form a complex 3D branching network inside the cell with the cytoskeleton. The association with the cytoskeleton determines mitochondrial shape, which can affect the function as well: different structures of the mitochondrial network may afford the population a variety of physical, chemical, and signalling advantages or disadvantages. Mitochondria in cells are always distributed along microtubules and the distribution of these organelles is also correlated with the endoplasmic reticulum. Recent evidence suggests that vimentin, one of the components of the cytoskeleton, is also critical to the association with the cytoskeleton.

Function

The most prominent roles of mitochondria are to produce the energy currency of the cell, ATP (i.e., phosphorylation of ADP), through respiration, and to regulate cellular metabolism. The central set of reactions involved in ATP production are collectively known as the citric acid cycle, or the Krebs cycle. However, the mitochondrion has many other functions in addition to the production of ATP.

Energy conversion

A dominant role for the mitochondria is the production of ATP, as reflected by the large number of proteins in the inner membrane for this task. This is done by oxidizing the major products of glucose: pyruvate, and NADH, which are produced in the cytosol. This type of cellular respiration known as aerobic respiration, is dependent on the presence of oxygen. When oxygen is limited, the glycolytic products will be metabolized by anaerobic fermentation, a process that is independent of the mitochondria. The production of ATP from glucose has an approximately 13-times higher yield during aerobic respiration compared to fermentation. Recently it has been shown that plant mitochondria can produce a limited amount of ATP without oxygen by using the alternate substrate nitrite. ATP crosses out through the inner membrane with the help of a specific protein, and across the outer membrane via porins. ADP returns via the same route.

Pyruvate and the citric acid cycle

Pyruvate molecules produced by glycolysis are actively transported across the inner mitochondrial membrane, and into the matrix where they can either be oxidized and combined with coenzyme A to form CO_2, acetyl-CoA, and NADH, or they can be carboxylated (by pyruvate carboxylase) to form oxaloacetate. This latter reaction "fills up" the amount of oxaloacetate in the citric acid cycle, and is therefore an anaplerotic reaction, increasing the cycle's capacity to metabolize acetyl-CoA when the tissue›s energy needs (e.g. in muscle) are suddenly increased by activity.

In the citric acid cycle, all the intermediates (e.g. citrate, iso-citrate, alpha-ketoglutarate, succinate, fumarate, malate and oxaloacetate) are regenerated during each turn of the cycle. Adding more of any of these intermediates to the mitochondrion therefore means that the additional amount is retained within the cycle, increasing all the other intermediates as one is converted into the other. Hence, the addition of any one of them to the cycle has an anaplerotic effect, and its removal has a cataplerotic effect. These anaplerotic and cataplerotic reactions will, during the course of the cycle, increase or decrease the amount of oxaloacetate available to combine with acetyl-CoA to form citric acid. This in turn increases or decreases the rate of ATP production by the mitochondrion, and thus the availability of ATP to the cell.

Acetyl-CoA, on the other hand, derived from pyruvate oxidation, or from the beta-oxidation of fatty acids, is the only fuel to enter the citric acid cycle. With each turn of the cycle one molecule of acetyl-CoA is consumed for every molecule of oxaloacetate present in the mitochondrial matrix, and is never regenerated. It is the oxidation of the acetate portion of acetyl-CoA that produces CO_2 and water, with the energy thus released captured in the form of ATP.

In the liver, the carboxylation of cytosolic pyruvate into intra-mitochondrial oxaloacetate is an early step in the gluconeogenic pathway, which converts lactate and de-aminated alanine into glucose, under the influence of high levels of glucagon and/or epinephrine in the blood.Here, the addition of oxaloacetate to the mitochondrion does not have a net anaplerotic effect, as another citric acid cycle intermediate (malate) is immediately removed from the mitochondrion to be converted into cytosolic oxaloacetate, which is ultimately converted into glucose, in a process that is almost the reverse of glycolysis.

The enzymes of the citric acid cycle are located in the mitochondrial matrix, with the exception of succinate dehydrogenase, which is bound to the inner mitochondrial membrane as part of Complex II. The citric acid cycle oxidizes the acetyl-CoA to carbon dioxide, and, in the process, produces reduced cofactors (three molecules of NADH and one molecule of $FADH_2$) that are a source of electrons for the *electron transport chain*, and a molecule of GTP (that is readily converted to an ATP).

NADH and $FADH_2$: the Electron Transport Chain

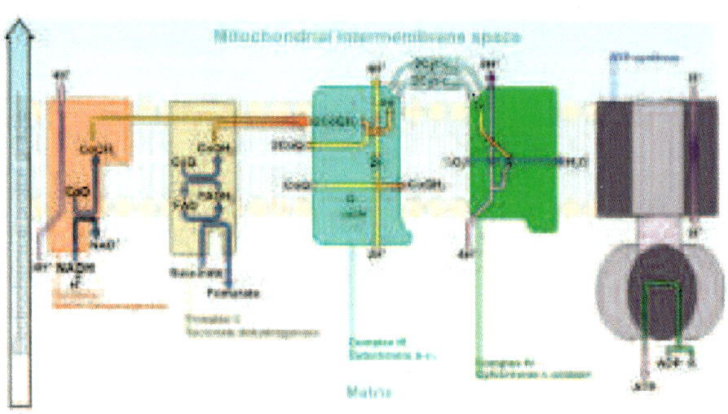

Diagram of the electron transport chain in the mitonchondrial intermembrane space

The redox energy from NADH and $FADH_2$ is transferred to oxygen (O_2) in several steps via the electron transport chain. These energy-rich molecules are produced within the matrix via the citric acid cycle but are also produced in the cytoplasm by glycolysis. Reducing equivalents from the cytoplasm can be imported via the malate-aspartate shuttle system of antiporter proteins or feed into the electron transport chain using a glycerol phosphate shuttle. Protein complexes in the inner membrane (NADH dehydrogenase (ubiquinone), cytochrome c reductase, and cytochrome c oxidase) perform the transfer and the incremental release of energy is used to pump protons (H^+) into the intermembrane space. This process is efficient, but a small percentage of electrons may prematurely reduce oxygen, forming reactive oxygen species such as superoxide. This can cause oxidative stress in the mitochondria and may contribute to the decline in mitochondrial function associated with the aging process.

As the proton concentration increases in the intermembrane space, a strong electrochemical gradient is established across the inner membrane. The protons can return to the matrix through the ATP synthase complex, and their potential energy is used to synthesize ATP from ADP and inorganic phosphate (P_i). This process is called chemiosmosis, and was first described by Peter Mitchell who was awarded the 1978 Nobel Prize in Chemistry for his work. Later, part of the 1997 Nobel Prize in Chemistry was awarded to Paul D. Boyer and John E. Walker for their clarification of the working mechanism of ATP synthase.

Heat Production

Under certain conditions, protons can re-enter the mitochondrial matrix without contributing to ATP synthesis. This process is known as *proton leak* or *mitochondrial uncoupling* and is due to the facilitated diffusion of protons into the matrix. The process results in the unharnessed potential energy of the proton electrochemical gradient being released as heat. The process is mediated by a proton channel called thermogenin, or UCP1. Thermogenin is a 33 kDa protein first discovered in 1973. Thermogenin is primarily found in brown adipose tissue, or brown fat, and is responsible for non-shivering thermogenesis. Brown adipose tissue is found in mammals, and is at its highest levels in early life and in hibernating animals. In humans, brown adipose tissue is present at birth and decreases with age.

Storage of Calcium Ions

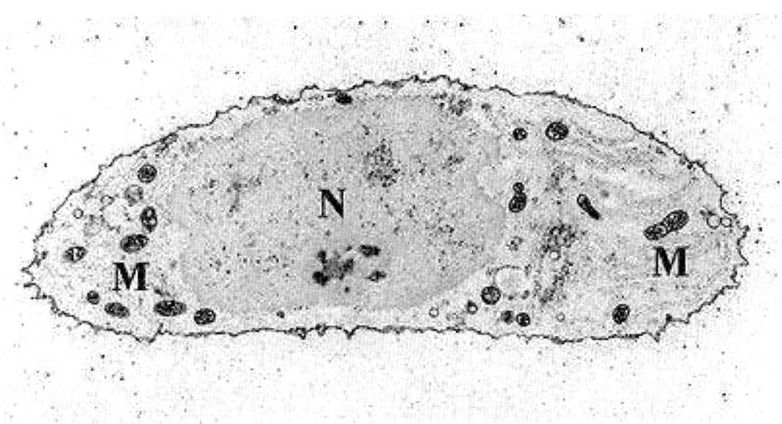

Transmission electron micrograph of a chondrocyte, stained for calcium, showing its nucleus (N) and mitochondria (M).

The concentrations of free calcium in the cell can regulate an array of reactions and is important for signal transduction in the cell. Mitochondria can transiently store calcium, a contributing process for the cell›s homeostasis of calcium. In fact, their ability to rapidly take in calcium for later release makes them very good cytosolic buffers for calcium. The endoplasmic reticulum (ER) is the most significant storage site of calcium,and there is a significant interplay between the mitochondrion and ER with regard to calcium. The calcium is taken up into the matrix by the mitochondrial calcium uniporter on the inner mitochondrial membrane. It is primarily driven by the mitochondrial membrane potential. Release of this calcium back into the cell›s interior can occur via a sodium-calcium exchange protein or via calcium-induced-calcium-release pathways. This can initiate calcium spikes or calcium waves with large changes in the membrane potential. These can activate a series of second messenger system proteins that can coordinate processes such as neurotransmitter release in nerve cells and release of hormones in endocrine cells.

Ca^{2+} influx to the mitochondrial matrix has recently been implicated as a mechanism to regulate respiratory bioenergetics by allowing the electrochemical potential across the membrane to transiently "pulse" from $\Delta\Psi$-dominated to pH-dominated, facilitating a reduction of oxidative stress. In neurons, concomitant increases in cytosolic and mitochondrial calcium act to synchronize neuronal activity with mitochondrial energy metabolism. Mitochondrial matrix calcium levels can reach the tens of micromolar levels, which is necessary for the activation of isocitrate dehydrogenase, one of the key regulatory enzymes of the Kreb›s cycle.

Additional Functions

Mitochondria play a central role in many other metabolic tasks, such as:

- Signaling through mitochondrial reactive oxygen species
- Regulation of the membrane potential
- Apoptosis-programmed cell death
- Calcium signaling (including calcium-evoked apoptosis)
- Regulation of cellular metabolism
- Certain heme synthesis reactions *(see also: porphyrin)*
- Steroid synthesis.
- Hormonal signaling Mitochondria are sensitive and responsive to hormones, in part by the action of mitochondrial estrogen receptors (mtERs). These receptors have been found in various tissues and cell types, including brain and heart

Some mitochondrial functions are performed only in specific types of cells. For example, mitochondria in liver cells contain enzymes that allow them to detoxify ammonia, a waste product of protein metabolism. A mutation in the genes regulating any of these functions can result in mitochondrial diseases.

CELLULAR PROLIFERATION REGULATION

The relationship between cellular proliferation and mitochondria has been investigated using cervical cancer HeLa cells. Tumor cells require an ample amount of ATP (Adenosine triphosphate) in order to synthesize bioactive compounds such as lipids, proteins, and nucleotides for rapid cell proliferation. The majority of ATP in tumor cells is generated via the oxidative phosphorylation pathway (OxPhos). Interference with OxPhos have shown to cause cell cycle arrest suggesting that mitochondria play a role in cell proliferation. Mitochondrial ATP production is also vital for cell division in addition to other basic functions in the cell including the regulation of cell volume, solute concentration, and cellular architecture. ATP levels differ at various stages of the cell cycle suggesting that there is a relationship between the abundance of ATP and the cell›s ability to enter a new cell cycle. ATP›s role in the basic functions of the cell make the cell cycle sensitive to changes in the availability of mitochondrial derived ATP. The variation in ATP levels at different stages of the cell cycle support the hypothesis that mitochondria play an important role in cell cycle regulation. Although the specific mechanisms between mitochondria and the cell cycle regulation is not well understood, studies have shown that low energy cell cycle checkpoints monitor the energy capability before committing to another round of cell division.

Genome

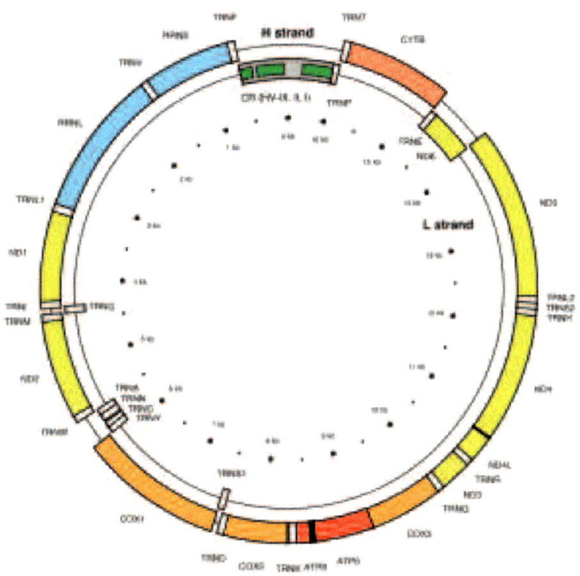

The 16,569 bp human mitochondrial genome encoding 37 genes, *i.e.,* 28 on the H-strand and 9 on the L-strand.

Main article: Mitochondrial DNA

Mitochondria contain their own genome, an indication that they are derived from bacteria through endoymbiosis (see Origin and evolution above). However, the ancestral endosymbiont genome has lost most of its genes so that the mitochondrial genome is one of the most reduced genomes across organisms.

The **human** mitochondrial genome is a circular DNA molecule of about 16 kilobases. It encodes 37 genes: 13 for subunits of respiratory complexes I, III, IV and V, 22 for mitochondrial tRNA (for the 20 standard amino acids, plus an extra gene for leucine and serine), and 2 for rRNA. One mitochondrion can contain two to ten copies of its DNA.

As in prokaryotes, there is a very high proportion of coding DNA and an absence of repeats. Mitochondrial genes are transcribed as multigenic transcripts, which are cleaved and polyadenylated to yield mature mRNAs. Not all proteins necessary for mitochondrial function are encoded by the mitochondrial genome; most are coded by genes in the cell nucleus and the corresponding proteins are imported into the mitochondrion. The exact number of genes encoded by the nucleus and the mitochondrial genome differs between species. Most mitochondrial genomes are circular, although exceptions have been reported. In general, mitochondrial DNA lacks introns, as is the case in the human mitochondrial genome;however, introns have been observed in some eukaryotic mitochondrial DNA, such as that of yeast and protists, including *Dictyostelium discoideum*. Between protein-coding regions, tRNAs are present. During transcription, the tRNAs acquire their characteristic L-shape that gets recognized and cleaved by specific enzymes. Mitochondrial tRNA genes have different sequences from the nuclear tRNAs but lookalikes of mitochondrial tRNAs have been found in the nuclear chromosomes with high sequence similarity.

In **animals**, the mitochondrial genome is typically a single circular chromosome that is approximately 16 kb long and has 37 genes. The genes, while highly conserved, may vary in location. Curiously, this pattern is not found in the human body louse (*Pediculus humanus*). Instead, this mitochondrial genome is arranged in 18 minicircular chromosomes, each of which is 3–4 kb long and has one to three genes. This pattern is also found in other sucking lice, but not in chewing lice. Recombination has been shown to occur between the minichromosomes. The reason for this difference is not known.

Alternative Genetic Code

While slight variations on the standard genetic code had been predicted earlier, none was discovered until 1979, when researchers studying human mitochondrial genes determined that they used an alternative code. However, the mitochondria of many other eukaryotes, including most plants, use the standard code. Many slight variants have been discovered since, including various alternative mitochondrial codes. Further, the AUA, AUC, and AUU codons are all allowable start codons.

Exceptions to the standard genetic code in mitochondria

Organism	Codon	Standard	Mitochondria
Mammals	AGA, AGG	Arginine	Stop codon
Invertebrates	AGA, AGG	Arginine	Serine
Fungi	CUA	Leucine	Threonine
All of the above	AUA	Isoleucine	Methionine
	UGA	Stop codon	Tryptophan

Some of these differences should be regarded as pseudo-changes in the genetic code due to the phenomenon of RNA editing, which is common in mitochondria. In higher plants, it was thought that CGG encoded for tryptophan and not arginine; however, the codon in the processed RNA was discovered to be the UGG codon, consistent with the standard genetic code for tryptophan. Of note, the arthropod mitochondrial genetic code has undergone parallel evolution within a phylum, with some organisms uniquely translating AGG to lysine.

Evolution and Diversity

Mitochondrial genomes have far fewer genes than the bacteria from which they are thought to be descended. Although some have been lost altogether, many have been transferred to the nucleus, such as the respiratory complex II protein subunits. This is thought to be relatively common over evolutionary time. A few organisms, such as the *Cryptosporidium*, actually have mitochondria that lack any DNA, presumably because all their genes have been lost or transferred. In *Cryptosporidium*, the mitochondria have an altered ATP generation system that renders the parasite resistant to many classical mitochondrial inhibitors such as cyanide, azide, and atovaquone.

Replication and Inheritance

Mitochondria divide by binary fission, similar to bacterial cell division. The regulation of this division differs between eukaryotes. In many single-celled eukaryotes, their growth and division is linked to the cell cycle. For example, a single mitochondrion may divide synchronously with the nucleus. This division and segregation process must be tightly controlled so that each daughter cell receives at least one mitochondrion. In other eukaryotes (in mammals for example), mitochondria may replicate their DNA and divide mainly in response to the energy needs of the cell, rather than in phase with the cell cycle. When the energy needs of a cell are high, mitochondria grow and divide. When the energy use is low, mitochondria are destroyed or become inactive. In such examples, and in contrast to the situation in many single celled eukaryotes, mitochondria are apparently randomly distributed to the daughter cells during the division of the cytoplasm. Understanding of mitochondrial dynamics, which is described as the balance between mitochondrial fusion and fission, has revealed that functional and structural alterations in mitochondrial morphology are important factors in pathologies associated with several disease conditions.

The hypothesis of mitochondrial binary fission has relied on the visualization by fluorescence microscopy and conventional transmission electron microscopy (TEM). The resolution of fluorescence microscopy(~200 nm) is insufficient to distinguish structural details, such as double mitochondrial membrane in mitochondrial division or even to distinguish individual mitochondria when several are close together. Conventional TEM has also some technical limitations in verifying mitochondrial division. Cryo-electron tomography was recently used to visualize mitochondrial division in frozen hydrated intact cells. It revealed that mitochondria divide by budding.

An individual›s mitochondrial genes are not inherited by the same mechanism as nuclear genes. Typically, the mitochondria are inherited from one parent only. In humans, when an egg cell is fertilized by a sperm, the egg nucleus and sperm nucleus each contribute equally to the genetic makeup of the zygote nucleus. In contrast, the mitochondria, and therefore the mitochondrial DNA, usually come from the egg only. The sperm›s mitochondria enter the egg, but do not contribute genetic information to the embryo. Instead, paternal mitochondria are marked with ubiquitin to select them for later destruction inside the embryo. The egg cell contains relatively few mitochondria, but it is these mitochondria that survive and divide to populate the cells of the adult organism. Mitochondria are, therefore, in most cases inherited only from mothers, a pattern known as maternal inheritance. This mode is seen in most organisms, including the majority of animals. However, mitochondria in some species can sometimes be inherited paternally. This is the norm among certain coniferous plants, although not in pine trees and yews. For Mytilids, paternal inheritance only occurs within males of the species. It has been suggested that it occurs at a very low level in humans. There is a recent suggestion that mitochondria that shorten male lifespan stay in the system because they are inherited only through the mother. By contrast, natural selection weeds out mitochondria that reduce female survival as such mitochondria are less likely to be passed on to the next generation. Therefore, it is suggested that human females and female animals tend to live longer than males. The authors claim that this is a partial explanation.

Uniparental inheritance leads to little opportunity for genetic recombination between different lineages of mitochondria, although a single mitochondrion can contain 2–10 copies of its DNA. For this reason, mitochondrial DNA is usually thought to reproduce by binary fission. What recombination does take place maintains genetic integrity rather than maintaining diversity. However, there are studies showing evidence of recombination in mitochondrial DNA. It is clear that the enzymes necessary for recombination are present in mammalian cells. Further, evidence suggests that animal mitochondria can undergo recombination. The data are a bit more controversial in humans, although indirect evidence of recombination exists. If recombination does not occur, the whole mitochondrial DNA sequence represents a single haplotype, which makes it useful for studying the evolutionary history of populations.

Entities undergoing uniparental inheritance and with little to no recombination may be expected to be subject to Muller›s ratchet, the inexorable accumulation of deleterious mutations until functionality is lost. Animal populations of mitochondria avoid this buildup through a developmental process known as the mtDNA bottleneck. The bottleneck exploits stochastic processes in the cell to increase in the cell-to-cell variability in mutant load as an organism develops: a single egg cell with some proportion of mutant mtDNA thus produces an embryo where different cells have different mutant loads. Cell-level selection may then act to remove those cells with more mutant mtDNA, leading to a stabilisation or reduction in mutant load between generations. The mechanism underlying the bottleneck is debated, with a recent mathematical and experimental metastudy providing evidence for a combination of random partitioning of mtDNAs at cell divisions and random turnover of mtDNA molecules within the cell.

POPULATION GENETIC STUDIES

The near-absence of genetic recombination in mitochondrial DNA makes it a useful source of information for scientists involved in population genetics and evolutionary biology. Because all the mitochondrial DNA is inherited as a single unit, or haplotype, the relationships between mitochondrial DNA from different individuals can be represented as a gene tree. Patterns in these gene trees can be used to infer the evolutionary history of populations. The classic example of this is in human evolutionary genetics, where the molecular clock can be used to provide a recent date for mitochondrial Eve. This is often interpreted as strong support for a recent modern human expansion out of Africa. Another human example is the sequencing of mitochondrial DNA from Neanderthal bones. The relatively large evolutionary distance between the mitochondrial DNA sequences of Neanderthals and living humans has been interpreted as evidence for the lack of interbreeding between Neanderthals and anatomically modern humans.

However, mitochondrial DNA reflects only the history of the females in a population and so may not represent the history of the population as a whole. This can be partially overcome by the use of paternal genetic sequences, such as the non-recombining region of the Y-chromosome. In a broader sense, only studies that also include nuclear DNA can provide a comprehensive evolutionary history of a population.

Recent measurements of the molecular clock for mitochondrial DNA reported a value of 1 mutation every 7884 years dating back to the most recent common ancestor of humans and apes, which is consistent with estimates of mutation rates of autosomal DNA (10^{-8} per base per generation).

DYSFUNCTION AND DISEASE

Mitochondrial Diseases

Damage and subsequent dysfunction in mitochondria is an important factor in a range of human diseases due to their influence in cell metabolism. Mitochondrial disorders often present themselves as neurological disorders, including autism. They can also manifest as myopathy, diabetes, multiple endocrinopathy, and a variety of other systemic disorders. Diseases caused by mutation in the mtDNA include Kearns-Sayre syndrome, MELAS syndrome and Leber›s hereditary optic neuropathy. In the vast majority of cases, these diseases are transmitted by a female to her children, as the zygote derives its mitochondria and hence its mtDNA from the ovum. Diseases such as Kearns-Sayre syndrome, Pearson syndrome, and progressive external ophthalmoplegia are thought to be due to large-scale mtDNA rearrangements, whereas other diseases such as MELAS syndrome, Leber›s hereditary optic neuropathy, myoclonic epilepsy with ragged red fibers (MERRF), and others are due to point mutations in mtDNA.

In other diseases, defects in nuclear genes lead to dysfunction of mitochondrial proteins. This is the case in Friedreich›s ataxia, hereditary spastic paraplegia, and Wilson›s disease. These diseases are inherited in a dominance relationship, as applies to most other genetic diseases. A variety of disorders can be caused by nuclear mutations of oxidative phosphorylation enzymes, such as coenzyme Q10 deficiency and Barth syndrome. Environmental influences may interact with hereditary predispositions and cause mitochondrial disease. For example, there may be a link between pesticide exposure and the later onset of Parkinson›s disease. Other pathologies with etiology involving mitochondrial dysfunction include schizophrenia, bipolar disorder, dementia, Alzheimer›s disease, Parkinson›s disease, epilepsy, stroke, cardiovascular disease, chronic fatigue syndrome, retinitis pigmentosa, and diabetes mellitus.

Mitochondria-mediated oxidative stress plays a role in cardiomyopathy in Type 2 diabetics. Increased fatty acid delivery to the heart increases fatty acid uptake by cardiomyocytes, resulting in increased fatty acid oxidation in these cells. This process increases the reducing equivalents available to the electron transport chain of the mitochondria, ultimately increasing reactive oxygen species (ROS) production. ROS increases uncoupling proteins (UCPs) and potentiate proton leakage through the adenine nucleotide translocator (ANT), the combination of which uncouples the mitochondria. Uncoupling then increases oxygen consumption by the mitochondria, compounding the increase in fatty acid oxidation. This creates a vicious cycle of uncoupling; furthermore, even though oxygen consumption increases, ATP synthesis does not increase proportionally because the mitochondria is uncoupled. Less ATP availability ultimately results in an energy deficit presenting as reduced cardiac efficiency and contractile dysfunction. To compound the problem, impaired sarcoplasmic reticulum calcium release and reduced mitochondrial reuptake limits peak cytosolic levels of the important signaling ion during muscle contraction. The decreased intra-mitochondrial calcium concentration increases dehydrogenase activation and ATP synthesis. So in addition to lower ATP synthesis due to fatty acid oxidation, ATP synthesis is impaired by poor calcium signaling as well, causing cardiac problems for diabetics.

Possible Relationships to Aging

Given the role of mitochondria as the cell's powerhouse, there may be some leakage of the high-energy electrons in the respiratory chain to form reactive oxygen species. This was thought to result in significant oxidative stress in the mitochondria with high mutation rates of mitochondrial DNA (mtDNA). Hypothesized links between aging and oxidative stress are not new and were proposed in 1956, which was later refined into the mitochondrial free radical theory of aging. A vicious cycle was thought to occur, as oxidative stress leads to mitochondrial DNA mutations, which can lead to enzymatic abnormalities and further oxidative stress.

A number of changes can occur to mitochondria during the aging process. Tissues from elderly patients show a decrease in enzymatic activity of the proteins of the respiratory chain. However, mutated mtDNA can only be found in about 0.2% of very old cells. Large deletions in the mitochondrial genome have been hypothesized to lead to high levels of oxidative stress and neuronal death in Parkinson›s disease.

In popular culture

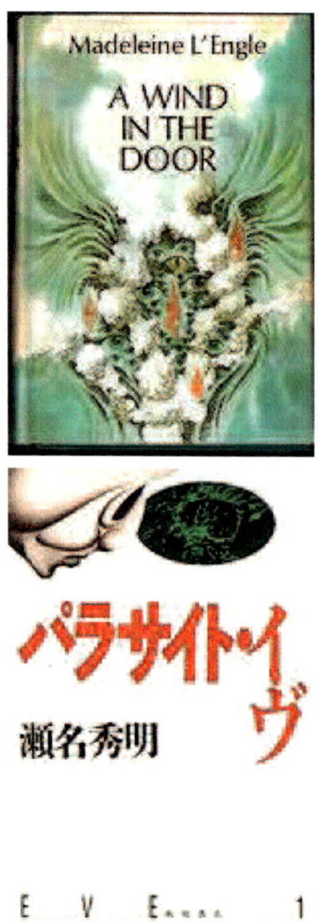

The science fiction novels *A Wind in the Door* and *Parasite Eve* include fictional depictions of mitochondria as major plot elements.

Madeleine L›Engle›s 1973 science fantasy novel *A Wind in the Door* prominently features the mitochondria of main character Charles Wallace Murry, as being inhabited by creatures known as the farandolae. The novel also features other characters travelling inside one of Murry›s mitochondria.

The 1995 horror fiction novel *Parasite Eve* by Hideaki Sena depicts mitochondria as having some consciousness and mind control abilities, attempting to use these to overtake eukaryotes as the dominant life form. This text was adapted into an eponymous film, video game, and video game sequel all involving a similar premise.

In the *Star Wars* franchise, microorganisms referred to as "midi-chlorians" give some characters the ability to sense and use the Force. George Lucas, director of the 1999 film *Star Wars: Episode I – The Phantom Menace*, in which midi-chlorians were introduced, described them as "a loose depiction of mitochondria". The non-fictional *Midichloria* genus of bacteria was later named after the midi-chlorians of *Star Wars*.

As a result of the mitochondrion's prominence in modern science education, the phrase "mitochondria is [sic] the powerhouse of the cell" became a popular Internet meme. The meme is used to imply that secondary education places an insufficient focus on life skills, compared to academic knowledge such as the role of the mitochondrion, which has been considered comparatively impractical.

REFERENCES

1. Henze K, Martin W; Martin, William (2003). Evolutionary biology: essence of mitochondria. Nature. 426 *(6963): 127–8. doi:10.1038/426127a. PMID 14614484.*
2. Karnkowska, Anna; Vacek, Vojtěch; Zubáčová, Zuzana; Treitli, Sebastian; Petrželková, Romana; Eme, Laura; Novák, Lukáš; Žárský, Vojtěch; Barlow, Lael; Herman, Emily; Soukal, Petr; Hroudová, Miluše; Doležal, Pavel; Stairs, Courtney; Roger, Andrew; Eliáš, Marek; Dacks, Joel; Vlček, Čestmír; Hampl, Vladimír. A Eukaryote without a Mitochondrial Organelle (PDF). Current Biology. 26: 1–11. doi:10.1016/j.cub.2016.03.053. Retrieved 16 May 2016.
3. mitochondria. Online Etymology Dictionary.
4. Campbell, Neil A.; Brad Williamson; Robin J. Heyden (2006). Biology: Exploring Life. Boston, Massachusetts: Pearson Prentice Hall. ISBN 0-13-250882-6.
5. Wiemerslage L, Lee D (2016). Quantification of mitochondrial morphology in neurites of dopaminergic neurons using multiple parameters.. J Neurosci Methods. doi:10.1016/j.jneumeth.2016.01.008. PMID 26777473.
6. McBride HM, Neuspiel M, Wasiak S (2006). Mitochondria: more than just a powerhouse. Curr. Biol. 16 *(14): R551–60. doi:10.1016/j.cub.2006.06.054. PMID 16860735.*
7. Valero T (2014). Mitochondrial biogenesis: pharmacological approaches. Curr. Pharm. Des. 20 *(35): 5507–9. doi:10.2174/138161282035140911142118. PMID 24606795. Mitochondrial biogenesis is therefore defined as the process via which cells increase their individual mitochondrial mass . ... Mitochondrial biogenesis occurs by growth and division of pre-existing organelles and is temporally coordinated with cell cycle events .*
8. Sanchis-Gomar F, García-Giménez JL, Gómez-Cabrera MC, Pallardó FV (2014).

Mitochondrial biogenesis in health and disease. Molecular and therapeutic approaches. Curr. Pharm. Des. 20 *(35): 5619–5633. doi:10.2174/1381612820666140306095106. PMID 24606801. Mitochondrial biogenesis (MB) is the essential mechanism by which cells control the number of mitochondria.*

9. Gardner A, Boles RG (2005). Is a ‹Mitochondrial Psychiatry› in the Future? A Review. Curr. Psychiatry Review. 1 *(3): 255–271. doi:10.2174/157340005774575064.*
10. Lesnefsky EJ, Moghaddas S, Tandler B, Kerner B, Hoppel CL (June 2001). Mitochondrial dysfunction in cardiac disease: ischemia—reperfusion. Journal of Molecular and Cellular Cardiology. 33 *(6): 1065–1089. doi:10.1006/jmcc.2001.1378. PMID 11444914.*
11. Dorn GW, Vega RB, Kelly DP (2015). Mitochondrial biogenesis and dynamics in the developing and diseased heart. Genes Dev. 29 *(19): 1981–91. doi:10.1101/gad.269894.115. PMC 4604339* . PMID 26443844.
12. Study Confirms Mitochondrial Deficits in Children with Autism. biosciencetechnology.com. May 2014
13. Alberts, Bruce; Alexander Johnson; Julian Lewis; Martin Raff; Keith Roberts; Peter Walter (1994). Molecular Biology of the Cell. New York: Garland Publishing Inc. ISBN 0-8153-3218-1.
14. Voet, Donald; Judith G. Voet; Charlotte W. Pratt (2006). Fundamentals of Biochemistry, 2nd Edition. John Wiley and Sons, Inc. pp. 547, 556. ISBN 0-471-21495-7.
15. Taylor SW, Fahy E, Zhang B, Glenn GM, Warnock DE, Wiley S, Murphy AN, Gaucher SP, Capaldi RA, Gibson BW, Ghosh SS (March 2003). Characterization of the human heart mitochondrial proteome. Nat. Biotechnol. 21 *(3): 281–6. doi:10.1038/nbt793. PMID 12592411.*
16. Zhang J, Li X, Mueller M, Wang Y, Zong C, Deng N, Vondriska TM, Liem DA, Yang J, Korge P, Honda H, Weiss JN, Apweiler R, Ping P (2008). Systematic Characterization of the Murine Mitochondrial Proteome Using Functionally Validated Cardiac Mitochondria. Proteomics. 8 *(8): 1564–1575. doi:10.1002/pmic.200700851. PMC 2799225* . PMID 18348319.
17. Zhang J, Liem DA, Mueller M, Wang Y, Zong C, Deng N, Vondriska TM, Yang J, Korge P, Drews O, Maclellan WR, Honda H, Weiss JN, Apweiler R, Ping P (2008). Altered Proteome Biology of Cardiac Mitochondria Under Stress Conditions. J. Proteome Res. 7 *(6): 2204–14. doi:10.1021/pr070371f. PMC 3805274* . PMID 18484766.
18. Andersson SG, Karlberg O, Canbäck B, Kurland CG (January 2003). On the origin of mitochondria: a genomics perspective. Philosophical Transactions of the Royal Society B. 358 *(1429): 165–77; discussion 177–9. doi:10.1098/rstb.2002.1193. PMC 1693097* . PMID 12594925.
19. Ernster, Lars; Schatz, Gottfried (1981). Mitochondria: a historical review(PDF). The Journal of Cell Biology. 91 *(3 Pt 2): 227s–255s. doi:10.1083/jcb.91.3.227s. PMC 2112799* . PMID 7033239.
20. Altmann, R. 1890 . Die Elementarorganismen und ihre Beziehungen zu den Zellen. Veit, Leipzig, .
21. Benda, C. 1898. Ueber die Spermatogenese der Vertebraten und höherer

Evertebraten. II. Theil: Die Histiogenese der Spermien. Arch. Anal. Physiol. 393-398, .

22. Ernster's citation Meves, Friedrich (May 1908). Die Chondriosomen als Träger erblicher Anlagen. Cytologische Studien am Hühnerembryo. Archiv für mikroskopische Anatomie. 72 *(1): 816–867. doi:10.1007/BF02982402.* is wrong, correct citation is Meves, Friedrich (1904). Über das Vorkommen von Mitochondrien bezw. Chondromiten in Pflanzenzellen. Ber. D. Deutsch. Bot. Ges. 22: 284–286., cited in Meves› 1908 paper and in Schmidt, Ernst Willy (1913). Pflanzliche Mitochondrien. Progressus rei botanicae. 4: 164–183. Retrieved 21 September 2012., with confirmation of Nymphaea alba

23. Siekevitz P (1957). Powerhouse of the cell. Scientific American. 197 *(1): 131–140. doi:10.1038/scientificamerican0757-131.*

24. Martin, William F.; Garg, Sriram; Zimorski, Verena (2015). Endosymbiotic theories for eukaryote origin. Philosophical Transaction of the Royal Society B. 370. doi:10.1098/rstb.2014.0330. PMID 20140330.

25. Margulis, Lynn; Sagan, Dorion (1986). Origins of Sex. Three Billion Years of Genetic Recombination. New Haven: Yale University Press. pp. 69–71, 87. ISBN 0 300 03340 0.

26. William F. Martin and Miklós Müller "Origin of mitochondria and hydrogenosomes", Springer Verlag, Heidelberg 2007.

27. Emelyanov VV (2003). Mitochondrial connection to the origin of the eukaryotic cell. Eu J Biochem. 270 *(8): 1599–1618. doi:10.1046/j.1432-1033.2003.03499.x. PMID 12694174.*

28. Gray MW, Burger G, Lang BF (March 1999). Mitochondrial evolution. Science. 283 *(5407): 1476–81. doi:10.1126/science.283.5407.1476. PMID 10066161.*

29. Thrash, J. Cameron; et al. (2011). Phylogenomic evidence for a common ancestor of mitochondria and the SAR11 clade. Scientific Reports. 1: 13. doi:10.1038/srep00013. PMID 22355532.

30. Ferla, M. P.; Thrash, J. C.; Giovannoni, S. J.; Patrick, W. M. (2013). New rRNA gene-based phylogenies of the Alphaproteobacteria provide perspective on major groups, mitochondrial ancestry and phylogenetic instability. PLoS ONE. 8 *(12): e83383. doi:10.1371/journal.pone.0083383. PMC 3859672* . PMID 24349502.

31. O›Brien TW (2003). Properties of human mitochondrial ribosomes. IUBMB Life. 55 *(9): 505–13. doi:10.1080/15216540310001626610. PMID 14658756.*

32. Lynn Sagan (1967). On the origin of mitosing cells. J Theor Biol. 14 *(3): 255–274. doi:10.1016/0022-5193(67)90079-3. PMID 11541392.*

33. Emelyanov VV (2001). Rickettsiaceae, rickettsia-like endosymbionts, and the origin of mitochondria. Biosci. Rep. 21 *(1): 1–17. doi:10.1023/A:1010409415723. PMID 11508688.*

34. Feng DF, Cho G, Doolittle RF (1997). Determining divergence times with a protein clock: Update and reevaluation. Proc. Natl. Acad. Sci. 94 *(24): 13028–13033. doi:10.1073/pnas.94.24.13028. PMC 24257* . PMID 9371794.

35. Cavalier-Smith T (1991). Archamoebae: the ancestral eukaryotes?. Biosystems. 25 *(1–2): 25–38. doi:10.1016/0303-2647(91)90010-I. PMID 1854912.*

36. Karnkowska, Anna; Vacek, Vojtěch; Zubáčová, Zuzana; Treitli, Sebastian C.; Petrželková, Romana; Eme, Laura; Novák, Lukáš; Žárský, Vojtěch; Barlow, Lael D. (2016-05-23). A Eukaryote without a Mitochondrial Organelle. Current Biology. 26 *(10): 1274–1284. doi:10.1016/j.cub.2016.03.053. ISSN 0960-9822.*

37. Mitochondrion – much more than an energy converter. British Society for Cell Biology. Retrieved 19 August 2013.

38. to:[a] [b] Herrmann JM, Neupert W (April 2000). Protein transport into mitochondria. Current Opinion in Microbiology. 3 *(2): 210–214. doi:10.1016/S1369-5274(00)00077-1. PMID 10744987.*

39. to:[a] [b] Chipuk JE, Bouchier-Hayes L, Green DR (2006). Mitochondrial outer membrane permeabilization during apoptosis: the innocent bystander scenario. Cell Death and Differentiation. 13 *(8): 1396–1402. doi:10.1038/sj.cdd.4401963. PMID 16710362.*

40. Hayashi T, Rizzuto R, Hajnoczky G, Su TP (February 2009). MAM: more than just a housekeeper. Trends Cell Biol. 19 *(2): 81–8. doi:10.1016/j.tcb.2008.12.002. PMC 2750097* . PMID 19144519.

41. McMillin JB, Dowhan W (December 2002). Cardiolipin and apoptosis. Biochim. Biophys. Acta. 1585 *(2–3): 97–107. doi:10.1016/S1388-1981(02)00329-3. PMID 12531542.*

42. Mannella CA (2006). Structure and dynamics of the mitochondrial inner membrane cristae. Biochimica et Biophysica Acta. 1763 *(5–6): 542–548. doi:10.1016/j.bbamcr.2006.04.006. PMID 16730811.*

43. Thar, R.; Kühl, Michael (2004). Propagation of electromagetic radiation in mitochondria?(PDF). J Theor Biol. 230 *(2): 261–270. doi:10.1016/j.jtbi.2004.05.021.*

44. Anderson S, Bankier AT, Barrell BG, de-Bruijn MH, Coulson AR, et al. (1981). Sequence and organization of the human mitochondrial genome. Nature. 290 *(5806): 427–465. doi:10.1038/290457a0. PMID 7219534.*

45. Rizzuto, R.; Marchi, Saverio; Bonora, Massimo; Aguiari, Paola; Bononi, Angela; De Stefani, Diego; Giorgi, Carlotta; Leo, Sara; Rimessi, Alessandro (2009). Ca^{2+} *transfer from the ER to mitochondria: when, how and* why. Biochim Biophys Acta. 1787 *(11): 1342–51. doi:10.1016/j.bbabio.2009.03.015. PMC 2730423* . PMID 19341702.

46. Hayashi, T.; Rizzuto, Rosario; Hajnoczky, Gyorgy; Su, Tsung-Ping (2009). MAM: more than just a housekeeper. Trends Cell Biol. 19 *(2): 81–88. doi:10.1016/j.tcb.2008.12.002. PMC 2750097* . PMID 19144519.

47. de Brito, OM; et al. (2010). An intimate liaison: spatial organization of the endoplasmic reticulum–mitochondria relationship. EMBO J. 29 *(16): 2715–2723. doi:10.1038/emboj.2010.177. PMC 2924651* . PMID 20717141.

48. to:[a] [b] Vance, JE; Shiao, YJ (1996). Intracellular trafficking of phospholipids: import of phosphatidylserine into mitochondria. Anticancer Research. 16 *(3B): 1333–9. PMID 8694499.*

49. to:[a] [b] [c] Lebiedzinska, M.; Szabadkai, György; Jones, Aleck W.E.; Duszynski, Jerzy; Wieckowski, Mariusz R. (2009). Interactions between the endoplasmic reticulum, mitochondria, plasma membrane and other subcellular organelles. Int J Biochem Cell Biol. 41 *(10): 1805–16. doi:10.1016/j.biocel.2009.02.017. PMID 19703651.*

50. Twig, G.; Elorza, Alvaro; Molina, Anthony J A; Mohamed, Hibo; Wikstrom, Jakob D; Walzer, Gil; Stiles, Linsey; Haigh, Sarah E; Katz, Steve (2008). Fission and selective fusion govern mitochondrial segregation and elimination by autophagy. The EMBO Journal. 27 *(2): 433–446. doi:10.1038/sj.emboj.7601963. PMC 2234339* . PMID 18200046.

51. Osman, C.; Voelker, D. R.; Langer, T. (2011). "Making heads or tails of phospholipids in mitochondria". J Cell Biol. 192 *(1): 7–16. doi:10.1083/jcb.201006159. PMC 3019561* . PMID 21220505.

52. Kornmann, B.; Currie, E.; Collins, S. R.; Schuldiner, M.; Nunnari, J.; Weissman, J. S.; Walter, P. (2009). An ER-Mitochondria Tethering Complex Revealed by a Synthetic Biology Screen. Science. 325 *(24): 477–481. doi:10.1126/science.1175088. PMC 2933203* . PMID 19556461.

53. Rusinol, A. E.; Cui, Z; Chen, MH; Vance, JE (1994). A Unique Mitochondria-associated Membrane Fraction from Rat Liver Has a High Capacity for Lipid Synthesis and Contains Pre-Golgi Secretory Proteins Including Nascent Lipoprotein. J Biol Chem. 269 *(44): 27494–27502. PMID 7961664.*

54. Santulli, Gaetano; Marks, Andrew (2015). Essential Roles of Intracellular Calcium Release Channels in Muscle, Brain, Metabolism, and Aging. Current Molecular Pharmacology. 8 *(2): 206–222. doi:10.2174/1874467208666150507105105. ISSN 1874-4672.*

55. Kopach, O.; Kruglikov, Illya; Pivneva, Tatyana; Voitenko, Nana; Fedirko, Nataliya (2008). Functional coupling between ryanodine receptors, mitochondria and Ca^{2+} *ATPases in rat submandibular acinar cells". Cell Calcium.* 43 *(5): 469–481. doi:10.1016/j.ceca.2007.08.001. PMID 17889347.*

56. Csordas, G.; Hajnóczky, G (2001). Sorting of calcium signals at the junctions of endoplasmic reticulum and mitochondria. Cell Calcium. 29 *(4): 249–262. doi:10.1054/ceca.2000.0191. PMID 11243933.*

57. Decuypere, J. P.; Monaco, Giovanni; Bultynck, Geert; Missiaen, Ludwig; De Smedt, Humbert; Parys, Jan B. (2011). The IP3 receptor–mitochondria connection in apoptosis and autophagy. Biochim Biophys Acta. 1813 *(5): 1003–13. doi:10.1016/j.bbamcr.2010.11.023. PMID 21146562.*

58. Hajnoczky, G.; Csordás, G; Yi, M (2011). Old players in a new role: mitochondria-associated membranes, VDAC, and ryanodine receptors as contributors to calcium signal propagation from endoplasmic reticulum to the mitochondria. Cell Calcium. 32 *(5–6): 363–377. doi:10.1016/S0143416002001872. PMID 12543096.*

59. Marriott, KS; Prasad, M; Thapliyal, V; Bose, HS (December 2012). σ-1 Receptor at the Mitochondrial-Associated Endoplasmic Reticulum Membrane Is Responsible for Mitochondrial Metabolic Regulation. The Journal of Pharmacology and Experimental Therapeutics. 343 *(3): 578–86. doi:10.1124/jpet.112.198168. PMC 3500540* . PMID 22923735.

60. Hayashi, T; Su, TP (Nov 2, 2007). Sigma-1 receptor chaperones at the ER-mitochondrion interface regulate Ca(2+) signaling and cell survival. Cell. 131 *(3): 596–610. doi:10.1016/j.cell.2007.08.036. PMID 17981125.*

61. The eukaryote Giardia lamblia, for example, does not contain mitochondria, but does have a mitochondrial-like gene, suggesting that it once included either mitochondria or

an endosymbiotic progenitor of it Roger, Andrew J.; Svärd, Staffan G.; Tovar, Jorge; Clark, C. Graham; Smith, Michael W. Smith, Gillin, Frances D., and Sogin, Mitchell L. (1998). A mitochondrial-like chaperonin 60 gene in Giardia lamblia: Evidence that diplomonads once harbored an endosymbiont related to the progenitor of mitochondria. Proceedings of the National Academy of Sciences. National Academy of Sciences. 95 *(1): 229–234. doi:10.1073/pnas.95.1.229. PMC 18184* . PMID 9419358.

62. das Neves RP, Jones NS, Andreu L, Gupta R, Enver T, Iborra FJ (2010). Weissman, Jonathan S, ed. Connecting Variability in Global Transcription Rate to Mitochondrial Variability. PLOS Biology. 8 *(12): e1000560. doi:10.1371/journal.pbio.1000560. PMC 3001896* . PMID 21179497.

63. Johnston IG, Gaal B, das Neves RP, Enver T, Iborra FJ, Jones NS (2012). Haugh, Jason M, ed. Mitochondrial Variability as a Source of Extrinsic Cellular Noise. PLOS Computational Biology. 8 *(3): e1002416. doi:10.1371/journal.pcbi.1002416. PMC 3297557* . PMID 22412363.

64. Rappaport L, Oliviero P, Samuel JL (1998). Cytoskeleton and mitochondrial morphology and function. Mol and Cell Biochem. 184: 101–105. doi:10.1023/A:1006843113166.

65. Hoitzing, H., Johnston, I.G. and Jones, N.S. (2015). What is the function of mitochondrial networks? A theoretical assessment of hypotheses and proposal for future research. BioEssays. 37 *(6): 687–700. doi:10.1002/bies.201400188.*

66. Soltys B. J.; Gupta R. S. (1992). Interrelationships of endoplasmic reticulum, mitochondria, intermediate filaments, and microtubules—a quadruple fluorescence labeling study. Biochem Cell Biol. 70 *(10–11): 1174–86. doi:10.1139/o92-163. PMID 1363623.*

67. Tang HL, Lung HL, Wu KC, Le AP, Tang HM, Fung MC (2007). Vimentin supports mitochondrial morphology and organization. Biochemical J. 410 *(1): 141–6. doi:10.1042/BJ20071072. PMID 17983357.*

68. Rich PR (2003). The molecular machinery of Keilin›s respiratory chain. Biochem. Soc. Trans. 31 *(Pt 6): 1095–105. doi:10.1042/BST0311095. PMID 14641005.*

69. Stoimenova M, Igamberdiev AU, Gupta KJ, Hill RD (July 2007). Nitrite-driven anaerobic ATP synthesis in barley and rice root mitochondria. Planta. 226 *(2): 465–74. doi:10.1007/s00425-007-0496-0. PMID 17333252.*

70. Stryer, Lubert (1995). Citric acid cycle.. In: Biochemistry. (Fourth ed.). New York: W.H. Freeman and Company. pp. 509–527, 569–579, 614–616, 638–641, 732–735, 739–748, 770–773. ISBN 0 7167 2009 4.

71. King A, Selak MA, Gottlieb E (2006). Succinate dehydrogenase and fumarate hydratase: linking mitochondrial dysfunction and cancer. Oncogene. 25 *(34): 4675–4682. doi:10.1038/sj.onc.1209594. PMID 16892081.*

72. Huang, K.; K. G. Manton (2004). The role of oxidative damage in mitochondria during aging: A review. Frontiers in Bioscience. 9: 1100–1117. doi:10.2741/1298. PMID 14977532.

73. Mitchell P, Moyle J (1967-01-14). Chemiosmotic hypothesis of oxidative phosphorylation. Nature. 213 *(5072): 137–9. doi:10.1038/213137a0. PMID 4291593.*

74. Mitchell P (1967-06-24). Proton current flow in mitochondrial systems. Nature. 25 *(5095): 1327–8. doi:10.1038/2141327a0. PMID 6056845.*

75. Nobel Foundation. Chemistry 1997. Retrieved 2007-12-16.

76. Mozo J, Emre Y, Bouillaud F, Ricquier D, Criscuolo F (November 2005). Thermoregulation: What Role for UCPs in Mammals and Birds?. Bioscience Reports. 25 *(3–4): 227–249. doi:10.1007/s10540-005-2887-4. PMID 16283555.*

77. Nicholls DG, Lindberg O (1973). Brown-adipose-tissue mitochondria. The influence of albumin and nucleotides on passive ion permeabilities. Eur. J. Biochem. 37 *(3): 523–30. doi:10.1111/j.1432-1033.1973.tb03014.x. PMID 4777251.*

78. Santulli, Gaetano; Xie, Wenjun; Reiken, Steven R.; Marks, Andrew R. (2015). Mitochondrial calcium overload is a key determinant in heart failure. Proceedings of the National Academy of Sciences. 112 *(36): 11389–11394. doi:10.1073/pnas.1513047112. ISSN 0027-8424.*

79. Siegel GJ, Agranoff BW, Fisher SK, Albers RW, Uhler MD, eds. (1999). Basic Neurochemistry (6 ed.). Lippincott Williams & Wilkins. ISBN 0-397-51820-X. Illustrations by Lorie M. Gavulic

80. Rossier MF (2006). T channels and steroid biosynthesis: in search of a link with mitochondria. Cell Calcium. 40 *(2): 155–64. doi:10.1016/j.ceca.2006.04.020. PMID 16759697.*

81. Brighton, Carl T.; Hunt, Robert M. (1974). Mitochondrial calcium and its role in calcification. Clinical Orthopaedics and Related Research. 100 *(5): 406–416. doi:10.1097/00003086-197405000-00057. PMID 4134194.*

82. Brighton, Carl T.; Hunt, Robert M. (1978). The role of mitochondria in growth plate calcification as demonstrated in a rachitic model. Journal of Bone and Joint Surgery. 60 *(5): 630–639. PMID 681381.*

83. Pizzo P, Pozzan T (2007). Mitochondria–endoplasmic reticulum choreography: structure and signaling dynamics. Trends Cell Biol. 17 *(10): 511–517. doi:10.1016/j.tcb.2007.07.011. PMID 17851078.*

84. Miller RJ (1998). Mitochondria – the kraken wakes!. Trends in Neurosci. 21 *(3): 95–97. doi:10.1016/S0166-2236(97)01206-X.*

85. Santulli, Gaetano; Pagano, Gennaro; Sardu, Celestino; Xie, Wenjun; Reiken, Steven; D'Ascia, Salvatore Luca; Cannone, Michele; Marziliano, Nicola; Trimarco, Bruno; Guise, Theresa A.; Lacampagne, Alain; Marks, Andrew R. (2015). Calcium release channel RyR2 regulates insulin release and glucose homeostasis. Journal of Clinical Investigation. 125 *(5): 1968–1978. doi:10.1172/JCI79273. ISSN 0021-9738.*

86. Schwarzlander M, Logan DC, Johnston IG, Jones NS, Meyer AJ, Fricker MD, Sweetlove LJ (2012). Pulsing of Membrane Potential in Individual Mitochondria: A Stress-Induced Mechanism to Regulate Respiratory Bioenergetics in Arabidopsis. Plant Cell. 24 *(3): 1188–201. doi:10.1105/tpc.112.096438. PMC 3336130* . PMID 22395486.

87. Ivannikov, M.; et al. (2013). Mitochondrial Free Ca^{2+} *Levels and Their Effects on Energy Metabolism in Drosophila Motor Nerve Terminals". Biophys. J.* 104 *(11): 2353–2361. doi:10.1016/j.bpj.2013.03.064. PMC 3672877* . PMID 23746507.

88. Li X, Fang P, Mai J, et al. (2013). Targeting mitochondrial reactive oxygen species as novel therapy for inflammatory diseases and cancers. J Hematol Oncol. 6 *(19): 19. doi:10.1186/1756-8722-6-19. PMC 3599349* . PMID 23442817.

89. Green DR (September 1998). Apoptotic pathways: the roads to ruin. Cell. 94 *(6): 695–8. doi:10.1016/S0092-8674(00)81728-6. PMID 9753316.*

90. Hajnóczky G, Csordás G, Das S, Garcia-Perez C, Saotome M, Sinha Roy S, Yi M (2006). Mitochondrial calcium signalling and cell death: approaches for assessing the role of mitochondrial Ca^{2+} *uptake in apoptosis". Cell Calcium.* 40 *(5–6): 553–60. doi:10.1016/j.ceca.2006.08.016. PMC 2692319* . PMID 17074387.

91. McBride HM, Neuspiel M, Wasiak S (2006). Mitochondria: more than just a powerhouse. Curr Biol. 16 *(14): R551–60. doi:10.1016/j.cub.2006.06.054. PMID 16860735.*

92. Oh-hama T (1997). Evolutionary consideration on 5-aminolevulinate synthase in nature. Orig Life Evol Biosph. 27 *(4): 405–12. doi:10.1023/A:1006583601341. PMID 9249985.*

93. Klinge, Carolyn (2008). Estrogenic Control of Mitochondrial Function and Biogenesis. J Cell Biochem. 105 *(6): 1342–1351. doi:10.1002/jcb.21936. PMC 2593138* . PMID 18846505.

94. *Álvarez-Delgado, Carolina (2010). Different expression of alpha and beta mitochondrial estrogen receptors in the aging rat brain: interaction with respiratory complex V.. Experimental Gerontology.* 45 *(7–8): 580–585. doi:10.1016/j.exger.2010.01.015.*

95. Pavón, Natalia (2012). Sexual hormones: effects on cardiac and mitochondrial activity after ischemia-reperfusion in adult rats. Gender difference.. J Steroid Biochem Mol Biol. 132 *(1–2): 135–146. doi:10.1016/j.jsbmb.2012.05.003.*

96. Weinberg, Frank; Chandel, Navdeep S. (2009). Mitochondrial Metabolism and Cancer. Annals of the New York Academy of Sciences. 1177 *(1): 66–73. doi:10.1111/j.1749-6632.2009.05039.x. PMID 19845608.*

97. Moreno-Sánchez, Rafael; Rodríguez-Enríquez, Sara; Marín-Hernández, Alvaro; Saavedra, Emma (2007). Energy metabolism in tumor cells. FEBS Journal. 274 *(6): 1393–1418. doi:10.1111/j.1742-4658.2007.05686.x. PMID 17302740.*

98. Pedersen, Peter L. (1994). ATP Synthase: The machine that makes ATP. Current Biology. 4 *(12): 1138–1141. doi:10.1016/S0960-9822(00)00257-8. PMID 7704582.*

99. Pattappa, Girish; Heywood, Hannah K.; de Bruijn, Joost D.; Lee, David A. (2011). The metabolism of human mesenchymal stem cells during proliferation and differentiation. Journal of Cellular Physiology. 226 *(10): 2562–2570. doi:10.1002/jcp.22605. PMID 21792913.*

100. Agarwal, Bhawana (2011). A role for anions in ATP synthesis and its molecular mechanistic interpretation. Journal of Bioenergetics and Biomembranes. 43 *(3): 299–310. doi:10.1007/s10863-011-9358-3. PMID 21647635.*

101. Sweet, S.; Singh, G. (1999). Changes in mitochondrial mass, membrane potential, and cellular adenosine triphosphate content during the cell cycle of human leukemic (HL-60) cells. Journal of Cellular Physiology. 180 *(1): 91–96. doi:10.1002/(SICI)1097-4652(199907)180:1<91::AID-JCP10>3.0.CO;2-6. PMID 10362021.*

102. McBride, Heidi M.; Neuspiel, Margaret; Wasiak, Sylwia (2006). Mitochondria: More Than Just a Powerhouse. Current Biology. 16 *(14): R551–R560. doi:10.1016/j.cub.2006.06.054. PMID 16860735.*

103. Chan DC (2006-06-30). Mitochondria: Dynamic Organelles in Disease, Aging, and Development. Cell. 125 *(7): 1241–1252. doi:10.1016/j.cell.2006.06.010. PMID 16814712.*

104. Wiesner RJ, Ruegg JC, Morano I (1992). Counting target molecules by exponential polymerase chain reaction, copy number of mitochondrial DNA in rat tissues. Biochemical and Biophysical Research Communications. 183 *(2): 553–559. doi:10.1016/0006-291X(92)90517-O. PMID 1550563.*

105. Fukuhara H, Sor F, Drissi R, Dinouël N, Miyakawa I, Rousset, and Viola AM (1993). Linear mitochondrial DNAs of yeasts: frequency of occurrence and general features. Mol Cell Biol. 13 *(4): 2309–2314. PMC 359551* . PMID 8455612.

106. Bernardi G (1978). Intervening sequences in the mitochondrial genome. Nature. 276 *(5688): 558–559. doi:10.1038/276558a0. PMID 214710.*

107. Hebbar SK, Belcher SM, Perlman PS (April 1992). A maturase-encoding group IIA intron of yeast mitochondria self-splices in vitro. Nucleic Acids Research. 20 *(7): 1747–54. doi:10.1093/nar/20.7.1747. PMC 312266* . PMID 1579468.

108. Gray MW, Lang BF, Cedergren R, Golding GB, Lemieux C, Sankoff D, et al. (1998). Genome structure and gene content in protist mitochondrial DNAs. Nucleic Acids Research. 26 *(4): 865–878. doi:10.1093/nar/26.4.865. PMC 147373* . PMID 9461442.

109. Gray MW, Lang BF, Burger G (2004). Mitochondria of protists. Annu Rev Genet. 38: 477–524. doi:10.1146/annurev.genet.37.110801.142526. PMID 15568984.

110. Telonis Aristeidis G.; et al. (2014). Nuclear and Mitochondrial tRNA-lookalikes in the Human Genome. Frontiers in Genetics. 5: 00344. doi:10.3389/fgene.2014.00344. PMC 4189335 . PMID 25339973.

111. Shao R, Kirkness EF, Barker SC (March 2009). The single mitochondrial chromosome typical of animals has evolved into 18 minichromosomes in the human body louse, Pediculus humanus. Genome Res. 19 *(5): 904–12. doi:10.1101/gr.083188.108. PMC 2675979* . PMID 19336451.

112. Crick, F. H. C. & Orgel, L. E. (1973). Directed panspermia (PDF). Icarus. 19: 341–346. Bibcode:1973Icar...19..341C. doi:10.1016/0019-1035(73)90110-3. p. 344: It is a little surprising that organisms with somewhat different codes do not coexist. Further discussion.

113. Barrell BG, Bankier AT, Drouin J (1979). A different genetic code in human mitochondria. Nature. 282 *(5735): 189–194. doi:10.1038/282189a0. PMID 226894.*

114. Mitochondrial Genetic Code in Taxonomy Tree. NCBI

115. Elzanowski, Andrzej and Ostell, Jim. The Genetic Codes. NCBI

116. Jukes TH, Osawa S (1990). The genetic code in mitochondria and chloroplasts. Experientia. 46 *(11–12): 1117–26. doi:10.1007/BF01936921. PMID 2253709.*

117. Hiesel R, Wissinger B, Schuster W, Brennicke A (2006). RNA editing in plant mitochondria. Science. 246 *(4937): 1632–4. doi:10.1126/science.2480644. PMID 2480644.*

118. Abascal F, Posada D, Knight RD, Zardoya R (2006). Parallel Evolution of the Genetic Code in Arthropod Mitochondrial Genomes. PLOS Biology. 4 *(5): 0711–0718. doi:10.1371/journal.pbio.0040127. PMC 1440934* . PMID 16620150.

119. Henriquez FL, Richards TA, Roberts F, McLeod R, Roberts CW (2005). The unusual mitochondrial compartment of Cryptosporidium parvum. Trends Parasitol. 21 *(2): 68–74. doi:10.1016/j.pt.2004.11.010. PMID 15664529.*

120. Pfeiffer, Ronald F. (2012). Parkinson›s Disease. CRC Press. p. 583.

121. Seo AY, Joseph AM, Dutta D, Hwang JC, Aris JP, Leeuwenburgh C (2010). New insights into the role of mitochondria in aging: mitochondrial dynamics and more. J. Cell. Sci. 123 *(15): 2533–42. doi:10.1242/jcs.070490. PMC 2912461* . PMID 20940129.

122. Hu GB (2014). Whole Cell Cryo-electron tomography suggest mitochondria divide by budding. Microsc and microanal. 20: 1–8. doi:10.1017/S1431927614001317.

123. Kimball, J.W. (2006) "Sexual Reproduction in Humans: Copulation and Fertilization,"Kimball›s Biology Pages (based on Biology, 6th ed., 1996)

124. Sutovsky, P.; et al. (1999). Ubiquitin tag for sperm mitochondria. Nature. 402 *(6760): 371–372. doi:10.1038/46466. PMID 10586873.*Discussed in Science News.

125. Mogensen HL (1996). The Hows and Whys of Cytoplasmic Inheritance in Seed Plants. American Journal of Botany. 83 *(3): 383–404. doi:10.2307/2446172. JSTOR 2446172.*

126. Zouros E (December 2000). The exceptional mitochondrial DNA system of the mussel family Mytilidae. Genes Genet. Syst. 75 *(6): 313–8. doi:10.1266/ggs.75.313. PMID 11280005.*

127. Sutherland B, Stewart D, Kenchington ER, Zouros E (1 January 1998). The fate of paternal mitochondrial DNA in developing female mussels, Mytilus edulis: implications for the mechanism of doubly uniparental inheritance of mitochondrial DNA. Genetics. 148 *(1): 341–7. PMC 1459795* . PMID 9475744.

128. Male and Female Mitochondrial DNA Lineages in the Blue Mussel (Mytilus edulis) Species Group by Donald T. Stewart, Carlos Saavedra, Rebecca R. Stanwood, Amy 0. Ball, and Eleftherios Zouros

129. Johns, D. R. (2003). Paternal transmission of mitochondrial DNA is (fortunately) rare. Annals of Neurology. 54 *(4): 422–4. doi:10.1002/ana.10771. PMID 14520651.*

130. Fruit flies offer DNA clue to why women live longer. BBC. 2 August 2012

131. Thyagarajan B, Padua RA, Campbell C (1996). Mammalian mitochondria possess homologous DNA recombination activity. J. Biol. Chem. 271 *(44): 27536–27543. doi:10.1074/jbc.271.44.27536. PMID 8910339.*

132. Lunt DB, Hyman BC (15 May 1997). Animal mitochondrial DNA recombination. Nature. 387 *(6630): 247. doi:10.1038/387247a0. PMID 9153388.*

133. Eyre-Walker A, Smith NH, Maynard Smith J (1999-03-07). How clonal are human mitochondria?. Proceedings of the Royal Society B. 266 *(1418): 477–483. doi:10.1098/rspb.1999.0662. PMC 1689787* . PMID 10189711.

134. Awadalla P, Eyre-Walker A, Maynard Smith J (1999). Linkage Disequilibrium and Recombination in Hominid Mitochondrial DNA. Science. 286 *(5449): 2524–2525. doi:10.1126/science.286.5449.2524. PMID 10617471.*

135. Cree, L.M., Samuels, D.C., de Sousa Lopes, S.C., Rajasimha, H.K., Wonnapinij, P., Mann, J.R., Dahl, H.H.M. and Chinnery, P.F. (2008). A reduction of mitochondrial DNA molecules during embryogenesis explains the rapid segregation of genotypes. Nature Genetics. 40 *(2): 249–254. doi:10.1038/ng.2007.63.*

136. Cao, L., Shitara, H., Horii, T., Nagao, Y., Imai, H., Abe, K., Hara, T., Hayashi, J.I. and Yonekawa, H. (2007). The mitochondrial bottleneck occurs without reduction of mtDNA content in female mouse germ cells. Nature Genetics. 39 *(3): 386–390. doi:10.1038/ng1970.*

137. Wai, T., Teoli, D. and Shoubridge, E.A. (2008). The mitochondrial DNA genetic bottleneck results from replication of a subpopulation of genomes. Nature Genetics. 40 *(12): 1484–1488. doi:10.1038/ng.258.*

138. Johnston, I.G., Burgstaller, J.P., Havlicek, V., Kolbe, T., Rülicke, T., Brem, G., Poulton, J. and Jones, N.S. (2015). Stochastic modelling, Bayesian inference, and new in vivo measurements elucidate the debated mtDNA bottleneck mechanism. Elife. 4: e07464. doi:10.7554/eLife.07464.

139. Castro JA, Picornell A, Ramon M (1998). Mitochondrial DNA: a tool for populational genetics studies. Int Microbiol. 1 *(4): 327–32. PMID 10943382.*

140. Cann RL, Stoneking M, Wilson AC (January 1987). Mitochondrial DNA and human evolution. Nature. 325 *(6099): 31–36. doi:10.1038/325031a0. PMID 3025745.*

141. Torroni A, Achilli A, Macaulay V, Richards M, Bandelt HJ (2006). Harvesting the fruit of the human mtDNA tree. Trends Genet. 22 *(6): 339–45. doi:10.1016/j.tig.2006.04.001. PMID 16678300.*

142. Garrigan D, Hammer MF (2006). Reconstructing human origins in the genomic era. Nature Reviews Genetics. 7 *(9): 669–80. doi:10.1038/nrg1941. PMID 16921345.*

143. Krings M, Stone A, Schmitz RW, Krainitzki H, Stoneking M, Pääbo S (1997). Neandertal DNA sequences and the origin of modern humans. Cell. 90 *(1): 19–30. doi:10.1016/S0092-8674(00)80310-4. PMID 9230299.*

144. Harding RM, Fullerton SM, Griffiths RC, Bond J, Cox MJ, Schneider JA, Moulin DS, Clegg JB (April 1997). Archaic African and Asian lineages in the genetic ancestry of modern humans. American Journal of Human Genetics. 60 *(4): 772–89. PMC 1712470* . PMID 9106523.

145. Soares, P;; et al. (June 2009). Correcting for Purifying Selection: An Improved Human Mitochondrial Molecular Clock. Am J Hum Genet. 84 *(6): 740–759. doi:10.1016/j.ajhg.2009.05.001. PMC 2694979* . PMID 19500773.

146. Michael W. Nachman and Susan L. Crowell Estimate of the Mutation Rate per Nucleotide in Humans Genetics, Vol. 156, 297–304, September 2000

147. Zeviani M, Di Donato S (2004). Mitochondrial disorders. Brain. 127 *(Pt 10): 2153–2172. doi:10.1093/brain/awh259. PMID 15358637.*

148. Taylor RW, Turnbull DM (2005). MITOCHONDRIAL DNA MUTATIONS IN HUMAN DISEASE. Nature Reviews Genetics. 6 *(5): 389–402. doi:10.1038/nrg1606. PMC 1762815* . PMID 15861210.

149. Chinnery PF, Schon EA (2003). Mitochondria. J. Neurol. Neurosurg. Psychiatr. 74 *(9): 1188–99. doi:10.1136/jnnp.74.9.1188. PMC 1738655* . PMID 12933917.

150. Sherer TB, Betarbet R, Greenamyre JT (2002). Environment, mitochondria, and Parkinson›s disease. The Neuroscientist. 8 *(3): 192–7. doi:10.1177/107385840200 8003004. PMID 12061498.*

151. Gomez C, Bandez MJ, Navarro A (2007). Pesticides and impairment of mitochondrial function in relation with the parkinsonian syndrome. Front. Biosci. 12: 1079–93. doi:10.2741/2128. PMID 17127363.

152. Lim Y. A.; Rhein, Virginie; Baysang, Ginette; Meier, Fides; Poljak, Anne; j. Raftery, Mark; Guilhaus, Michael; Ittner, Lars M.; Eckert, Anne (2010). Abeta and human amylin share a common toxicity pathway via mitochondrial dysfunction. Proteomics. 10 *(8): 1621–33. doi:10.1002/pmic.200900651. PMID 20186753.*

153. Schapira AH (2006). Mitochondrial disease. Lancet. 368 *(9529): 70–82. doi:10.1016/ S0140-6736(06)68970-8. PMID 16815381.*

154. Pieczenik SR, Neustadt J (2007). Mitochondrial dysfunction and molecular pathways of disease. Exp. Mol. Pathol. 83 *(1): 84–92. doi:10.1016/j.yexmp.2006.09.008. PMID 17239370.*

155. Bugger H, Abel ED (2010). Mitochondria in the diabetic heart. Cardiovascular Research. 88 *(2): 229–240. doi:10.1093/cvr/cvq239. PMC 2952534* . PMID 20639213.

156. Richter C, Park J, Ames BN (September 1988). Normal oxidative damage to mitochondrial and nuclear DNA is extensive. PNAS. 85 *(17): 6465–6467. doi:10.1073/ pnas.85.17.6465. PMC 281993* . PMID 3413108.

157. Harman D (1956). Aging: a theory based on free radical and radiation chemistry. J. Gerontol. 11 *(3): 298–300. doi:10.1093/geronj/11.3.298. PMID 13332224.*

158. Harman, D (1972). A biologic clock: the mitochondria?. Journal of the American Geriatrics Society. 20 *(4): 145–147. PMID 5016631.*

159. Mitochondria and Aging. circuitblue.co.

160. Boffoli D, Scacco SC, Vergari R, Solarino G, Santacroce G, Papa S (1994). Decline with age of the respiratory chain activity in human skeletal muscle. Biochim. Biophys. Acta. 1226 *(1): 73–82. doi:10.1016/0925-4439(94)90061-2. PMID 8155742.*

161. de Grey, Aubrey (2004). Mitochondrial Mutations in Mammalian Aging: An Over-Hasty About-Turn? (PDF). Rejuvenation Res. 7 *(3): 171–4. doi:10.1089/rej.2004.7.171. PMID 15588517.*

162. Bender A, Krishnan KJ, Morris CM, Taylor GA, Reeve AK, Perry RH, Jaros E, Hersheson JS, Betts J, Klopstock T, Taylor RW, Turnbull DM (2006). High levels of mitochondrial DNA deletions in substantia nigra neurons in aging and Parkinson disease. Nat. Genet. 38 *(5): 515–517. doi:10.1038/ng1769. PMID 16604074.*

163. Narcisse, Evan (August 10, 2010). 20,000 Per Cell: Why Midi-chlorians Suck. Time. Retrieved June 19, 2016.

164. Byrne, Nathan (May 13, 2016). So, Mitochondria Aren›t Actually The Powerhouse Of The Cell After All. Newsy. Retrieved June 19, 2016.

165. Woodward, Nathaniel (January 20, 2016). Education: Mitochondria is the powerhouse of the cell. The Eagle Online. Retrieved June 19, 2016.

CHAPTER 5

Endosymbiont

Chapter Contents

Endosymbionts are smaller symbiotic partners living inside a host organism, establishing endosymbiosis. The symbionts may live within the host's cells (intracellular) or outside cells (extracellular) in multicellular hosts.

INTRODUCTION

An endosymbiont or endobiont is any organism that lives within the body or cells of another organism, i.e. forming an endosymbiosis . Examples are nitrogen-fixingbacteria (called rhizobia), which live in root nodules on legume roots, single-cell algae inside reef-building corals, and bacterial endosymbion ts that provide essential nutrients to about 10–15% of insects.

Many instances of endosymbiosis are obligate; that is, either the endosymbiont or the host cannot survive without the other, such as the gutless marine worms of the genus *Riftia*, which get nutrition from their endosymbiotic bacteria. The most common examples of obligate endosymbioses are mitochondria and chloroplasts. Some human parasites, e.g. *Wuchereria bancrofti* and *Mansonella perstans*, thrive in their intermediate insect hosts because of an obligate endosymbiosis with *Wolbachia spp*. They can both be eliminated from said hosts by treatments that target this bacterium.

However, not all endosymbioses are obligate. Also, some endosymbioses can be harmful to either of the organisms involved.

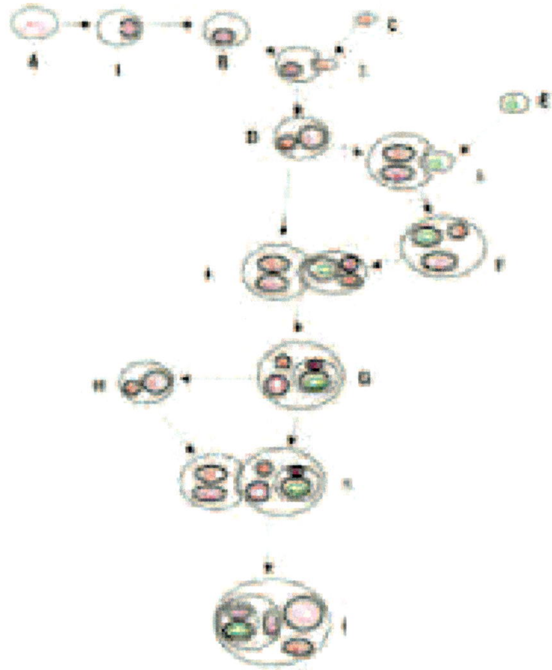

A representation of the endosymbiotic theory

Two major types of organelle in eukaryotic cells, mitochondria and plastids such as chloroplasts, originated as bacterial endosymbionts. Such symbiogenesis was first suggested in 1905, and articulated by the Russian botanist Konstantin Mereschkowski in 1910. It was reintroduced by Lynn Margulis in the 1960s.

SYMBIOGENESIS AND ORGANELLES

Symbiogenesis explains the origins of organelles such as mitochondria and chloroplasts in eukaryotic cells. The theory proposes that chloroplasts and mitochondria evolved from certain types of bacteria that eukaryotic cells engulfed through endophagocytosis. These cells and the bacteria trapped inside them entered a symbiotic relationship, a close association between different types of organisms over an extended time. However, to be specific, the relationship was endosymbiotic, meaning that one of the organisms (the bacteria) lived within the other (the eukaryotic cells).

The theory suggests that an anaerobic cell probably ingested an aerobic bacterium but failed to digest it. The aerobic bacterium flourished within the cell because the cell's cytoplasm was abundant in half-digested food molecules. The bacterium digested these molecules with oxygen and gained great amounts of energy. Because the bacterium had so much energy, it probably leaked some of it as adenosine triphosphate into the cell's cytoplasm. This benefited the anaerobic cell because it was now able to breathe aerobically, which means more potential for energy gain. Eventually, the aerobic bacterium could no longer live independently from the cell, and it, therefore, became a mitochondrion. The origin of the chloroplast is very similar to that of the mitochondrion. A cell must have captured a photosynthetic cyanobacterium and failed to digest it. The cyanobacterium thrived in the cell and eventually evolved into the first chloroplast. Other eukaryotic organelles may have also evolved through endosymbiosis; it has been proposed that cilia, flagella, centrioles, and microtubules may have originated from a symbiosis between a Spirochaete bacterium and an early eukaryotic cell, but this is not widely accepted among biologists.

There are several lines of evidence that support endosymbiosis theory. Mitochondria and chloroplasts contain their own small supply of DNA, which may be remnants of the genome the organelles had when they were independent aerobic bacteria. The single most convincing evidence of the descent of organelles from bacteria is the position of mitochondria and plastid DNA sequences in phylogenetic trees of bacteria. Mitochondria have sequences that clearly indicate origin from a group of bacteria called the alphaproteobacteria. Plastids have DNA sequences that indicate origin from the cyanobacteria (blue-green algae). In addition, there are organisms alive today, called living intermediates, that are in a similar endosymbiotic condition to the prokaryotic cells and the aerobic bacteria. Living intermediates show that the evolution proposed by the endosymbiont theory is possible. For example, the giant amoeba *Pelomyxa* lacks mitochondria but has aerobic bacteria that carry out a similar role. A variety of corals, clams, snails, and one species of *Paramecium* permanently host algae in their cells. Many of the insect endosymbionts have been shown to have ancient associations with their hosts, involving strictly vertical inheritance. In addition, these insect symbionts have similar patterns of genome evolution to those found in true organelles: genome reduction, rapid rates of gene evolution, and bias in nucleotide base composition favoring adenine and thymine, at the expense of guanine and cytosine.

Further evidence of endosymbiosis is provided by the prokaryotic ribosomes found within chloroplasts and mitochondria as well as the double-membrane enclosing them. It used to be widely assumed that the inner membrane is the original membrane of the once independent prokaryote, while the outer one is the food vacuole (phagosomal membrane) it was enclosed in initially. However, this view neglects the fact that i) both modern cyanobacteria and alpha-proteobacteria are Gram-negative bacteria, which are surrounded by double membranes; ii) the outer membranes of the endosymbiotic organelles (chloroplasts and mitochondria) are very similar to those of these bacteria in their lipid and protein compositions. Accumulating biochemical data strongly suggests that the double-membrane-enclosing chloroplasts and mitochondria derived from those of the ancestral bacteria, and the phagosomal membrane disappeared during organelle evolution. Triple or quadruple membranes are found among certain algae, probably resulting from repeated endosymbiosis (although little else was retained of the engulfed cell).

These modern organisms with endosymbiotic relationships with aerobic bacteria have verified the endosymbiotic theory, which explains the origin of mitochondria and chloroplasts from bacteria. Researchers in molecular and evolutionary biology no longer question this theory, although some of the details, such as the mechanisms for loss of genes from organelles to host nuclear genomes, are still being worked out.

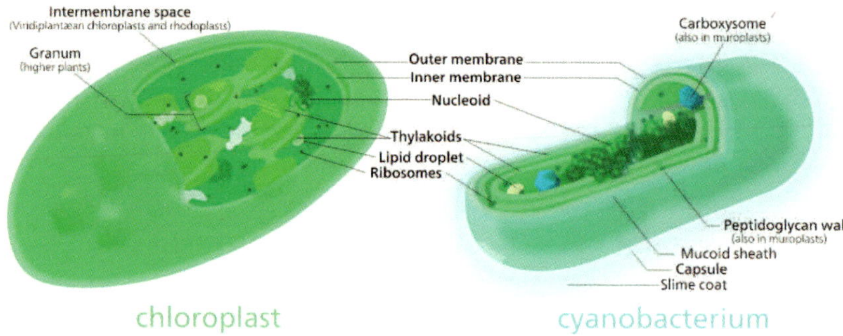

Comparison of chloroplasts and cyanobacteria emphasizing their similarities.

Bacterial endosymbionts in marine invertebrates

Extracellular endosymbionts are also represented in all four extant classes of Echinodermata (Crinoidea, Ophiuroidea, Echinoidea, and Holothuroidea). Little is known of the nature of the association (mode of infection, transmission, metabolic requirements, etc.) but phylogenetic analysis indicates that these symbionts belong to the alpha group of the class Proteobacteria, relating them to *Rhizobium* and *Thiobacillus*. Other studies indicate that these subcuticular bacteria may be both abundant within their hosts and widely distributed among the Echinoderms in general.

Some marine oligochaeta (e.g., Olavius or Inanidrillus) have obligate extracellular endosymbionts that fill the entire body of their host. These marine worms are nutritionally dependent on their symbiotic chemoautotrophic bacteria lacking any digestive or excretory system (no gut, mouth, or nephridia).

Symbiodinium dinoflagellate endosymbionts in marine metazoa and protists

Dinoflagellate endosymbionts of the genus *Symbiodinium*, commonly known as zooxanthellae, are found in corals, mollusks (esp. giant clams, the *Tridacna*), sponges, and foraminifera. These endosymbionts drive the formation of coral reefs by capturing sunlight and providing their hosts with energy for carbonate deposition.

Previously thought to be a single species, molecular phylogenetic evidence over the past couple decades has shown there to be great diversity in *Symbiodinium*. In some cases, there is specificity between host and *Symbiodinium* clade. More often, however, there is an ecological distribution of *Symbiodinium*, the symbionts switching between hosts with apparent ease. When reefs become environmentally stressed, this distribution of symbionts is related to the observed pattern of coral bleaching and recovery. Thus, the distribution of *Symbiodinium* on coral reefs and its role in coral bleaching presents one of the most complex and interesting current problems in reef ecology.

IN PROTISTS

Mixotricha paradoxa is a protozoan that lacks mitochondria. However, spherical bacteria live inside the cell and serve the function of the mitochondria. *Mixotricha* also has three other species of symbionts that live on the surface of the cell.

Paramecium bursaria, a species of ciliate, has a mutualistic symbiotic relationship with green alga called Zoochlorella. The algae live inside the cell, in the cytoplasm.

Paulinella chromatophora is a freshwater amoeboid which has recently (evolutionarily speaking) taken on a cyanobacterium as an endosymbiont.

BACTERIAL ENDOSYMBIONTS IN INSECTS

Scientists classify insect endosymbionts in two broad categories, 'Primary' and 'Secondary'. Primary endosymbionts (sometimes referred to as P-endosymbionts) have been associated with their insect hosts for many millions of years (from 10 to several hundred million years in some cases). They form obligate associations (see below), and display cospeciation with their insect hosts. Secondary endosymbionts exhibit a more recently developed association, are sometimes horizontally transferred between hosts, live in the hemolymph of the insects (not specialized bacteriocytes, see below), and are not obligate.

Among primary endosymbionts of insects, the best-studied are the pea aphid (*Acyrthosiphon pisum*) and its endosymbiont *Buchnera sp.* APS, the tsetse fly*Glossina morsitans morsitans* and its endosymbiont *Wigglesworthia glossinidia brevipalpis* and the endosymbiotic protists in lower termites. As with endosymbiosis in other insects, the symbiosis is obligate in that neither the bacteria nor the insect is viable without the other. Scientists have been unable to cultivate the bacteria in lab conditions outside of the insect. With special nutritionally-enhanced diets, the insects can survive, but are unhealthy, and at best survive only a few generations.

In some insect groups, these endosymbionts live in specialized insect cells called bacteriocytes (also called *mycetocytes*), and are maternally-transmitted, i.e. the mother transmits her endosymbionts to her offspring. In some cases, the bacteria are transmitted in the egg, as in *Buchnera*; in others like *Wigglesworthia*, they are transmitted via milk to the developing insect embryo. In termites, the endosymbionts reside within the hindguts and are transmitted through trophallaxis among colony members.

The primary endosymbionts are thought to help the host either by providing nutrients that the host cannot obtain itself or by metabolizing insect waste products into safer forms. For example, the putative primary role of *Buchnera* is to synthesize essential amino acids that the aphid cannot acquire from its natural diet of plant sap. Likewise, the primary role of *Wigglesworthia*, it is presumed, is to synthesize vitamins that the tsetse fly does not get from the blood that it eats. In lower termites, the endosymbiotic protists play a major role in the digestion of lignocellulosic materials that constitute a bulk of the termites' diet.

Bacteria benefit from the reduced exposure to predators and competition from other bacterial species, the ample supply of nutrients and relative environmental stability inside the host.

Genome sequencing reveals that obligate bacterial endosymbionts of insects have among the smallest of known bacterial genomes and have lost many genes that are commonly found in closely related bacteria. Several theories have been put forth to explain the loss of genes. It is presumed that some of these genes are not needed in the environment of the host insect cell. A complementary theory suggests that the relatively small numbers of bacteria inside each insect decrease the efficiency of natural selection in 'purging' deleterious mutations and small mutations from the population, resulting in a loss of genes over many millions of years. Research in which a parallel phylogeny of bacteria and insects was inferred supports the belief that the primary endosymbionts are transferred only vertically (i.e., from the mother), and not horizontally (i.e., by escaping the host and entering a new host).

Attacking obligate bacterial endosymbionts may present a way to control their insect hosts, many of which pests or carriers of human disease. For example, aphids are crop pests and the tsetse fly carries the organism *Trypanosoma brucei* that causes African sleeping sickness. Other motivations for their study is to understand symbiosis, and to understand how bacteria with severely depleted genomes are able to survive, thus improving our knowledge of genetics and molecular biology.

Less is known about secondary endosymbionts. The pea aphid (*Acyrthosiphon pisum*) is known to contain at least three secondary endosymbionts, *Hamiltonella defensa*, *Regiella insecticola*, and *Serratia symbiotica*. *H. defensa* aids in defending the insect from parasitoids. *Sodalis glossinidius* is a secondary endosymbiont of tsetse flies that lives inter- and intracellularly in various host tissues, including the midgut and hemolymph. Phylogenetic studies have not indicated a correlation between evolution of *Sodalis* and tsetse. Unlike tsetse's P-symbiont *Wigglesworthia*, though, *Sodalis* has been cultured *in vitro*.

VIRAL ENDOSYMBIONTS AND ENDOGENOUS RETROVIRUS

During pregnancy in viviparous mammals, endogenous retroviruses (ERVs) are activated and produced in high quantities during the implantation of the embryo. On one hand, they are hypothesized to act as immunosuppressors, perhaps protecting the embryo from the immune system of the mother; on the other hand viral fusion proteins cause the formation of the placental syncytium in order to limit the exchange of migratory cells between the developing embryo and the body of the mother, where an epithelium will not be adequate because certain blood cells are specialized to be able to insert themselves between adjacent epithelial cells. The ERV is an endogenized form of what was once an infectious retrovirus. The immunodepressive action was important for the infection of the original virus. The fusion proteins may have been a way to spread the infection to other cells by simply merging them with the infected one. It is believed that the ancestors of modern viviparous mammals evolved after an infection of an ancestor with this virus, perhaps improving the ability of the fetus to survive the immune system of the mother.

The human genome project found several thousand ERVs, which are organized into 24 families.

ORGANISM

"Biological form" redirects here. For the informal taxonomic term in botany, see Race (biology) § Physiological race.

"Form of life" redirects here. For the philosophical concept, see Form of life (philosophy).

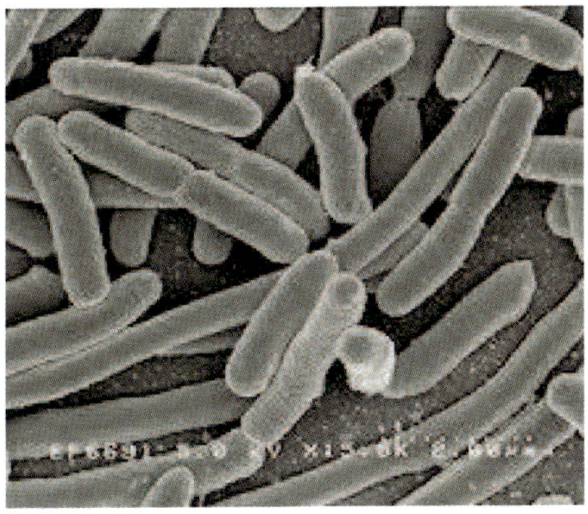

Escherichia coli is a prokaryote.

Polypores are often parasitic on hosts such as trees.

In biology, an **organism** is any contiguous living system, such as an animal, plant, fungus, archaeon, or bacterium. All known types of organisms are capable of some degree of response to stimuli, reproduction with variation, growth and development and homeostasis. An organism consists of one or more cells; when it has one cell it is known as a unicellular organism; and when it has more than one it is known as a multicellular organism. Most unicellular organisms are of microscopic scale and are thus loosely described as microorganisms. Humans are multicellular organisms composed of many trillions of cells grouped into specialized tissues and organs.

An organism may be either a prokaryote or a eukaryote. Prokaryotes are represented by two separate domains—bacteria and archaea. Eukaryotic organisms are characterized by the presence of a membrane-bound cell nucleus and contain additional membrane-bound compartments called organelles (such as mitochondria in animals and plants and plastids in plants and algae, all generally considered to be derived from endosymbiotic bacteria). Fungi, animals and plants are examples of kingdoms of organisms within the eukaryotes.

Estimates on the number of Earth›s current species range from 10 million to 14 million, of which only about 1.2 million have been documented. More than 99% of all species, amounting to over five billion species, that ever lived are estimated to be extinct. In 2016, a set of 355 genes from the last universal common ancestor (LUCA) of all living organisms living was identified.

Etymology

The term "organism" (from Greek, *organismos*, from, *organon*, i.e. "instrument, implement, tool, organ of sense or apprehension") first appeared in the English language in 1703 and took on its current definition by 1834 (Oxford English Dictionary). It is directly related to the term "organization". There is a long tradition of defining organisms as self-organizing beings, going back at least to Immanuel Kant's 1790 *Critique of Judgment.*

An organism may be defined as an assembly of molecules functioning as a more or less stable whole that exhibits the properties of life. Dictionary definitions can be broad, using phrases such as any living structure, such as a plant, animal, fungus or bacterium, capable of growth and reproduction. Many definitions exclude viruses and possible man-made non-organic life forms, as viruses are dependent on the biochemical machinery of a host cell for reproduction. A superorganism is an organism consisting of many individuals working together as a single functional or social unit.

There has been controversy about the best way to define the organism and indeed about whether or not such a definition is necessary. Several contributions are responses to the suggestion that the category of organism may well not be adequate in biology.

Non-cellular Life

Viruses are not typically considered to be organisms because they are incapable of autonomous reproduction, growth or metabolism. This controversy is problematic because some cellular organisms are also incapable of independent survival (but are capable of independent metabolism and procreation) and live as obligatory intracellular parasites. Although viruses have a few enzymes and molecules characteristic of living organisms, they have no metabolism of their own; they cannot synthesize and organize the organic compounds from which they are formed. Naturally, this rules out autonomous reproduction: they can only be passively replicated by the machinery of the host cell. In this sense they are similar to inanimate matter. While viruses sustain no independent metabolism, and thus are usually not classified as organisms, they do have their own genes, and they do evolve by mechanisms similar to the evolutionary mechanisms of organisms.

The most common argument in support of viruses as living organisms is their ability to undergo evolution and replicate through self-assembly. Some scientists argue that viruses neither evolve, nor self- reproduce. In fact, viruses are evolved by their host cells, meaning that there was co-evolution of viruses and host cells. If host cells did not exist, viral evolution would be impossible. This is not true for cells. If viruses did not exist, the direction of cellular evolution could be different, but cells would nevertheless be able to evolve. As for the reproduction, viruses totally rely on hosts› machinery to replicate. The discovery of viral megagenomes with genes coding for energy metabolism and protein synthesis fueled the debate about whether viruses belong in the tree of life. The presence of these genes suggested that viruses were once able to metabolize.

However, it was found later that the genes coding for energy and protein metabolism have cellular origin. Most likely, these genes were acquired through horizontal gene transfer from viral hosts.

Chemistry

Organisms are complex chemical systems, organized in ways that promote reproduction and some measure of sustainability or survival. The same laws that govern non-living chemistry govern the chemical processes of life. It is generally the phenomena of entire organisms that determine their fitness to an environment and therefore the survivability of their DNA-based genes.

Organisms clearly owe their origin, metabolism, and many other internal functions to chemical phenomena, especially the chemistry of large organic molecules. Organisms are complex systems of chemical compounds that, through interaction and environment, play a wide variety of roles.

Organisms are semi-closed chemical systems. Although they are individual units of life (as the definition requires), they are not closed to the environment around them. To operate they constantly take in and release energy. Autotrophs produce usable energy (in the form of organic compounds) using light from the sun or inorganic compounds while heterotrophs take in organic compounds from the environment.

The primary chemical element in these compounds is carbon. The chemical properties of this element such as its great affinity for bonding with other small atoms, including other carbon atoms, and its small size making it capable of forming multiple bonds, make it ideal as the basis of organic life. It is able to form small three-atom compounds (such as carbon dioxide), as well as large chains of many thousands of atoms that can store data (nucleic acids), hold cells together, and transmit information (protein).

Macromolecules

Compounds that make up organisms may be divided into macromolecules and other, smaller molecules. The four groups of macromolecule are nucleic acids, proteins, carbohydrates and lipids. Nucleic acids (specifically deoxyribonucleic acid, or DNA) store genetic data as a sequence of nucleotides. The particular sequence of the four different types of nucleotides (adenine, cytosine, guanine, and thymine) dictate many characteristics that constitute the organism. The sequence is divided up into codons, each of which is a particular sequence of three nucleotides and corresponds to a particular amino acid. Thus a sequence of DNA codes for a particular protein that, due to the chemical properties of the amino acids it is made from, folds in a particular manner and so performs a particular function.

These protein functions have been recognized:

- Enzymes, which catalyze all of the reactions of metabolism
- Structural proteins, such as tubulin, or collagen
- Regulatory proteins, such as transcription factors or cyclins that regulate the cell cycle
- Signaling molecules or their receptors such as some hormones and their receptors
- Defensive proteins, which can include everything from antibodies of the immune system, to toxins (e.g., dendrotoxins of snakes), to proteins that include unusual amino acids like canavanine

A bilayer of phospholipids makes up the membrane of cells that constitutes a barrier, containing everything within the cell and preventing compounds from freely passing into, and out of, the cell. Due to the selective permeability of the phospholipid membraine only specific compounds can pass through it. In some multicellular organisms they serve as a storage of energy and mediate communication between cells. Carbohydrates are more easily broken down than lipids and yield more energy to compare to lipids and proteins. In fact, carbohydrates are the number one source of energy for all living organisms.

Structure

All organisms consist of monomeric units called cells; some contain a single cell (unicellular) and others contain many units (multicellular). Multicellular organisms are able to specialize cells to perform specific functions. A group of such cells is a tissue, and in animals these occur as four basic types, namely epithelium, nervous tissue, muscle tissue, and connective tissue. Several types of tissue work together in the form of an organ to produce a particular function (such as the pumping of the blood by the heart, or as a barrier to the environment as the skin). This pattern continues to a higher level with several organs functioning as an organ system to allow for reproduction, digestion, etc. Many multicellular organisms consist of several organ systems, which coordinate to allow for life.

Cell

The cell theory, first developed in 1839 by Schleiden and Schwann, states that all organisms are composed of one or more cells; all cells come from preexisting cells; and cells contain the hereditary information necessary for regulating cell functions and for transmitting information to the next generation of cells.

There are two types of cells, eukaryotic and prokaryotic. Prokaryotic cells are usually singletons, while eukaryotic cells are usually found in multicellular organisms. Prokaryotic cells lack a nuclear membrane so DNA is unbound within the cell; eukaryotic cells have nuclear membranes.

All cells, whether prokaryotic or eukaryotic, have a membrane, which envelops the cell, separates its interior from its environment, regulates what moves in and out, and maintains the electric potential of the cell. Inside the membrane, a saltycytoplasm takes up most of the cell volume. All cells possess DNA, the hereditary material of genes, and RNA, containing the information necessary to build various proteins such as enzymes, the cell›s primary machinery. There are also other kinds of biomolecules in cells.

All cells share several similar characteristics of:

- Reproduction by cell division (binary fission, mitosis or meiosis).
- Use of enzymes and other proteins coded by DNA genes and made via messenger RNA intermediates and ribosomes.
- Metabolism, including taking in raw materials, building cell components, converting energy, molecules and releasing by-products. The functioning of a cell depends upon its ability to extract and use chemical energy stored in organic molecules. This energy is derived from metabolic pathways.
- Response to external and internal stimuli such as changes in temperature, pH or nutrient levels.

- Cell contents are contained within a cell surface membrane that contains proteins and a lipid bilayer.

Evolution

Last universal common ancestor

Precambrian stromatolites in the Siyeh Formation, Glacier National Park. In 2002, a paper in the scientific journal *Nature* suggested that these 3.5 Gya (billion years old) geological formations contain fossilized cyanobacteria microbes. This suggests they are evidence of one of the earliest known life forms on Earth.

Main article: Last universal common ancestor

Further information: Timeline of the evolutionary history of life

The last universal common ancestor (LUCA) is the most recent organism from which all organisms now living on Earth descend. Thus it is the most recent common ancestor of all current life on Earth. The LUCA is estimated to have lived some 3.5 to 3.8 billion years ago (sometime in the Paleoarchean era). The earliest evidences for life on Earth are graphite found to be biogenic in 3.7 billion-year-old metasedimentary rocks discovered in Western Greenland and microbial mat fossils found in 3.48 billion-year-old sandstone discovered in Western Australia.Although more than 99 percent of all species that ever lived on the planet are estimated to be extinct, there are currently 10–14 million species of life on Earth.

Information about the early development of life includes input from many different fields, including geology and planetary science. These sciences provide information about the history of the Earth and the changes produced by life. However, a great deal of information about the early Earth has been destroyed by geological processes over the course of time.

All organisms are descended from a common ancestor or ancestral gene pool. Evidence for common descent may be found in traits shared between all living organisms. In Darwin›s day, the evidence of shared traits was based solely on visible observation of morphologic similarities, such as the fact that all birds have wings, even those that do not fly.

There is strong evidence from genetics that all organisms have a common ancestor. For example, every living cell makes use of nucleic acids as its genetic material, and uses the same twenty amino acids as the building blocks for proteins. All organisms use the same genetic code (with some extremely rare and minor deviations) to translate nucleic acid sequences into proteins. The universality of these traits strongly suggests common ancestry, because the selection of many of these traits seems arbitrary. Horizontal gene transfer makes it more difficult to study the last universal ancestor. However, the universal use of the same genetic code, same nucleotides, and same amino acids makes the existence of such an ancestor overwhelmingly likely.

Reproduction

Sexual reproduction is widespread among current eukaryotes, and was likely present in the last common ancestor. This is suggested by the finding of a core set of genes for meiosis in the descendants of lineages that diverged early form the eukaryotic evolutionary tree. and Malik et al. It is further supported by evidence that eukaryotes previously regarded as "ancient asexuals", such as *Amoeba*, were likely sexual in the past, and that most present day asexual amoeboid lineages likely arose recently and independently.

In prokaryotes, natural bacterial transformation involves the transfer of DNA from one bacterium to another and integration of the donor DNA into the recipient chromosome by recombination. Natural bacterial transformation is considered to be a primitive sexual process and occurs in both bacteria and archaea, although it has been studied mainly in bacteria. Transformation is clearly a bacterial adaptation and not an accidental occurrence, because it depends on numerous gene products that specifically interact with each other to enter a state of natural competence to perform this complex process.Transformation is a common mode of DNA transfer among prokaryotes.

Horizontal Gene Transfer

The ancestry of living organisms has traditionally been reconstructed from morphology, but is increasingly supplemented with phylogenetics—the reconstruction of phylogenies by the comparison of genetic (DNA) sequence.

Sequence comparisons suggest recent horizontal transfer of many genes among diverse species including across the boundaries of phylogenetic "domains". Thus determining the phylogenetic history of a species can not be done conclusively by determining evolutionary trees for single genes.

Biologist Gogarten suggests the original metaphor of a tree no longer fits the data from recent genome research, therefore biologists (should) use the metaphor of a mosaic to describe the different histories combined in individual genomes and use (the) metaphor of a net to visualize the rich exchange and cooperative effects of HGT among microbes.

Future of Life (Cloning and Synthetic Organisms)

Modern biotechnology is challenging traditional concepts of organism and species. Cloning is the process of creating a new multicellular organism, genetically identical to another, with the potential of creating entirely new species of organisms. Cloning is the subject of much ethical debate.

In 2008 the J. Craig Venter Institute assembled a synthetic bacterial genome, *Mycoplasma genitalium*, by using recombination in yeast of 25 overlapping DNA fragments in a single step. *The use of yeast recombination greatly simplifies the assembly of large DNA molecules from both synthetic and natural fragments.* Other companies, such as Synthetic Genomics, have already been formed to take advantage of the many commercial uses of custom designed genomes.

CHLOROPLAST

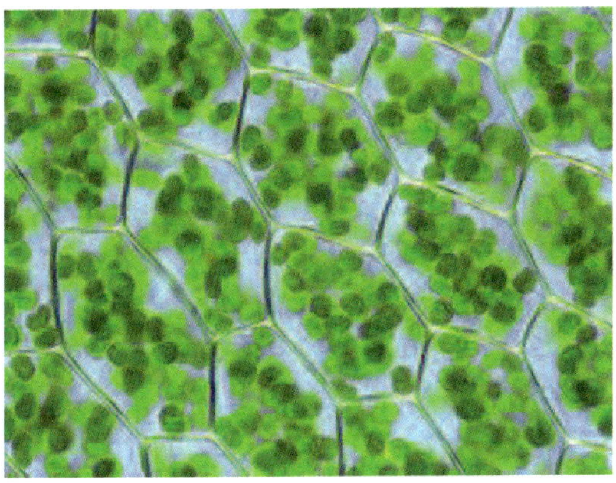

Chloroplasts visible in the cells of *Plagiomnium affine*, the many-fruited thyme moss

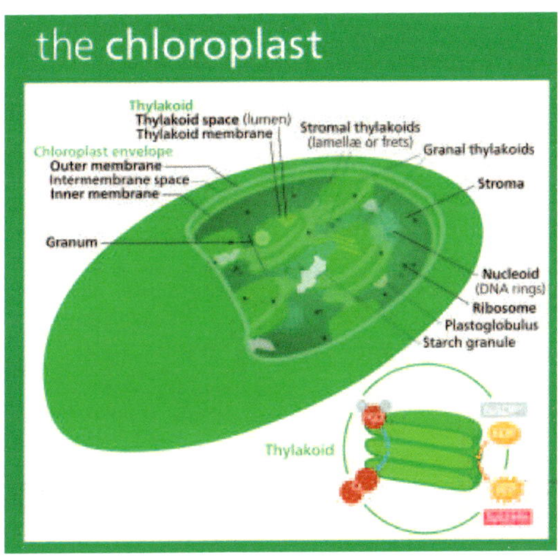

Structure of a typical higher-plant chloroplast

Chloroplasts are organelles, specialized subunits, in plant and algal cells. Their discovery inside plant cells is usually credited to Julius von Sachs (1832–1897), an influential botanist and author of standard botanical textbooks – sometimes called "The Father of Plant Physiology".

Chloroplasts› main role is to conduct photosynthesis, where the photosynthetic pigment chlorophyll captures the energy from sunlight and converts it and stores it in the energy-storage molecules ATP and NADPH while freeing oxygen from water. They then use the ATP and NADPH to make organic molecules from carbon dioxide in a process known as the Calvin cycle. Chloroplasts carry out a number of other functions, including fatty acid synthesis, much amino acid synthesis, and the immune response in plants. The number of chloroplasts per cell varies from one, in unicellular algae, up to 100 in plants like Arabidopsis and wheat.

A chloroplast is a type of organelle known as a plastid, characterized by its high concentration of chlorophyll, other plastid types, such as the leucoplast and the chromoplast, contain little chlorophyll and do not carry out photosynthesis.

Chloroplasts are highly dynamic—they circulate and are moved around within plant cells, and occasionally pinch in two to reproduce. Their behavior is strongly influenced by environmental factors like light color and intensity. Chloroplasts, like mitochondria, contain their own DNA, which is thought to be inherited from their ancestor—a photosynthetic cyanobacterium that was engulfed by an early eukaryotic cell. Chloroplasts cannot be made by the plant cell and must be inherited by each daughter cell during cell division.

With one exception (the amoeboid *Paulinella chromatophora*), all chloroplasts can probably be traced back to a single endosymbiotic event, when a cyanobacterium was engulfed by the eukaryote. Despite this, chloroplasts can be found in an extremely wide set of organisms, some not even directly related to each other—a consequence of many secondary and even tertiary endosymbiotic events.

The word *chloroplast* is derived from the Greek words *chloros*, which means green, and *plastes*, which means "the one who forms".

Discovery

The first definitive description of a chloroplast (*Chlorophyllkörnen*, "grain of chlorophyll") was given by Hugo von Mohl in 1837 as discrete bodies within the green plant cell. In 1883, A. F. W. Schimper would name these bodies as "chloroplastids" (*Chloroplastiden*). In 1884, Eduard Strasburger adopted the term "chloroplasts (*Chloroplasten*).

Chloroplast lineages and evolution

Chloroplasts are one of many types of organelles in the plant cell. They are considered to have originated from cyanobacteria through endosymbiosis—when a eukaryotic cell engulfed a photosynthesizing cyanobacterium that became a permanent resident in the cell. Mitochondria are thought to have come from a similar event, where an aerobic prokaryote was engulfed.This origin of chloroplasts was first suggested by the Russian biologist Konstantin Mereschkowski in 1905 after Andreas Schimper observed in 1883 that chloroplasts closely resemble cyanobacteria. Chloroplasts are only found in plants, algae, and the amoeboid *Paulinella chromatophora*.

Cyanobacterial Ancestor

Cyanobacteria are considered the ancestors of chloroplasts. They are sometimes called blue-green algae even though they are prokaryotes. They are a diverse phylum of bacteria capable of carrying out photosynthesis, and are gram-negative, meaning that they have two cell membranes. Cyanobacteria also contain a peptidoglycan cell wall, which is thicker than in other gram-negative bacteria, and which is located between their two cell membranes. Like chloroplasts, they have thylakoids within. On the thylakoid membranes are photosynthetic pigments, including chlorophyll *a*. Phycobilins are also common cyanobacterial pigments, usually organized into hemispherical phycobilisomes attached to the outside of the thylakoid membranes (phycobilins are not shared with all chloroplasts though).

Both chloroplasts and cyanobacteria have a double membrane, DNA, ribosomes, and thylakoids. Both the chloroplast and cyanobacterium depicted are idealized versions (the chloroplast is that of a higher plant)—a lot of diversity exists among chloroplasts and cyanobacteria.

Primary Endosymbiosis

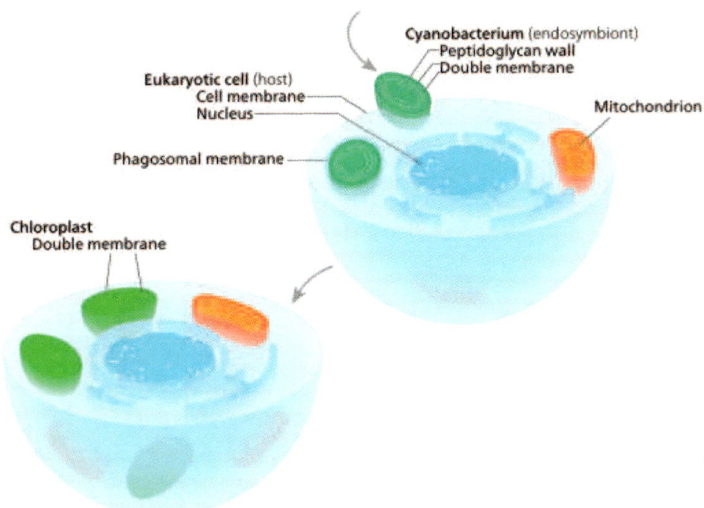

Primary Endosymbiosis

A eukaryote with mitochondria engulfed a cyanobacterium in an event of serial primary endosymbiosis, creating a lineage of cells with both organelles. It is important to note that the cyanobacterial endosymbiont already had a double membrane—the phagosomal vacuole-derived membrane was lost.

Somewhere around a billion years ago, a free-living cyanobacterium entered an early eukaryotic cell, either as food or as an internal parasite, but managed to escape the phagocytic vacuole it was contained in. The two innermost lipid-bilayer membranes that surround all chloroplasts correspond to the outer and inner membranes of the ancestral cyanobacterium's gram negative cell wall, and not the phagosomal membrane from the host, which was probably lost.The new cellular resident quickly became an advantage, providing food for the eukaryotic host, which allowed it to live within it. Over time, the cyanobacterium was assimilated, and many of its genes were lost or transferred to the nucleus of the host. Some of its proteins were then synthesized in the cytoplasm of the host cell, and imported back into the chloroplast (formerly the cyanobacterium). Separately, somewhere around 100 million years ago, it happened again and led to the amoeboid *Paulinella chromatophora*.

This event is called *endosymbiosis*, or "cell living inside another cell". The cell living inside the other cell is called the *endosymbiont*; the endosymbiont is found inside the *host cell*.

Chloroplasts are believed to have arisen after mitochondria, since all eukaryotes contain mitochondria, but not all have chloroplasts. This is called *serial endosymbiosis*—an early eukaryote engulfing the mitochondrion ancestor, and some descendants of it then engulfing the chloroplast ancestor, creating a cell with both chloroplasts and mitochondria.

Whether or not chloroplasts came from a single endosymbiotic event, or many independent engulfments across various eukaryotic lineages, has been long debated. It is now generally held that most organisms with chloroplasts either share a single ancestor or obtained their chloroplast from organisms that share a common ancestor that took in a cyanobacterium 600–1600 million years ago. The exception is the amoeboid *Paulinella chromatophora*, which descends from an ancestor that took in a cyanobacterium 90–140 million years ago.

These chloroplasts, which can be traced back directly to a cyanobacterial ancestor, are known as *primary plastids* (*plastid* in this context means almost the same thing as chloroplast). All primary chloroplasts belong to one of four chloroplast lineages—the glaucophyte chloroplast lineage, the amoeboid *Paulinella chromatophora* lineage, the rhodophyte (red algal) chloroplast lineage, or the chloroplastidan (green) chloroplast lineage. The rhodophyte and chloroplastidan lineages are the largest, with chloroplastidan (green) being the one that contains the land plants.

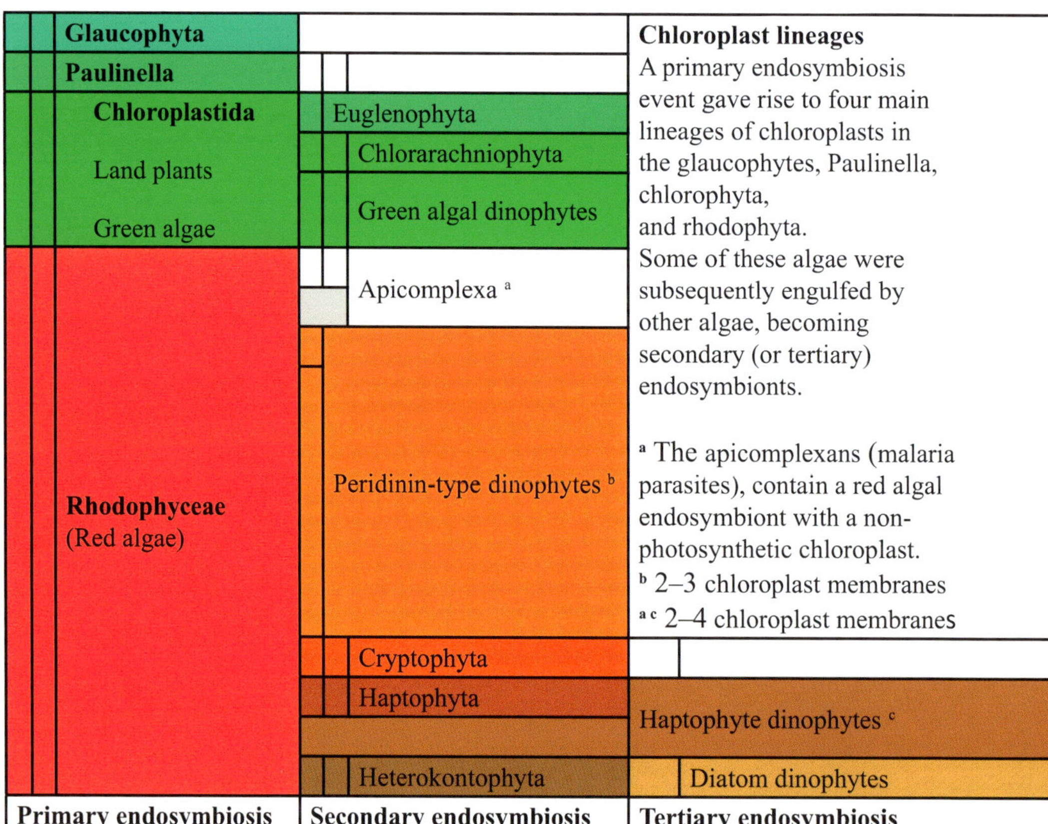

Glaucophyta

The alga *Cyanophora*, a glaucophyte, is thought to be one of the first organisms to contain a chloroplast. The glaucophyte chloroplast group is the smallest of the three primary chloroplast lineages, being found in only 13 species, and is thought to be the one that branched off the earliest.Glaucophytes have chloroplasts that retain a peptidoglycan wall between their double membranes, like their cyanobacterial parent. For this reason, glaucophyte chloroplasts are also known as *muroplasts*. Glaucophyte chloroplasts also contain concentric unstacked thylakoids, which surround a carboxysome - an icosahedral structure that glaucophyte chloroplasts and cyanobacteria keep their carbon fixation enzyme rubisco in. The starch that they synthesize collects outside the chloroplast. Like cyanobacteria, glaucophyte chloroplast thylakoids are studded with light collecting structures called phycobilisomes. For these reasons, glaucophyte chloroplasts are considered a primitive intermediate between cyanobacteria and the more evolved chloroplasts in red algae and plants.

Diversity of red algae Clockwise from top left: *Bornetia secundiflora*, *Peyssonnelia squamaria*, *Cyanidium*, *Laurencia*, *Callophyllis laciniata*. Red algal chloroplasts are characterized by phycobilin pigments which often give them their reddish color.

Rhodophyceae (red algae)

The rhodophyte, or red algal chloroplast group is another large and diverse chloroplast lineage.Rhodophyte chloroplasts are also called *rhodoplasts*, literally "red chloroplasts".

Rhodoplasts have a double membrane with an intermembrane space and phycobilin pigments organized into phycobilisomes on the thylakoid membranes, preventing their thylakoids from stacking. Some contain pyrenoids.Rhodoplasts have chlorophyll *a* and phycobilins for photosynthetic pigments; the phycobilin phycoerytherin is responsible for giving many red algae their distinctive red color.However, since they also contain the blue-green chlorophyll *a* and other pigments, many are reddish to purple from the combination. The red phycoerytherin pigment is an adaptation to help red algae catch more sunlight in deep water—as such, some red algae that live in shallow water have less phycoerytherin in their rhodoplasts, and can appear more greenish. Rhodoplasts synthesize a form of starch called floridean starch, which collects into granules outside the rhodoplast, in the cytoplasm of the red alga.

Chloroplastida (green algae and plants)

Diversity of green algae Clockwise from top left: *Scenedesmus*, *Micrasterias*, *Hydrodictyon*, *Stigeoclonium*, *Volvox*. Green algal chloroplasts are characterized by their pigments chlorophyll *a* and chlorophyll *b* which give them their green color.

The chloroplastidan chloroplasts, or green chloroplasts, are another large, highly diverse primary chloroplast lineage. Their host organisms are commonly known as the green algae and land plants. They differ from glaucophyte and red algal chloroplasts in that they have lost their phycobilisomes, and contain chlorophyll *b* instead. Most green chloroplasts are (obviously) green, though some aren›t, like some forms of *Hæmatococcus pluvialis*, due to accessory pigments that override the chlorophylls' green colors. Chloroplastidan chloroplasts have lost the peptidoglycan wall between their double membrane, leaving an intermembrane space.Some plants seem to have kept the genes for the synthesis of the peptidoglycan layer, though they›ve been repurposed for use in chloroplast division instead.

Most of the chloroplasts depicted in this article are green chloroplasts.

Green algae and plants keep their starch *inside* their chloroplasts, and in plants and some algae, the chloroplast thylakoids are arranged in grana stacks. Some green algal chloroplasts contain a structure called a pyrenoid, which is functionally similar to the glaucophyte carboxysome in that it is where rubisco and CO_2 are concentrated in the chloroplast.

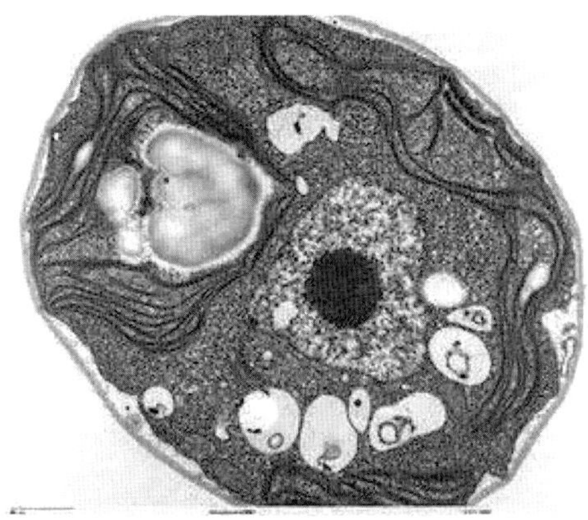

Transmission electron micrograph of *Chlamydomonas reinhardtii*, a green alga that contains a pyrenoid surrounded by starch.

Helicosporidium

Helicosporidium is a genus of nonphotosynthetic parasitic green algae that is thought to contain a vestigial chloroplast. Genes from a chloroplastand nuclear genes indicating the presence of a chloroplast have been found in Helicosporidium even if nobody's seen the chloroplast itself.

Secondary and Tertiary Endosymbiosis

Many other organisms obtained chloroplasts from the primary chloroplast lineages through secondary endosymbiosis—engulfing a red or green alga that contained a chloroplast. These chloroplasts are known as secondary plastids.

While primary chloroplasts have a double membrane from their cyanobacterial ancestor, secondary chloroplasts have additional membranes outside of the original two, as a result of the secondary endosymbiotic event, when a nonphotosynthetic eukaryote engulfed a chloroplast-containing alga but failed to digest it—much like the cyanobacterium at the beginning of this story.The engulfed alga was broken down, leaving only its chloroplast, and sometimes its cell membrane and nucleus, forming a chloroplast with three or four membranes—the two cyanobacterial membranes, sometimes the eaten alga›s cell membrane, and the phagosomal vacuole from the host's cell membrane.

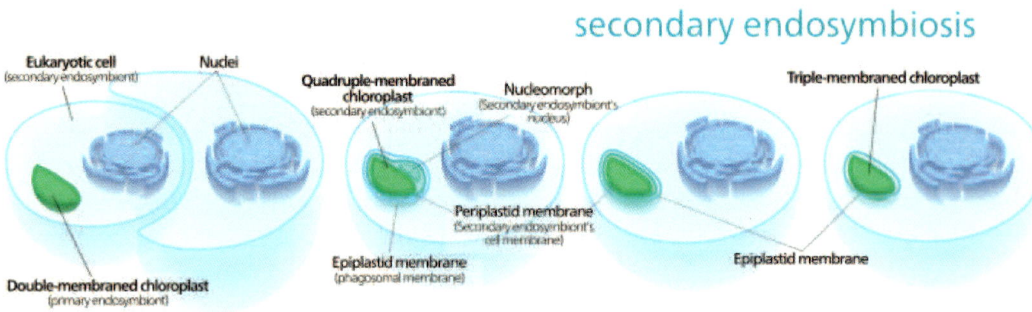

Secondary endosymbiosis consisted of a eukaryotic alga being engulfed by another eukaryote, forming a chloroplast with three or four membranes.

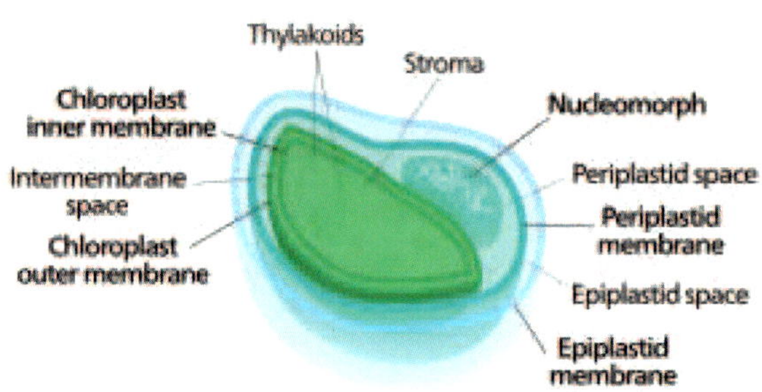

Diagram of a four membraned chloroplast containing a nucleomorph.

The genes in the phagocytosed eukaryote's nucleus are often transferred to the secondary host's nucleus. Cryptomonads and chlorarachniophytes retain the phagocytosed eukaryote's nucleus, an object called a nucleomorph, located between the second and third membranes of the chloroplast.

All secondary chloroplasts come from green and red algae—no secondary chloroplasts from glaucophytes have been observed, probably because glaucophytes are relatively rare in nature, making them less likely to have been taken up by another eukaryote.

Green Algal Derived Chloroplasts

Green algae have been taken up by the euglenids, chlorarachniophytes, a lineage of dinoflagellates, and possibly the ancestor of the chromalveolates in three or four separate engulfments. Many green algal derived chloroplasts contain pyrenoids, but unlike chloroplasts in their green algal ancestors, starch collects in granules outside the chloroplast.

Euglena, a euglenophyte, contains secondary chloroplasts from green algae.

Euglenophytes

Euglenophytes are a group of common flagellated protists that contain chloroplasts derived from a green alga. Euglenophyte chloroplasts have three membranes—it is thought that the membrane of the primary endosymbiont was lost, leaving the cyanobacterial membranes, and the secondary host›s phagosomal membrane.Euglenophyte chloroplasts have a pyrenoid and thylakoids stacked in groups of three. Starch is stored in the form of paramylon, which is contained in membrane-bound granules in the cytoplasm of the euglenophyte.

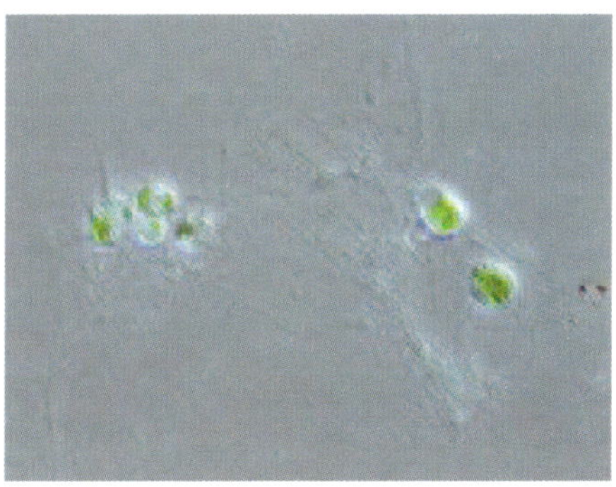

Chlorarachnion reptans is a chlorarachniophyte. Chlorarachniophytes replaced their original red algal endosymbiont with a green alga.

Chlorarachniophytes

Chlorarachniophytes are a rare group of organisms that also contain chloroplasts derived from green algae, though their story is more complicated than that of the euglenophytes. The ancestor of chlorarachniophytes is thought to have been a chromalveolate, a eukaryote with a *red* algal derived chloroplast. It is then thought to have lost its first red algal chloroplast, and later engulfed a green alga, giving it its second, green algal derived chloroplast.

Chlorarachniophyte chloroplasts are bounded by four membranes, except near the cell membrane, where the chloroplast membranes fuse into a double membrane. Their thylakoids are arranged in loose stacks of three. Chlorarachniophytes have a form of starch called chrysolaminarin, which they store in the cytoplasm, often collected around the chloroplast pyrenoid, which bulges into the cytoplasm.

Chlorarachniophyte chloroplasts are notable because the green alga they are derived from has not been completely broken down—its nucleus still persists as a nucleomorph found between the second and third chloroplast membranes—the periplastid space, which corresponds to the green alga›s cytoplasm.

Early Chromalveolates

Recent research has suggested that the ancestor of the chromalveolates acquired a green algal prasinophyte endosymbiont. The green algal derived chloroplast was lost and replaced with a red algal derived chloroplast, but not before contributing some of its genes to the early chromalveolate's nucleus. The presence of both green algal and red algal genes in chromalveolates probably helps them thrive under fluctuating light conditions.

Red algal Derived Chloroplasts *(chromalveolate chloroplasts)*

Like green algae, red algae have also been taken up in secondary endosymbiosis, though it is thought that all red algal derived chloroplasts are descended from a single red alga that was engulfed by an early chromalveolate, giving rise to the chromalveolates, some of which, like the ciliates, subsequently lost the chloroplast. This is still debated though.

Pyrenoids and stacked thylakoids are common in chromalveolate chloroplasts, and the outermost membrane of many are continuous with the rough endoplasmic reticulum and studded with ribosomes. They have lost their phycobilisomes and exchanged them for chlorophyll *c*, which isn't found in primary red algal chloroplasts themselves.

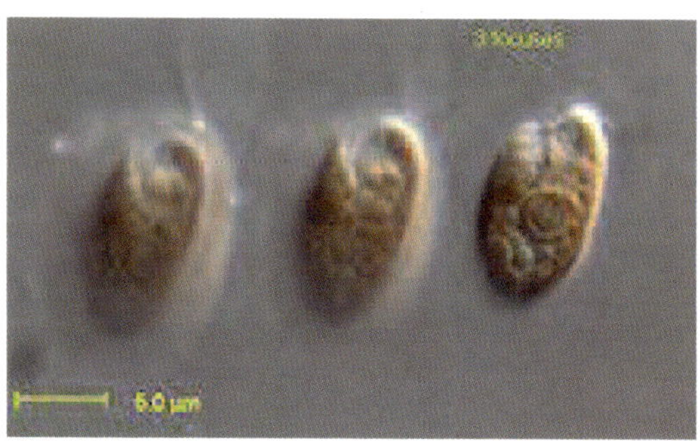

Rhodomonas salina is a cryptophyte.

Cryptophytes

Cryptophytes, or cryptomonads are a group of algae that contain a red-algal derived chloroplast. Cryptophyte chloroplasts contain a nucleomorph that superficially resembles that of the chlorarachniophytes. Cryptophyte chloroplasts have four membranes, the outermost of which is continuous with the rough endoplasmic reticulum. They synthesize ordinary starch, which is stored in granules found in the periplastid space—outside the original double membrane, in the place that corresponds to the red alga's cytoplasm. Inside cryptophyte chloroplasts is a pyrenoid and thylakoids in stacks of two.

Their chloroplasts do not have phycobilisomes, but they do have phycobilin pigments which they keep in their thylakoid space, rather than anchored on the outside of their thylakoid membranes.

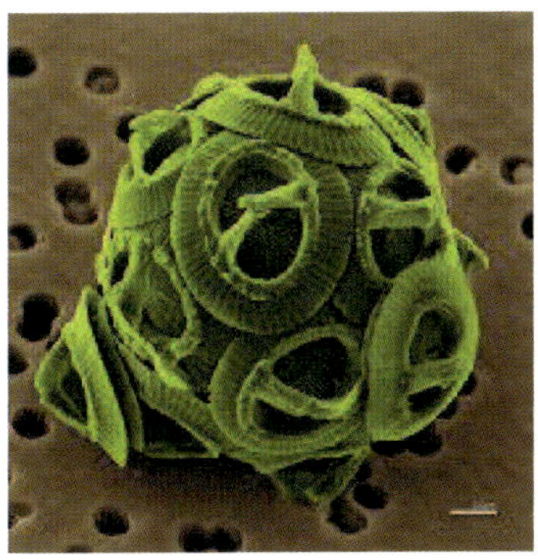

Scanning electron micrograph of *Gephyrocapsa oceanica*, a haptophyte.

Haptophytes

Haptophytes are similar and closely related to cryptophytes, and are thought to be the first chromalveolates to branch off. Their chloroplasts lack a nucleomorph, their thylakoids are in stacks of three, and they synthesize chrysolaminarin sugar, which they store completely outside of the chloroplast, in the cytoplasm of the haptophyte.

Heterokontophytes (stramenopiles)

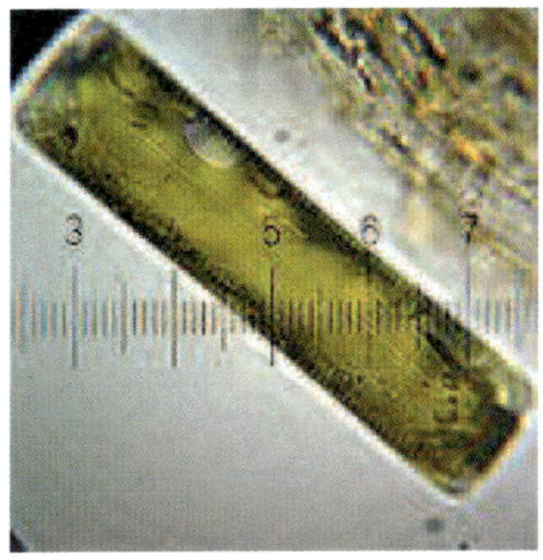

The photosynthetic pigments present in their chloroplasts give diatoms a greenish-brown color.

The heterokontophytes, also known as the stramenopiles, are a very large and diverse group of algae that also contain red algal derived chloroplasts.Heterokonts include the diatoms and the brown algae, golden algae, and yellow-green algae.

Heterokont chloroplasts are very similar to haptophyte chloroplasts, containing a pyrenoid, triplet thylakoids, and with some exceptions, having an epiplastid membrane connected to the endoplasmic reticulum. Like haptophytes, heterokontophytes store sugar in chrysolaminarin granules in the cytoplasm.Heterokontophyte chloroplasts contain chlorophyll *a* and with a few exceptions chlorophyll *c*, but also have carotenoids which give them their many colors.

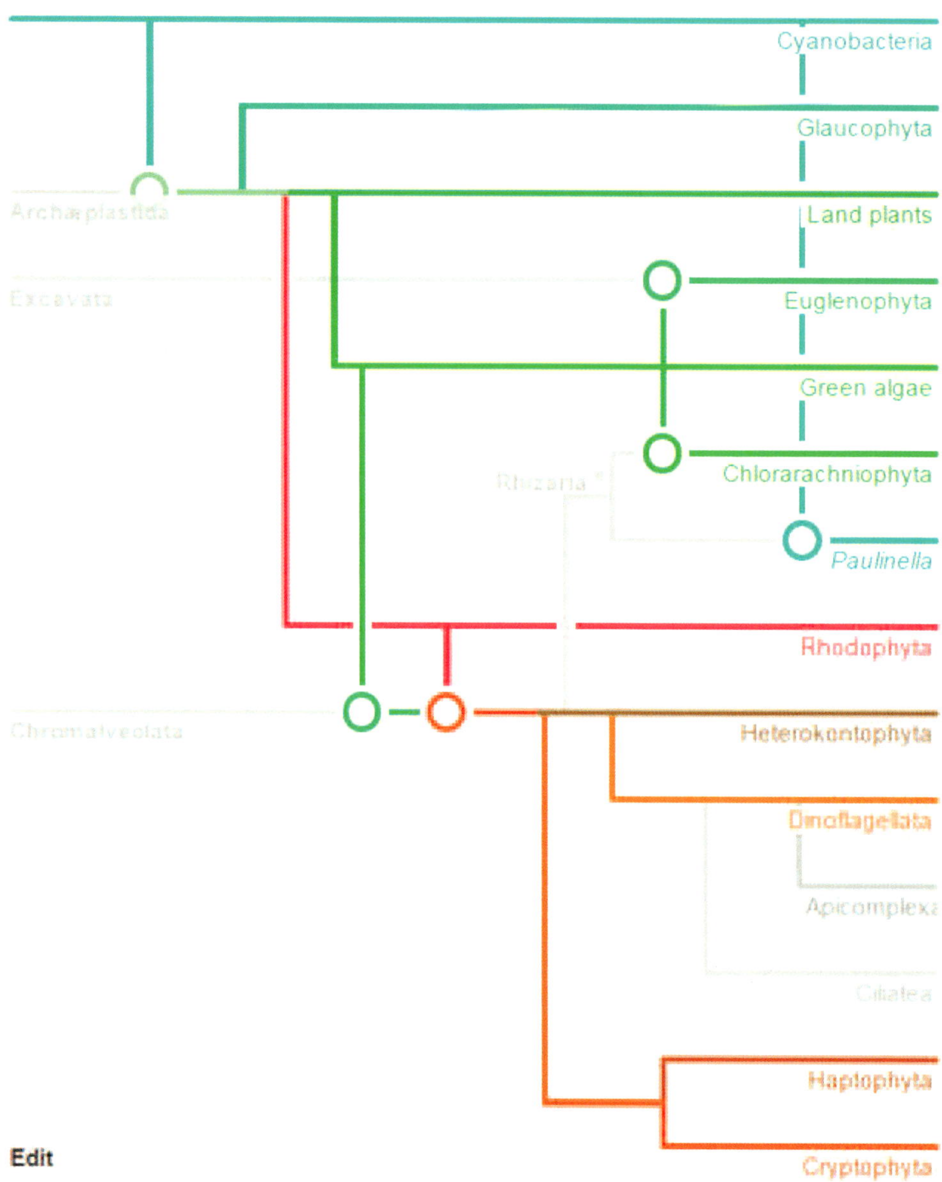

Edit

Possible cladogram of chloroplast evolution

Circles represent endosymbiotic events. For clarity, dinophyte tertiary endosymbioses and many nonphotosynthetic lineages have been omitted.

Apicomplexans

Apicomplexans are another group of chromalveolates. Like the helicosproidia, they›re parasitic, and have a nonphotosynthetic chloroplast. They were once thought to be related to the helicosproidia, but it is now known that the helicosproida are green algae rather than chromalveolates. The apicomplexans include *Plasmodium*, the malaria parasite. Many apicomplexans keep a vestigial red algal derived chloroplast called an apicoplast, which they inherited from their ancestors. Other apicomplexans like *Cryptosporidium* have lost the chloroplast completely. Apicomplexans store their energy in amylopectin starch granules that are located in their cytoplasm, even though they are nonphotosynthetic.

Apicoplasts have lost all photosynthetic function, and contain no photosynthetic pigments or true thylakoids. They are bounded by four membranes, but the membranes are not connected to the endoplasmic reticulum. The fact that apicomplexans still keep their nonphotosynthetic chloroplast around demonstrates how the chloroplast carries out important functions other than photosynthesis. Plant chloroplasts provide plant cells with many important things besides sugar, and apicoplasts are no different—they synthesize fatty acids, isopentenyl pyrophosphate, iron-sulfur clusters, and carry out part of the heme pathway. This makes the apicoplast an attractive target for drugs to cure apicomplexan-related diseases. The most important apicoplast function is isopentenyl pyrophosphate synthesis—in fact, apicomplexans die when something interferes with this apicoplast function, and when apicomplexans are grown in an isopentenyl pyrophosphate-rich medium, they dump the organelle.

Dinophytes

The dinoflagellates are yet another very large and diverse group of protists, around half of which are (at least partially) photosynthetic.

Most dinophyte chloroplasts are secondary red algal derived chloroplasts, like other chromalveolate chloroplasts. Many other dinophytes have lost the chloroplast (becoming the nonphotosynthetic kind of dinoflagellate), or replaced it though *tertiary* endosymbiosis—the engulfment of another chromalveolate containing a red algal derived chloroplast. Others replaced their original chloroplast with a green algal derived one.

Most dinophyte chloroplasts contain at least the photosynthetic pigments chlorophyll *a*, chlorophyll c_2, *beta*-carotene, and at least one dinophyte-unique xanthophyll (peridinin, dinoxanthin, or diadinoxanthin), giving many a golden-brown color. All dinophytes store starch in their cytoplasm, and most have chloroplasts with thylakoids arranged in stacks of three.

Peridinin-Containing Dinophyte Chloroplast

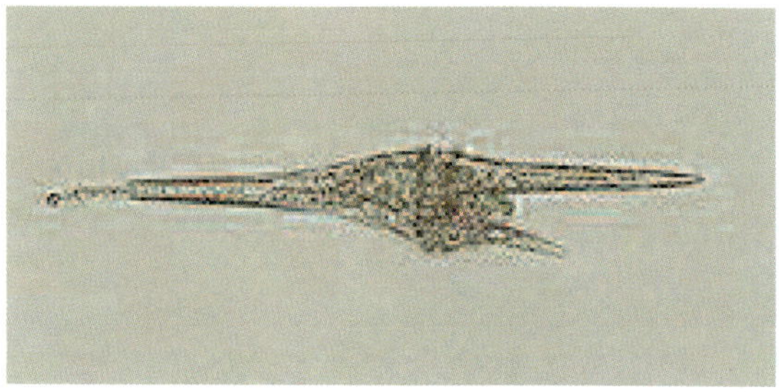

Ceratium furca, a peridinin-containing dinophyte

The most common dinophyte chloroplast is the peridinin-type chloroplast, characterized by the carotenoid pigment peridinin in their chloroplasts, along with chlorophyll *a* and chlorophyll c_2. Peridinin is not found in any other group of chloroplasts. The peridinin chloroplast is bounded by three membranes (occasionally two), having lost the red algal endosymbiont's original cell membrane. The outermost membrane is not connected to the endoplasmic reticulum. They contain a pyrenoid, and have triplet-stacked thylakoids. Starch is found outside the chloroplast. An important feature of these chloroplasts is that their chloroplast DNA is highly reduced and fragmented into many small circles. Most of the genome has migrated to the nucleus, and only critical photosynthesis-related genes remain in the chloroplast.

The peridinin chloroplast is thought to be the dinophytes› original chloroplast, which has been lost, reduced, replaced, or has company in several other dinophyte lineages.

Fucoxanthin-Containing Dinophyte Chloroplasts *(haptophyte endosymbionts)*

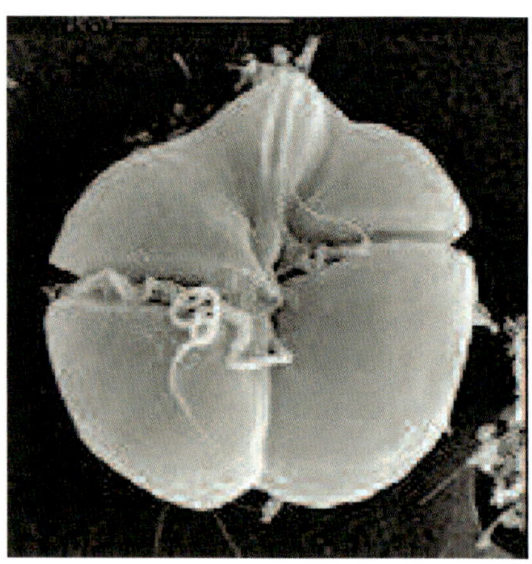

Karenia brevis is a fucoxanthin-containing dynophyte responsible for algal blooms called "red tides.

The fucoxanthin dinophyte lineages (including *Karlodinium* and *Karenia*) lost their original red algal derived chloroplast, and replaced it with a new chloroplast derived from a haptophyte endosymbiont. *Karlodinium* and *Karenia* probably took up different heterokontophytes. Because the haptophyte chloroplast has four membranes, tertiary endosymbiosis would be expected to create a six membraned chloroplast, adding the haptophyte's cell membrane and the dinophyte's phagosomal vacuole. However, the haptophyte was heavily reduced, stripped of a few membranes and its nucleus, leaving only its chloroplast (with its original double membrane), and possibly one or two additional membranes around it.

Fucoxanthin-containing chloroplasts are characterized by having the pigment fucoxanthin (actually 19′-hexanoyloxy-fucoxanthin and/or 19′-butanoyloxy-fucoxanthin) and no peridinin. Fucoxanthin is also found in haptophyte chloroplasts, providing evidence of ancestry.

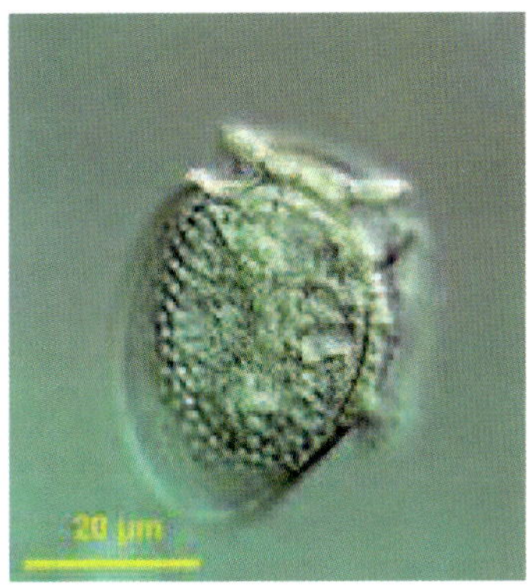

Dinophysis acuminata has chloroplasts taken from a cryptophyte.

Cryptophyte Derived Dinophyte Chloroplast

Members of the genus *Dinophysis* have a phycobilin-containing chloroplast taken from a cryptophyte. However, the cryptophyte is not an endosymbiont—only the chloroplast seems to have been taken, and the chloroplast has been stripped of its nucleomorph and outermost two membranes, leaving just a two-membraned chloroplast. Cryptophyte chloroplasts require their nucleomorph to maintain themselves, and *Dinophysis* species grown in cell culture alone cannot survive, so it is possible (but not confirmed) that the *Dinophysis* chloroplast is a kleptoplast—if so, *Dinophysis* chloroplasts wear out and *Dinophysis* species must continually engulf cryptophytes to obtain new chloroplasts to replace the old ones.]

Diatom Derived Dinophyte Chloroplasts

Some dinophytes, like *Kryptoperidinium* and *Durinskia* have a diatom (heterokontophyte) derived chloroplast. These chloroplasts are bounded by up to *five* membranes, (depending on whether you count the entire diatom endosymbiont as the chloroplast, or just the red algal derived chloroplast inside it). The diatom endosymbiont has been reduced relatively little—it still retains its original mitochondria, and has endoplasmic reticulum, ribosomes, a nucleus, and of course, red algal derived chloroplasts—practically a complete cell, all inside the host›s endoplasmic reticulum lumen. However the diatom endosymbiont can›t store its own food—its starch is found in granules in the dinophyte host›s cytoplasm instead. The diatom endosymbiont›s nucleus is present, but it probably can›t be called a nucleomorph because it shows no sign of genome reduction, and might have even been *expanded*. Diatoms have been engulfed by dinoflagellates at least three times.

The diatom endosymbiont is bounded by a single membrane, inside it are chloroplasts with four membranes. Like the diatom endosymbiont's diatom ancestor, the chloroplasts have triplet thylakoids and pyrenoids.

In some of these genera, the diatom endosymbiont's chloroplasts aren't the only chloroplasts in the dinophyte. The original three-membraned peridinin chloroplast is still around, converted to an eyespot.

Prasinophyte (green algal) Derived Dinophyte Chloroplast

Lepidodinium viride and its close relatives are dinophytes that lost their original peridinin chloroplast and replaced it with a green algal derived chloroplast (more specifically, a prasinophyte). *Lepidodinium* is the only dinophyte that has a chloroplast that's not from the rhodoplast lineage. The chloroplast is surrounded by two membranes and has no nucleomorph—all the nucleomorph genes have been transferred to the dinophyte nucleus. The endosymbiotic event that led to this chloroplast was serial secondary endosymbiosis rather than tertiary endosymbiosis—the endosymbiont was a green alga containing a primary chloroplast (making a secondary chloroplast).

Chromatophores

While most chloroplasts originate from that first set of endosymbiotic events, *Paulinella chromatophora* is an exception that acquired a photosynthetic cyanobacterial endosymbiont more recently. It is not clear whether that symbiont is closely related to the ancestral chloroplast of other eukaryotes. Being in the early stages of endosymbiosis, *Paulinella chromatophora* can offer some insights into how chloroplasts evolved. *Paulinella* cells contain one or two sausage shaped blue-green photosynthesizing structures called chromatophores, descended from the cyanobacterium *Synechococcus*. Chromatophores cannot survive outside their host. Chromatophore DNA is about a million base pairs long, containing around 850 protein encoding genes—far less than the three million base pair *Synechococcus* genome, but much larger than the approximately 150,000 base pair genome of the more assimilated chloroplast. Chromatophores have transferred much less of their DNA to the nucleus of their host. About 0.3–0.8% of the nuclear DNA in *Paulinella* is from the chromatophore, compared with 11–14% from the chloroplast in plants.

Kleptoplastidy

In some groups of mixotrophic protists, like some dinoflagellates, chloroplasts are separated from a captured alga or diatom and used temporarily. These klepto chloroplasts may only have a lifetime of a few days and are then replaced.

Chloroplast DNA

Chloroplasts have their own DNA, often abbreviated as ctDNA, or **cpDNA**. It is also known as the plastome. Its existence was first proved in 1962, and first sequenced in 1986—when two Japanese research teams sequenced the chloroplast DNA of liverwort and tobacco. Since then, hundreds of chloroplast DNAs from various species have been sequenced, but they›re mostly those of land plants and green algae—glaucophytes, red algae, and other algal groups are extremely underrepresented, potentially introducing some bias in views of "typical" chloroplast DNA structure and content.

Molecular Structure

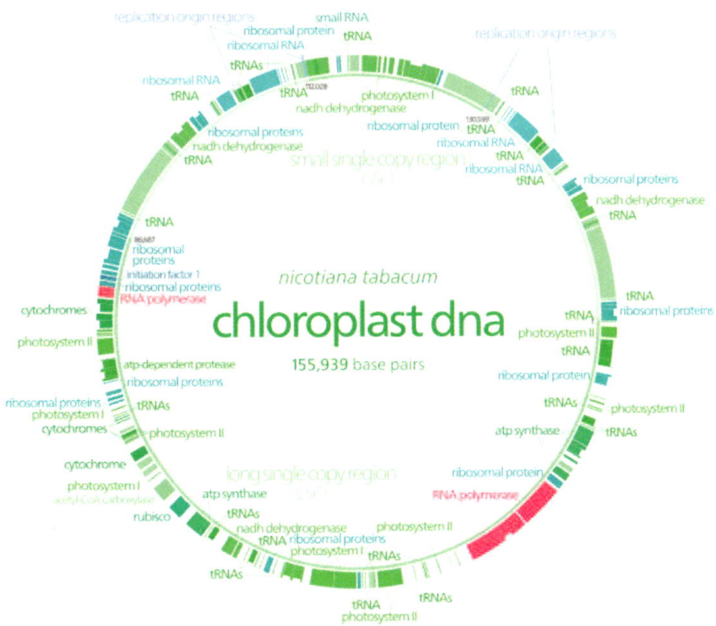

Chloroplast DNA Interactive gene map of chloroplast DNA from *Nicotiana tabacum*. Segments with labels on the inside reside on the B strand of DNA, segments with labels on the outside are on the A strand. Notches indicate introns.

With few exceptions, most chloroplasts have their entire chloroplast genome combined into a single large circular DNA molecule, typically 120,000–170,000 base pairs long. They can have a contour length of around 30–60 micrometers, and have a mass of about 80–130 million daltons.

While usually thought of as a circular molecule, there is some evidence that chloroplast DNA molecules more often take on a linear shape.

Inverted Repeats

Many chloroplast DNAs contain two *inverted repeats*, which separate a long single copy section (LSC) from a short single copy section (SSC). While a given pair of inverted repeats are rarely completely identical, they are always very similar to each other, apparently resulting from concerted evolution.

The inverted repeats vary wildly in length, ranging from 4,000 to 25,000 base pairs long each and containing as few as four or as many as over 150 genes. Inverted repeats in plants tend to be at the upper end of this range, each being 20,000–25,000 base pairs long.

The inverted repeat regions are highly conserved among land plants, and accumulate few mutations. Similar inverted repeats exist in the genomes of cyanobacteria and the other two chloroplast lineages (glaucophyta and rhodophyceae), suggesting that they predate the chloroplast, though some chloroplast DNAs have since lost or flipped the inverted repeats (making them direct repeats). It is possible that the inverted repeats help stabilize the rest of the chloroplast genome, as chloroplast DNAs which have lost some of the inverted repeat segments tend to get rearranged more.

Nucleoids

New chloroplasts may contain up to 100 copies of their DNA, though the number of chloroplast DNA copies decreases to about 15–20 as the chloroplasts age. They are usually packed into nucleoids, which can contain several identical chloroplast DNA rings. Many nucleoids can be found in each chloroplast. In primitive red algae, the chloroplast DNA nucleoids are clustered in the center of the chloroplast, while in green plants and green algae, the nucleoids are dispersed throughout the stroma.

Though chloroplast DNA is not associated with true histones, in red algae, similar proteins that tightly pack each chloroplast DNA ring into a nucleoid have been found.

DNA Replication

The Leading Model of cpDNA Replication

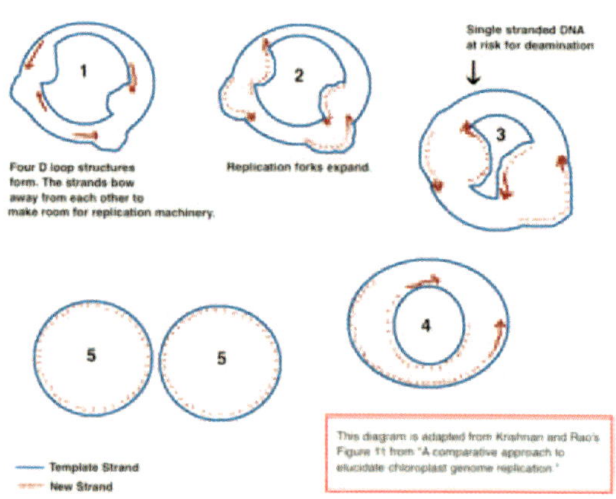

Chloroplast DNA replication via multiple D loop mechanisms. Adapted from Krishnan NM, Rao BJ's paper "A comparative approach to elucidate chloroplast genome replication."

The mechanism for chloroplast DNA (cpDNA) replication has not been conclusively determined, but two main models have been proposed. Scientists have attempted to observe chloroplast replication via electron microscopy since the 1970s. The results of the microscopy experiments led to the idea that chloroplast DNA replicates using a double displacement loop (D-loop). As the D-loop moves through the circular DNA, it adopts a theta intermediary form, also known as a Cairns replication intermediate, and completes replication with a rolling circle mechanism. Transcription starts at specific points of origin. Multiple replication forks open up, allowing replication machinery to transcribe the DNA. As replication continues, the forks grow and eventually converge. The new cpDNA structures separate, creating daughter cpDNA chromosomes.

In addition to the early microscopy experiments, this model is also supported by the amounts of deamination seen in cpDNA. Deamination occurs when an amino group is lost and is a mutation that often results in base changes. When adenine is deaminated, it becomes hypoxanthine. Hypoxanthine can bind to cytosine, and when the XC base pair is replicated, it becomes a GC (thus, an A $\rightarrow$ G base change).

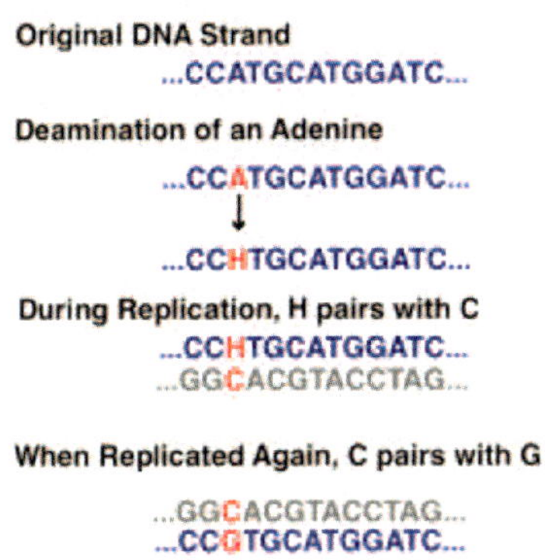

Over time, base changes in the DNA sequence can arise from deamination mutations. When adenine is deaminated, it becomes hypoxanthine, which can pair with cytosine. During replication, the cytosine will pair with guanine, causing an A --> G base change.

Deamination

In cpDNA, there are several A $\rightarrow$ G deamination gradients. DNA becomes susceptible to deamination events when it is single stranded. When replication forks form, the strand not being copied is single stranded, and thus at risk for A $\rightarrow$ G deamination. Therefore, gradients in deamination indicate that replication forks were most likely present and the direction that they initially opened (the highest gradient is most likely nearest the start site because it was single stranded for the longest amount of time). This mechanism is still the leading theory today; however, a second theory suggests that most cpDNA is actually linear and replicates through homologous recombination. It further contends that only a minority of the genetic material is kept in circular chromosomes while the rest is in branched, linear, or other complex structures.

Alternative Model of Replication

One of competing model for cpDNA replication asserts that most cpDNA is linear and participates in homologous recombination and replication structures similar to bacteriophage T4. It has been established that some plants have linear cpDNA, such as maize, and that more species still contain complex structures that scientists do not yet understand. When the original experiments on cpDNA were performed, scientists did notice linear structures; however, they attributed these linear forms to broken circles. If the branched and complex structures seen in cpDNA experiments are real and not artifacts of concatenated circular DNA or broken circles, then a D-loop mechanism of replication is insufficient to explain how those structures would replicate. At the same time, homologous recombination does not expand the multiple A --> G gradients seen in plastomes. Because of the failure to explain the deamination gradient as well as the numerous plant species that have been shown to have circular cpDNA, the predominant theory continues to hold that most cpDNA is circular and most likely replicates via a D loop mechanism.

Gene content and Protein Synthesis

The chloroplast genome most commonly includes around 100 genes that code for a variety of things, mostly to do with the protein pipeline and photosynthesis. As in prokaryotes, genes in chloroplast DNA are organized into operons.Interestingly though, unlike prokaryotic DNA molecules, chloroplast DNA molecules contain introns (plant mitochondrial DNAs do too, but not human mtDNAs).

Among land plants, the contents of the chloroplast genome are fairly similar.

Chloroplast Genome Reduction and Gene Transfer

Over time, many parts of the chloroplast genome were transferred to the nuclear genome of the host, a process called *endosymbiotic gene transfer*. As a result, the chloroplast genome is heavily reduced compared to that of free-living cyanobacteria. Chloroplasts may contain 60–100 genes whereas cyanobacteria often have more than 1500 genes in their genome. Recently, a plastid without a genome was found, demonstrating chloroplasts can lose their genome during endosymbiotic the gene transfer process.

Endosymbiotic gene transfer is how we know about the lost chloroplasts in many chromalveolate lineages. Even if a chloroplast is eventually lost, the genes it donated to the former host's nucleus persist, providing evidence for the lost chloroplast's existence. For example, while diatoms (a heterokontophyte) now have a red algal derived chloroplast, the presence of many green algal genes in the diatom nucleus provide evidence that the diatom ancestor (probably the ancestor of all chromalveolates too) had a green algal derived chloroplast at some point, which was subsequently replaced by the red chloroplast.

In land plants, some 11–14% of the DNA in their nuclei can be traced back to the chloroplast, up to 18% in *Arabidopsis*, corresponding to about 4,500 protein-coding genes. There have been a few recent transfers of genes from the chloroplast DNA to the nuclear genome in land plants.

Of the approximately 3000 proteins found in chloroplasts, some 95% of them are encoded by nuclear genes. Many of the chloroplast's protein complexes consist of subunits from both the chloroplast genome and the host's nuclear genome. As a result, protein synthesis must be coordinated between the chloroplast and the nucleus. The chloroplast is mostly under nuclear control, though chloroplasts can also give out signals regulating gene expression in the nucleus, called *retrograde signaling*.

Protein Synthesis

Protein synthesis within chloroplasts relies on two RNA polymerases. One is coded by the chloroplast DNA, the other is of nuclear origin. The two RNA polymerases may recognize and bind to different kinds of promoters within the chloroplast genome. The ribosomes in chloroplasts are similar to bacterial ribosomes.

Protein Targeting and Import

Because so many chloroplast genes have been moved to the nucleus, many proteins that would originally have been translated in the chloroplast are now synthesized in the cytoplasm of the plant cell. These proteins must be directed back to the chloroplast, and imported through at least two chloroplast membranes.

Curiously, around half of the protein products of transferred genes aren›t even targeted back to the chloroplast. Many became exaptations, taking on new functions like participating in cell division, protein routing, and even disease resistance. A few chloroplast genes found new homes in the mitochondrial genome—most became nonfunctional pseudogenes, though a few tRNA genes still work in the mitochondrion. Some transferred chloroplast DNA protein products get directed to the secretory pathway though it should be noted that many secondary plastids are bounded by an outermost membrane derived from the host's cell membrane, and therefore topologically outside of the cell, because to reach the chloroplast from the cytosol, you have to cross the cell membrane, just like if you were headed for the extracellular space. In those cases, chloroplast-targeted proteins do initially travel along the secretory pathway.

Because the cell acquiring a chloroplast already had mitochondria (and peroxisomes, and a cell membrane for secretion), the new chloroplast host had to develop a unique protein targeting system to avoid having chloroplast proteins being sent to the wrong organelle.

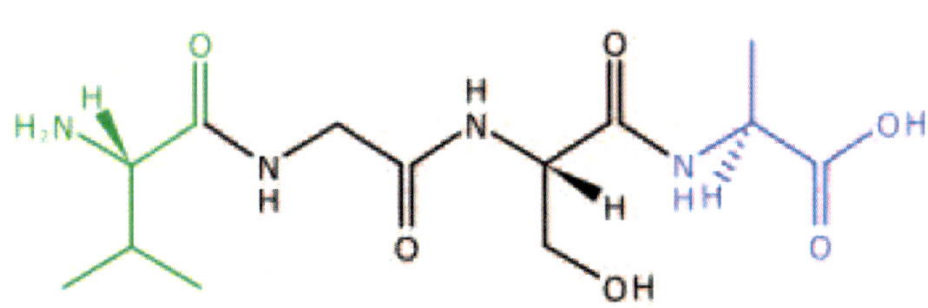

The two ends of a polypeptide are called the N-terminus, or *amino end*, and the C-terminus, or *carboxyl end*.This polypeptide has four amino acids linked together. At the left is the N-terminus, with its amino (H_2**N**) group in green. The blue C-terminus, with its carboxyl group (**C**O_2H) is at the right.

In most, but not all cases, nuclear-encoded chloroplast proteins are translated with a *cleavable transit peptide* that's added to the N-terminus of the protein precursor. Sometimes the transit sequence is found on the C-terminus of the protein, or within the functional part of the protein.

Transport Proteins and Membrane Translocons

After a chloroplast polypeptide is synthesized on a ribosome in the cytosol, an enzyme specific to chloroplast proteinsphosphorylates, or adds a phosphate group to many (but not all) of them in their transit sequences. Phosphorylation helps many proteins bind the polypeptide, keeping it from folding prematurely. This is important because it prevents chloroplast proteins from assuming their active form and carrying out their chloroplast functions in the wrong place—the cytosol. At the same time, they have to keep just enough shape so that they can be recognized by the chloroplast. These proteins also help the polypeptide get imported into the chloroplast.

From here, chloroplast proteins bound for the stroma must pass through two protein complexes—the TOC complex, or *translocon on the outer chloroplast membrane*, and the TIC translocon, or *translocon on the inner chloroplast membrane translocon*. Chloroplast polypeptide chains probably often travel through the two complexes at the same time, but the TIC complex can also retrieve preproteins lost in the intermembrane space.

Structure

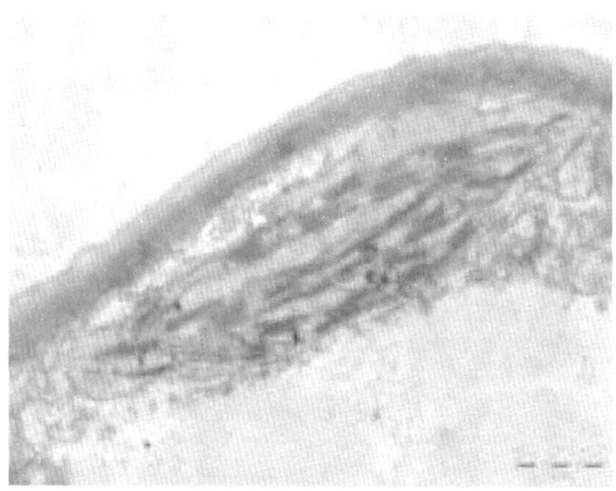

Transmission electron microscope image of a chloroplast. Grana of thylakoids and their connecting lamellae are clearly visible.

In land plants, chloroplasts are generally lens-shaped, 5–8 μm in diameter and 1–3 μm thick. Greater diversity in chloroplast shapes exists among the algae, which often contain a single chloroplast that can be shaped like a net (e.g., *Oedogonium*), a cup (e.g., *Chlamydomonas*), a ribbon-like spiral around the edges of the cell (e.g., *Spirogyra*), or slightly twisted bands at the cell edges (e.g., *Sirogonium*). Some algae have two chloroplasts in each cell; they are star-shaped in *Zygnema*, or may follow the shape of half the cell in order Desmidiales. In some algae, the chloroplast takes up most of the cell, with pockets for the nucleus and other organelles, for example, some species of *Chlorella* have a cup-shaped chloroplast that occupies much of the cell.

All chloroplasts have at least three membrane systems—the outer chloroplast membrane, the inner chloroplast membrane, and the thylakoid system. Chloroplasts that are the product of secondary endosymbiosis may have additional membranes surrounding these three. Inside the outer and inner chloroplast membranes is the chloroplast stroma, a semi-gel-like fluid that makes up much of a chloroplast›s volume, and in which the thylakoid system floats.

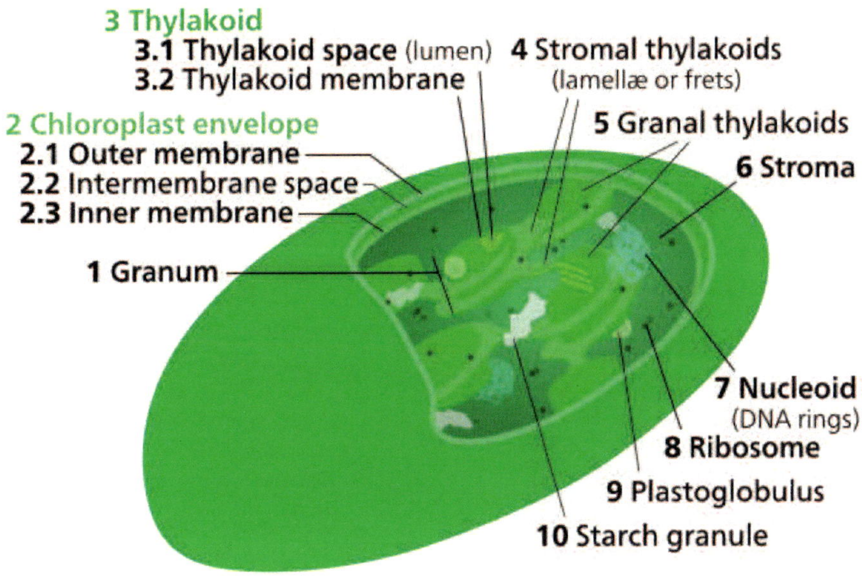

Chloroplast ultrastructure *(interactive diagram)* Chloroplasts have at least three distinct membrane systems, and a variety of things can be found in their stroma.

There are some common misconceptions about the outer and inner chloroplast membranes. The fact that chloroplasts are surrounded by a double membrane is often cited as evidence that they are the descendants of endosymbiotic cyanobacteria. This is often interpreted as meaning the outer chloroplast membrane is the product of the host›s cell membrane infolding to form a vesicle to surround the ancestral cyanobacterium—which is not true—both chloroplast membranes are homologous to the cyanobacterium's original double membranes.

The chloroplast double membrane is also often compared to the mitochondrial double membrane. This is not a valid comparison—the inner mitochondria membrane is used to run proton pumps and carry out oxidative phosphorylation across to generate ATP energy. The only chloroplast structure that can considered analogous to it is the internal thylakoid system. Even so, in terms of "in-out", the direction of chloroplast H^+ ion flow is in the opposite direction compared to oxidative phosphorylation in mitochondria. In addition, in terms of function, the inner chloroplast membrane, which regulates metabolite passage and synthesizes some materials, has no counterpart in the mitochondrion.

Outer Chloroplast Membrane

The outer chloroplast membrane is a semi-porous membrane that small molecules and ions can easily diffuse across.However, it is not permeable to larger proteins, so chloroplast polypeptides being synthesized in the cell cytoplasm must be transported across the outer chloroplast membrane by the TOC complex, or *translocon on the outer chloroplast* membrane.

The chloroplast membranes sometimes protrude out into the cytoplasm, forming a stromule, or **stroma**-containing tub**ule**. Stromules are very rare in chloroplasts, and are much more common in other plastids like chromoplasts and amyloplasts in petals and roots, respectively. They may exist to increase the chloroplast's surface area for cross-membrane transport, because they are often branched and tangled with the endoplasmic reticulum. When they were first observed in 1962, some plant biologists dismissed the structures as artifactual, claiming that stromules were just oddly shaped chloroplasts with constricted regions or dividing chloroplasts. However, there is a growing body of evidence that stromules are functional, integral features of plant cell plastids, not merely artifacts.

Intermembrane Space and Peptidoglycan Wall

Instead of an intermembrane space, glaucophyte algae have a peptidoglycan wall between their inner and outer chloroplast membranes.

Usually, a thin intermembrane space about 10–20 nanometers thick exists between the outer and inner chloroplast membranes.

Glaucophyte algal chloroplasts have a peptidoglycan layer between the chloroplast membranes. It corresponds to the peptidoglycan cell wall of their cyanobacterial ancestors, which is located between their two cell membranes. These chloroplasts are called *muroplasts* (from Latin *mura*, meaning "wall"). Other chloroplasts have lost the cyanobacterial wall, leaving an intermembrane space between the two chloroplast envelope membranes.

Inner Chloroplast Membrane

The inner chloroplast membrane borders the stroma and regulates passage of materials in and out of the chloroplast. After passing through the TOC complex in the outer chloroplast membrane, polypeptides must pass through the TIC complex *(translocon on the inner chloroplast membrane)* which is located in the inner chloroplast membrane.

In addition to regulating the passage of materials, the inner chloroplast membrane is where fatty acids, lipids, and carotenoids are synthesized.

Peripheral reticulum

Some chloroplasts contain a structure called the chloroplast peripheral reticulum. It is often found in the chloroplasts of C_4plants, though it has also been found in some C_3 angiosperms, and even some gymnosperms. The chloroplast peripheral reticulum consists of a maze of membranous tubes and vesicles continuous with the inner chloroplast membrane that extends into the internal stromal fluid of the chloroplast. Its purpose is thought to be to increase the chloroplast's surface area for cross-membrane transport between its stroma and the cell cytoplasm. The small vesicles sometimes observed may serve as transport vesicles to shuttle stuff between the thylakoids and intermembrane space.

Stroma

The protein-rich, alkaline, aqueous fluid within the inner chloroplast membrane and outside of the thylakoid space is called the stroma, which corresponds to the cytosol of the original cyanobacterium. Nucleoids of chloroplast DNA, chloroplast ribosomes, the thylakoid system with plastoglobuli, starch granules, and many proteins can be found floating around in it. The Calvin cycle, which fixes CO_2 into sugar takes place in the stroma.

Chloroplast Ribosomes

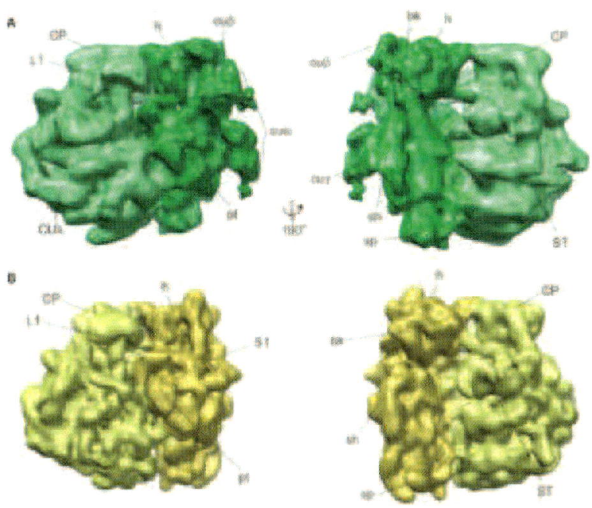

Chloroplast ribosomes Comparison of a chloroplast ribosome (green) and a bacterial ribosome (yellow). Important features common to both ribosomes and chloroplast-unique features are labeled.

Chloroplasts have their own ribosomes, which they use to synthesize a small fraction of their proteins. Chloroplast ribosomes are about two-thirds the size of cytoplasmic ribosomes (around 17 nm vs 25 nm). They take mRNAs transcribed from the chloroplast DNA and translate them into protein. While similar to bacterial ribosomes, chloroplast translation is more complex than in bacteria, so chloroplast ribosomes include some chloroplast-unique features. Small subunit ribosomal RNAs in several Chlorophyta and euglenid chloroplasts lack motifs for shine-dalgarno sequence recognition, which is considered essential for translation initiation in most chloroplasts and prokaryotes. Such loss is also rarely observed in other plastids and prokaryotes.

Plastoglobuli

Plastoglobuli (**singular** *plastoglobulus*, sometimes spelled *plastoglobule(s)*), are spherical bubbles of lipids and proteins about 45–60 nanometers across. They are surrounded by a lipid monolayer. Plastoglobuli are found in all chloroplasts, but become more common when the chloroplast is under oxidative stress, or when it ages and transitions into a gerontoplast. Plastoglobuli also exhibit a greater size variation under these conditions. They are also common in etioplasts, but decrease in number as the etioplasts mature into chloroplasts.

Plastoglubuli contain both structural proteins and enzymes involved in lipid synthesis and metabolism. They contain many types of lipids including plastoquinone, vitamin E, carotenoids and chlorophylls.

Plastoglobuli were once thought to be free-floating in the stroma, but it is now thought that they are permanently attached either to a thylakoid or to another plastoglobulus attached to a thylakoid, a configuration that allows a plastoglobulus to exchange its contents with the thylakoid network. In normal green chloroplasts, the vast majority of plastoglobuli occur singularly, attached directly to their parent thylakoid. In old or stressed chloroplasts, plastoglobuli tend to occur in linked groups or chains, still always anchored to a thylakoid.

Plastoglobuli form when a bubble appears between the layers of the lipid bilayer of the thylakoid membrane, or bud from existing plastoglubuli—though they never detach and float off into the stroma. Practically all plastoglobuli form on or near the highly curved edges of the thylakoid disks or sheets. They are also more common on stromal thylakoids than on granal ones.

Starch Granules

Starch granules are very common in chloroplasts, typically taking up 15% of the organelle›s volume, though in some other plastids like amyloplasts, they can be big enough to distort the shape of the organelle. Starch granules are simply accumulations of starch in the stroma, and are not bounded by a membrane.

Starch granules appear and grow throughout the day, as the chloroplast synthesizes sugars, and are consumed at night to fuel respiration and continue sugar export into the phloem, though in mature chloroplasts, it is rare for a starch granule to be completely consumed or for a new granule to accumulate.

Starch granules vary in composition and location across different chloroplast lineages. In red algae, starch granules are found in the cytoplasm rather than in the chloroplast. In C_4 plants, mesophyll chloroplasts, which do not synthesize sugars, lack starch granules.

Rubisco

Rubisco, shown here in a space-filling model, is the main enzyme responsible for carbon fixation in chloroplasts.

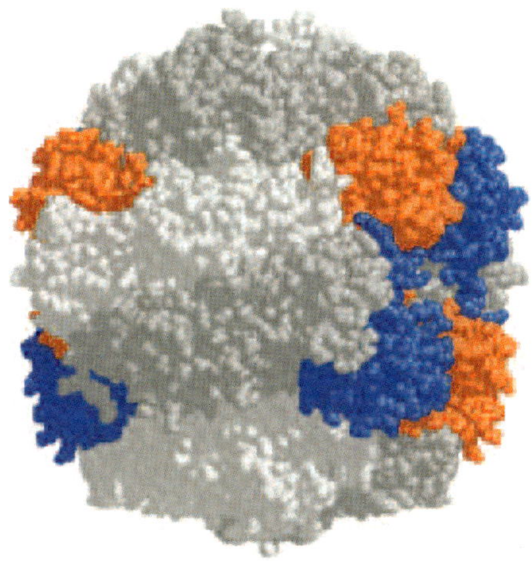

The chloroplast stroma contains many proteins, though the most common and important is Rubisco, which is probably also the most abundant protein on the planet. Rubisco is the enzyme that fixes CO_2 into sugar molecules. In C_3 plants, rubisco is abundant in all chloroplasts, though in C_4plants, it is confined to the bundle sheath chloroplasts, where the Calvin cycle is carried out in C_4 plants.

Pyrenoids

The chloroplasts of some hornworts and algae contain structures called pyrenoids. They are not found in higher plants. Pyrenoids are roughly spherical and highly refractive bodies which are a site of starch accumulation in plants that contain them. They consist of a matrix opaque to electrons, surrounded by two hemispherical starch plates. The starch is accumulated as the pyrenoids mature. In algae with carbon concentrating mechanisms, the enzyme rubisco is found in the pyrenoids. Starch can also accumulate around the pyrenoids when CO_2 is scarce. Pyrenoids can divide to form new pyrenoids, or be produced de novo.

Thylakoid System

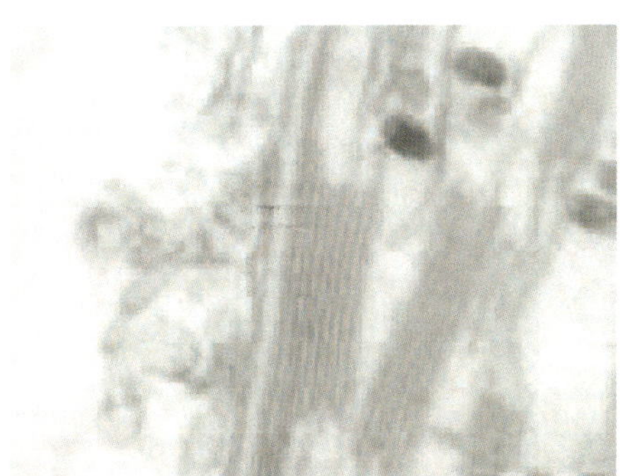

Transmission electron microscope image of some thylakoids arranged in grana stacks and lamellæ. Plastoglobuli (dark blobs) are also present.

Suspended within the chloroplast stroma is the thylakoid system, a highly dynamic collection of membranous sacks called thylakoids where chlorophyll is found and the light reactions of photosynthesis happen. In most vascular plant chloroplasts, the thylakoids are arranged in stacks called grana, though in certain C_4 plant chloroplasts and some algal chloroplasts, the thylakoids are free floating.

Granal Structure

Using a light microscope, it is just barely possible to see tiny green granules—which were named grana. With electron microscopy, it became possible to see the thylakoid system in more detail, revealing it to consist of stacks of flat thylakoids which made up the grana, and long interconnecting stromal thylakoids which linked different grana. In the transmission electron microscope, thylakoid membranes appear as alternating light-and-dark bands, 8.5 nanometers thick.

For a long time, the three-dimensional structure of the thylakoid system has been unknown or disputed. One model has the granum as a stack of thylakoids linked by helical stromal thylakoids; the other has the granum as a single folded thylakoid connected in a "hub and spoke" way to other grana by stromal thylakoids. While the thylakoid system is still commonly depicted according to the folded thylakoid model, it was determined in 2011 that the stacked and helical thylakoids model is correct.

Granum structure The prevailing model for granal structure is a stack of granal thylakoids linked by helical stromal thylakoids that wrap around the grana stacks and form large sheets that connect different grana.

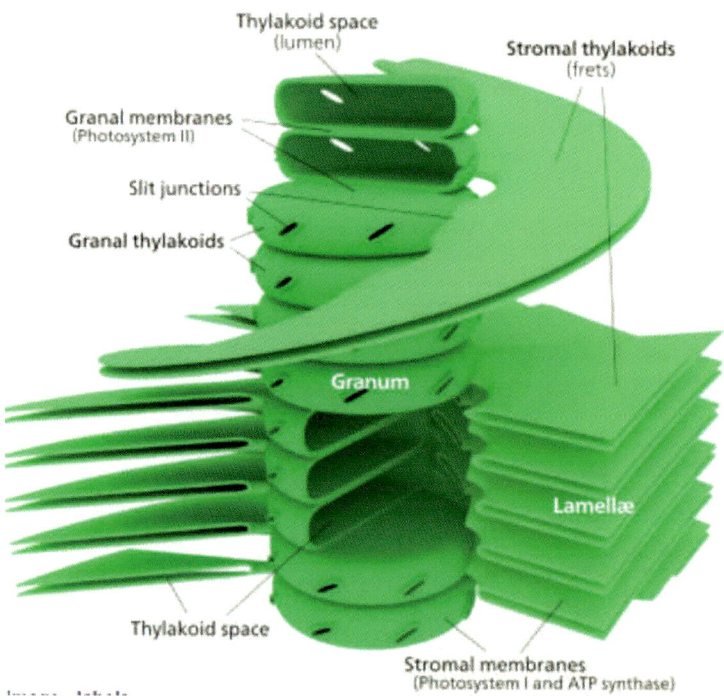

In the helical thylakoid model, grana consist of a stack of flattened circular granal thylakoids that resemble pancakes. Each granum can contain anywhere from two to a hundred thylakoids, though grana with 10–20 thylakoids are most common. Wrapped around the grana are helicoid stromal thylakoids, also known as frets or lamellar thylakoids. The helices ascend at an angle of 20–25°, connecting to each granal thylakoid at a bridge-like slit junction. The helicoids may extend as large sheets that link multiple grana, or narrow to tube-like bridges betwcen grana. While different parts of the thylakoid system contain different membrane proteins, the thylakoid membranes are continuous and the thylakoid space they enclose form a single continuous labyrinth.

Thylakoids

Thylakoids (sometimes spelled *thylakoïds*), are small interconnected sacks which contain the membranes that the light reactions of photosynthesis take place on. The word *thylakoid* comes from the Greek word *thylakos* which means "sack".

Embedded in the thylakoid membranes are important protein complexes which carry out the light reactions of photosynthesis. Photosystem II and photosystem I contain light-harvesting complexes with chlorophyll and carotenoids that absorb light energy and use it to energize electrons. Molecules in the thylakoid membrane use the energized electrons to pump hydrogen ions into the thylakoid space, decreasing the pH and turning it acidic. ATP synthase is a large protein complex that harnesses the concentration gradient of the hydrogen ions in the thylakoid space to generate ATP energy as the hydrogen ions flow back out into the stroma—much like a dam turbine.

There are two types of thylakoids—granal thylakoids, which are arranged in grana, and stromal thylakoids, which are in contact with the stroma. Granal thylakoids are pancake-shaped circular disks about 300–600 nanometers in diameter. Stromal thylakoids are helicoid sheets that spiral around grana. The flat tops and bottoms of granal thylakoids contain only the relatively flat photosystem II protein complex. This allows them to stack tightly, forming grana with many layers of tightly appressed membrane, called granal membrane, increasing stability and surface area for light capture.

In contrast, photosystem I and ATP synthase are large protein complexes which jut out into the stroma. They can't fit in the appressed granal membranes, and so are found in the stromal thylakoid membrane—the edges of the granal thylakoid disks and the stromal thylakoids. These large protein complexes may act as spacers between the sheets of stromal thylakoids.

The number of thylakoids and the total thylakoid area of a chloroplast is influenced by light exposure. Shaded chloroplasts contain larger and more grana with more thylakoid membrane area than chloroplasts exposed to bright light, which have smaller and fewer grana and less thylakoid area. Thylakoid extent can change within minutes of light exposure or removal.

Pigments and Chloroplast Colors

Inside the photosystems embedded in chloroplast thylakoid membranes are various photosynthetic pigments, which absorb and transfer light energy. The types of pigments found are different in various groups of chloroplasts, and are responsible for a wide variety of chloroplast colorations.

Paper chroma-tography of some spinach leaf extract shows the various pigments present in their chloroplasts.

Chlorophylls

Chlorophyll *a* is found in all chloroplasts, as well as their cyanobacterial ancestors. Chlorophyll *a* is a blue-green pigment partially responsible for giving most cyanobacteria and chloroplasts their color. Other forms of chlorophyll exist, such as the accessory pigments chlorophyll *b*, chlorophyll *c*, chlorophyll *d*, and chlorophyll *f*.

Chlorophyll *b* is an olive green pigment found only in the chloroplasts of plants, green algae, any secondary chloroplasts obtained through the secondary endosymbiosis of a green alga, and a few cyanobacteria. It is the chlorophylls *a* and *b* together that make most plant and green algal chloroplasts green.

Chlorophyll *c* is mainly found in secondary endosymbiotic chloroplasts that originated from a red alga, although it is not found in chloroplasts of red algae themselves. Chlorophyll *c* is also found in some green algae and cyanobacteria.

Chlorophylls *d* and *f* are pigments found only in some cyanobacteria.

Carotenoids

Delesseria sanguinea, a red alga, has chloroplasts that contain red pigments like phycoerytherin that mask their blue-green chlorophyll *a*.

In addition to chlorophylls, another group of yellow–orange pigments called carotenoids are also found in the photosystems. There are about thirty photosynthetic carotenoids. They help transfer and dissipate excess energy, and their bright colors sometimes override the chlorophyll green, like during the fall, when the leaves of some land plants change color. β-carotene is a bright red-orange carotenoid found in nearly all chloroplasts, like chlorophyll *a*. Xanthophylls, especially the orange-red zeaxanthin, are also common. Many other forms of carotenoids exist that are only found in certain groups of chloroplasts.

Phycobilins

Phycobilins are a third group of pigments found in cyanobacteria, and glaucophyte, red algal, and cryptophyte chloroplasts. Phycobilins come in all colors, though phycoerytherin is one of the pigments that makes many red algae red. Phycobilins often organize into relatively large protein complexes about 40 nanometers across called phycobilisomes. Like photosystem I and ATP synthase, phycobilisomes jut into the stroma, preventing thylakoid stacking in red algal chloroplasts. Cryptophyte chloroplasts and some cyanobacteria don›t have their phycobilin pigments organized into phycobilisomes, and keep them in their thylakoid space instead.

Photosynthetic pigments Table of the presence of various pigments across chloroplast groups. Colored cells represent pigment presence.

	Chlorophyll *a*	Chlorophyll *b*	Chlorophyll *c*	Chlorophyll *d* and *f*	Xanthophylls	α-carotene	β-carotene	Phycobilins
Land plants	■	■			■	■	■	
Green algae	■	■	■		■	■	■	
Euglenophytes and **Chlorarachniophytes**	■	■			■		■	
Multicellular red algae	■				■	■	■	■
Unicellular red algae	■				■		■	■
Haptophytes and **Dinophytes**	■		■		■		■	
Cryptophytes	■		■		■	■		■
Glaucophytes	■				■		■	■
Cyanobacteria	■	■	■	■	■	■	■	■

Specialized Chloroplasts in C_4 Plants

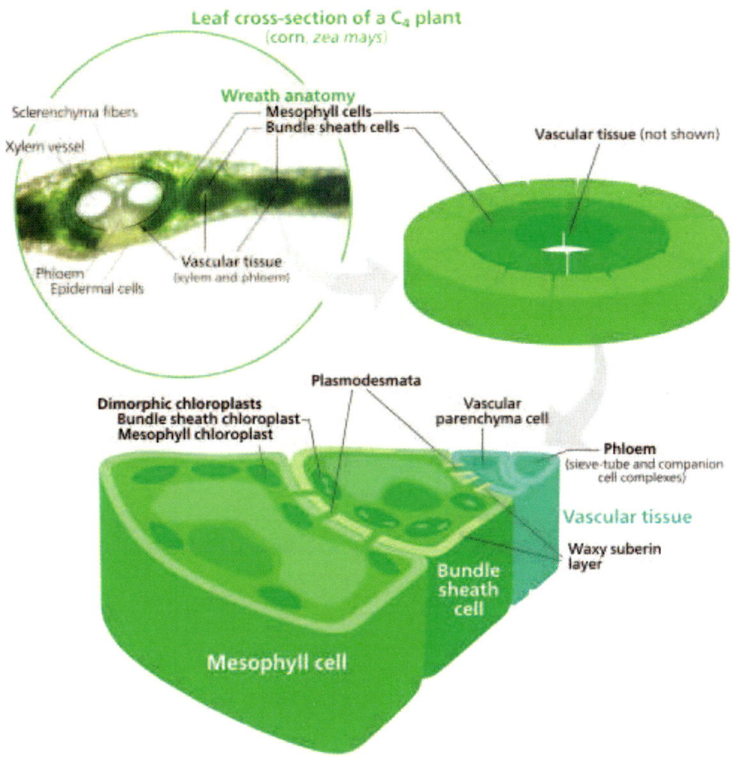

Many C_4 plants have their mesophyll cells and bundle sheath cells arranged radially around their leaf veins. The two types of cells contain different types of chloroplasts specialized for a particular part of photosynthesis.

To fix carbon dioxide into sugar molecules in the process of photosynthesis, chloroplasts use an enzyme called rubisco. Rubisco has a problem—it has trouble distinguishing between carbon dioxide and oxygen, so at high oxygen concentrations, rubisco starts accidentally adding oxygen to sugar precursors. This has the end result of ATP energy being wasted and CO_2 being released, all with no sugar being produced. This is a big problem, since O_2 is produced by the initial light reactions of photosynthesis, causing issues down the line in the Calvin cycle which uses rubisco.

C_4 plants evolved a way to solve this—by spatially separating the light reactions and the Calvin cycle. The light reactions, which store light energy in ATP and NADPH, are done in the mesophyll cells of a C_4 leaf. The Calvin cycle, which uses the stored energy to make sugar using rubisco, is done in the bundle sheath cells, a layer of cells surrounding a vein in a leaf.

As a result, chloroplasts in C_4 mesophyll cells and bundle sheath cells are specialized for each stage of photosynthesis. In mesophyll cells, chloroplasts are specialized for the light reactions, so they lack rubisco, and have normal grana and thylakoids, which they use to make ATP and NADPH, as well as oxygen. They store CO_2 in a four-carbon compound, which is why the process is called *C_4 photosynthesis*. The four-carbon compound is then transported to the bundle sheath chloroplasts, where it drops off CO_2 and returns to the mesophyll. Bundle sheath chloroplasts do not carry out the light reactions, preventing oxygen from building up in them and disrupting rubisco activity. Because of this, they lack thylakoids organized into grana stacks—though bundle sheath chloroplasts still have free-floating thylakoids in the stroma where they still carry out cyclic electron flow, a light-driven method of synthesizing ATP to power the Calvin cycle without generating oxygen. They lack photosystem II, and only have photosystem I—the only protein complex needed for cyclic electron flow.Because the job of bundle sheath chloroplasts is to carry out the Calvin cycle and make sugar, they often contain large starch grains.

Both types of chloroplast contain large amounts of chloroplast peripheral reticulum, which they use to get more surface area to transport stuff in and out of them. Mesophyll chloroplasts have a little more peripheral reticulum than bundle sheath chloroplasts.

Location

Distribution in a Plant

Not all cells in a multicellular plant contain chloroplasts. All green parts of a plant contain chloroplasts—the chloroplasts, or more specifically, the chlorophyll in them are what make the photosynthetic parts of a plant green. The plant cells which contain chloroplasts are usually parenchyma cells, though chloroplasts can also be found in collenchyma tissue. A plant cell which contains chloroplasts is known as a chlorenchyma cell. A typical chlorenchyma cell of a land plant contains about 10 to 100 chloroplasts.

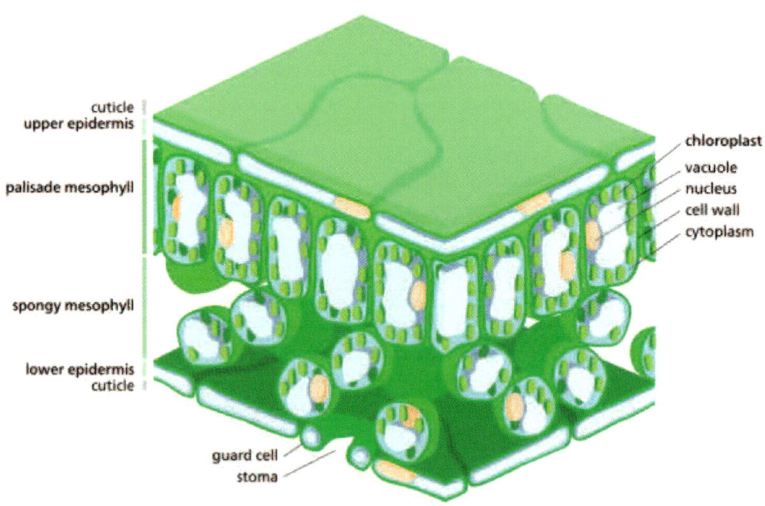

A cross section of a leaf, showing chloroplasts in its mesophyll cells. Stomal guard cells also have chloroplasts, though much fewer than mesophyll cells.

In some plants such as cacti, chloroplasts are found in the stems, though in most plants, chloroplasts are concentrated in the leaves. One square millimeter of leaf tissue can contain half a million chloroplasts. Within a leaf, chloroplasts are mainly found in the mesophyll layers of a leaf, and the guard cells of stomata. Palisade mesophyll cells can contain 30–70 chloroplasts per cell, while stomatal guard cells contain only around 8–15 per cell, as well as much less chlorophyll. Chloroplasts can also be found in the bundle sheath cells of a leaf, especially in C_4 plants, which carry out the Calvin cycle in their bundle sheath cells. They are often absent from the epidermis of a leaf.

Cellular Location

Chloroplast Movement

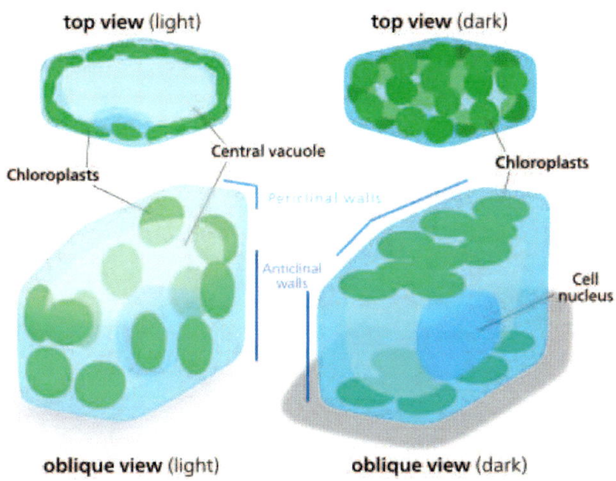

When chloroplasts are exposed to direct sunlight, they stack along the anticlinal cell walls to minimize exposure. In the dark they spread out in sheets along the periclinal walls to maximize light absorption.

The chloroplasts of plant and algal cells can orient themselves to best suit the available light. In low-light conditions, they will spread out in a sheet—maximizing the surface area to absorb light. Under intense light, they will seek shelter by aligning in vertical columns along the plant cell›s cell wall or turning sideways so that light strikes them edge-on. This reduces exposure and protects them from photooxidative damage.This ability to distribute chloroplasts so that they can take shelter behind each other or spread out may be the reason why land plants evolved to have many small chloroplasts instead of a few big ones. Chloroplast movement is considered one of the most closely regulated stimulus-response systems that can be found in plants.Mitochondria have also been observed to follow chloroplasts as they move.

In higher plants, chloroplast movement is run by phototropins, blue light photoreceptors also responsible for plant phototropism. In some algae, mosses, ferns, and flowering plants, chloroplast movement is influenced by red light in addition to blue light, though very long red wavelengths inhibit movement rather than speeding it up. Blue light generally causes chloroplasts to seek shelter, while red light draws them out to maximize light absorption.

Studies of *Vallisneria gigantea*, an aquatic flowering plant, have shown that chloroplasts can get moving within five minutes of light exposure, though they don't initially show any net directionality. They may move along microfilament tracks, and the fact that the microfilament mesh changes shape to form a honeycomb structure surrounding the chloroplasts after they have moved suggests that microfilaments may help to anchor chloroplasts in place.

Function and Chemistry

Guard cell Chloroplasts

Unlike most epidermal cells, the guard cells of plant stomata contain relatively well-developed chloroplasts. However, exactly what they do is controversial.

Plant Innate Immunity

Plants lack specialized immune cells—all plant cells participate in the plant immune response. Chloroplasts, along with the nucleus, cell membrane, and endoplasmic reticulum, are key players in pathogen defense. Due to its role in a plant cell's immune response, pathogens frequently target the chloroplast.

Plants have two main immune responses—the hypersensitive response, in which infected cells seal themselves off and undergo programmed cell death, and systemic acquired resistance, where infected cells release signals warning the rest of the plant of a pathogen›s presence. Chloroplasts stimulate both responses by purposely damaging their photosynthetic system, producing reactive oxygen species. High levels of reactive oxygen species will cause the hypersensitive response. The reactive oxygen species also directly kill any pathogens within the cell. Lower levels of reactive oxygen species initiate systemic acquired resistance, triggering defense-molecule production in the rest of the plant.

In some plants, chloroplasts are known to move closer to the infection site and the nucleus during an infection.

Chloroplasts can serve as cellular sensors. After detecting stress in a cell, which might be due to a pathogen, chloroplasts begin producing molecules like salicylic acid, jasmonic acid, nitric oxide and reactive oxygen species which can serve as defense-signals. As cellular signals, reactive oxygen species are unstable molecules, so they probably don't leave the chloroplast, but instead pass on their signal to an unknown second messenger molecule. All these molecules initiate retrograde signaling—signals from the chloroplast that regulate gene expression in the nucleus.

In addition to defense signaling, chloroplasts, with the help of the peroxisomes, help synthesize an important defense molecule, jasmonate. Chloroplasts synthesize all the fatty acids in a plant cell—linoleic acid, a fatty acid, is a precursor to jasmonate.

Photosynthesis

One of the main functions of the chloroplast is its role in photosynthesis, the process by which light is transformed into chemical energy, to subsequently produce food in the form of sugars. Water (H_2O) and carbon dioxide (CO_2) are used in photosynthesis, and sugar and oxygen (O_2) is made, using light energy. Photosynthesis is divided into two stages—the light reactions, where water is split to produce oxygen, and the dark reactions, or Calvin cycle, which builds sugar molecules from carbon dioxide. The two phases are linked by the energy carriers adenosine triphosphate (ATP) and nicotinamide adenine dinucleotide phosphate ($NADP^+$).

Light Reactions

The light reactions of photosynthesis take place across the thylakoid membranes.

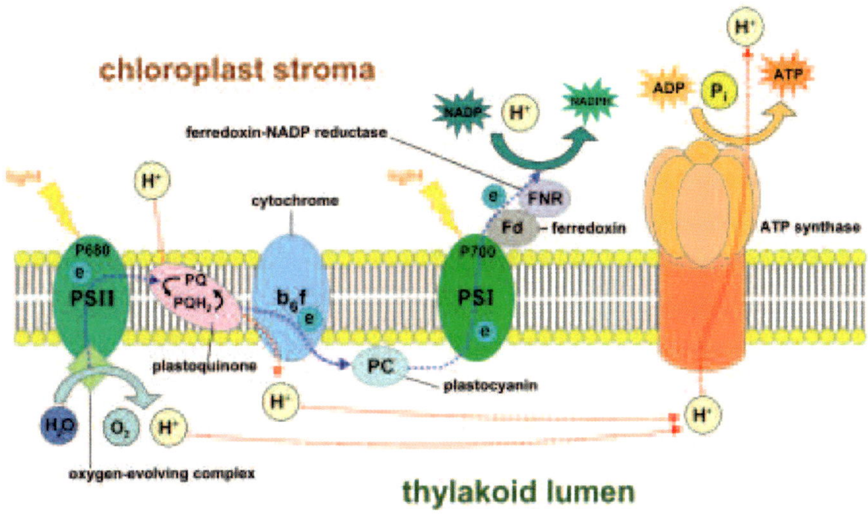

The light reactions take place on the thylakoid membranes. They take light energy and store it in NADPH, a form of $NADP^+$, and ATP to fuel the dark reactions.

Energy carriers

ATP is the phosphorylated version of adenosine diphosphate (ADP), which stores energy in a cell and powers most cellular activities. ATP is the energized form, while ADP is the (partially) depleted form. $NADP^+$ is an electron carrier which ferries high energy electrons. In the light reactions, it gets reduced, meaning it picks up electrons, becoming NADPH.

Photophosphorylation

Like mitochondria, chloroplasts use the potential energy stored in an H^+, or hydrogen ion gradient to generate ATP energy. The two photosystems capture light energy to energize electrons taken from water, and release them down an electron transport chain. The molecules between the photosystems harness the electrons' energy to pump hydrogen ions into the thylakoid space, creating a concentration gradient, with more hydrogen ions (up to a thousand times as many) inside the thylakoid system than in the stroma. The hydrogen ions in the thylakoid space then diffuse back down their concentration gradient, flowing back out into the stroma through ATP synthase. ATP synthase uses the energy from the flowing hydrogen ions to phosphorylate adenosine diphosphate into adenosine triphosphate, or ATP. Because chloroplast ATP synthase projects out into the stroma, the ATP is synthesized there, in position to be used in the dark reactions.

NADP+ Reduction

Electrons are often removed from the electron transport chains to charge $NADP^+$ with electrons, reducing it to NADPH. Like ATP synthase, ferredoxin-$NADP^+$ reductase, the enzyme that reduces $NADP^+$, releases the NADPH it makes into the stroma, right where it is needed for the dark reactions.

Because $NADP^+$ reduction removes electrons from the electron transport chains, they must be replaced—the job of photosystem II, which splits water molecules (H_2O) to obtain the electrons from its hydrogen atoms.

Cyclic Photophosphorylation

While photosystem II photolyzes water to obtain and energize new electrons, photosystem I simply reenergizes depleted electrons at the end of an electron transport chain. Normally, the reenergized electrons are taken by $NADP^+$, though sometimes they can flow back down more H^+-pumping electron transport chains to transport more hydrogen ions into the thylakoid space to generate more ATP. This is termed cyclic photophosphorylation because the electrons are recycled. Cyclic photophosphorylation is common in C_4 plants, which need more ATP than NADPH.

Dark Reactions

The Calvin cycle *(Interactive diagram)* The Calvin cycle incorporates carbon dioxide into sugar molecules.

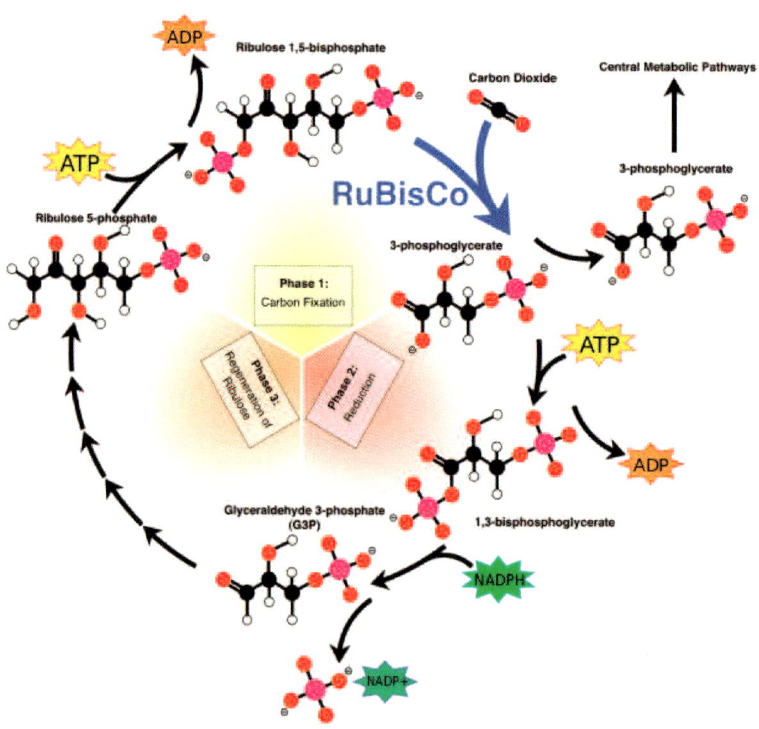

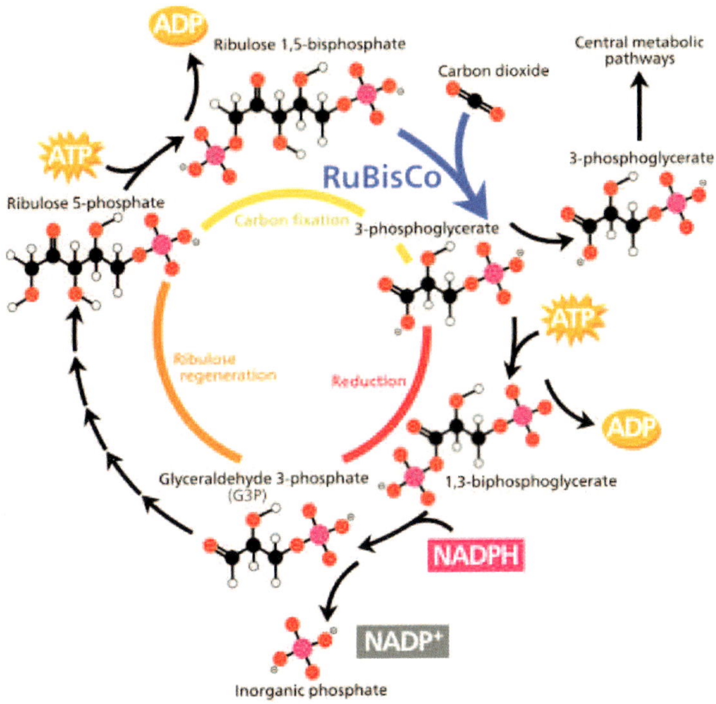

The Calvin cycle, also known as the dark reactions, is a series of biochemical reactions that fixes CO_2 into G3P sugar molecules and uses the energy and electrons from the ATP and NADPH made in the light reactions. The Calvin cycle takes place in the stroma of the chloroplast.

While named *"the dark reactions"*, in most plants, they take place in the light, since the dark reactions are dependent on the products of the light reactions.

Carbon Fixation and G3P Synthesis

The Calvin cycle starts by using the enzyme Rubisco to fix CO_2 into five-carbon Ribulose bisphosphate (RuBP) molecules. The result is unstable six-carbon molecules that immediately break down into three-carbon molecules called 3-phosphoglyceric acid, or 3-PGA. The ATP and NADPH made in the light reactions is used to convert the 3-PGA into glyceraldehyde-3-phosphate, or G3P sugar molecules. Most of the G3P molecules are recycled back into RuBP using energy from more ATP, but one out of every six produced leaves the cycle—the end product of the dark reactions.

Sugars and Starches

Glyceraldehyde-3-phosphate can double up to form larger sugar molecules like glucose and fructose. These molecules are processed, and from them, the still larger sucrose, a disaccharide commonly known as table sugar, is made, though this process takes place outside of the chloroplast, in the cytoplasm.

Sucrose is made up of a glucose monomer (left), and a fructose monomer (right).

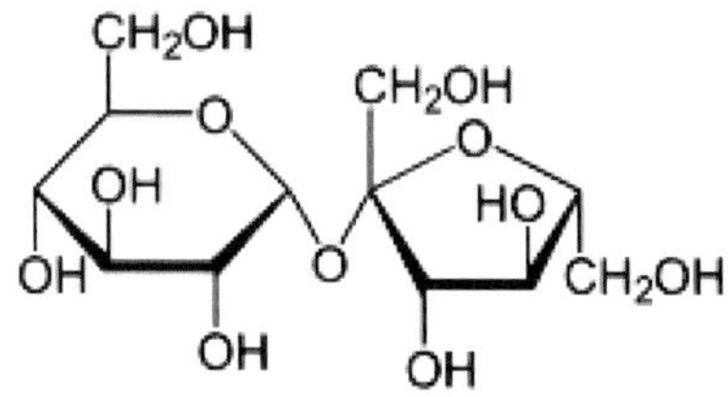

Alternatively, glucose monomers in the chloroplast can be linked together to make starch, which accumulates into the starch grains found in the chloroplast. Under conditions such as high atmospheric CO_2 concentrations, these starch grains may grow very large, distorting the grana and thylakoids. The starch granules displace the thylakoids, but leave them intact. Waterlogged roots can also cause starch buildup in the chloroplasts, possibly due to less sucrose being exported out of the chloroplast (or more accurately, the plant cell). This depletes a plant›s free phosphate supply, which indirectly stimulates chloroplast starch synthesis. While linked to low photosynthesis rates, the starch grains themselves may not necessarily interfere significantly with the efficiency of photosynthesis, and might simply be a side effect of another photosynthesis-depressing factor.

Photorespiration

Photorespiration can occur when the oxygen concentration is too high. Rubisco cannot distinguish between oxygen and carbon dioxide very well, so it can accidentally add O_2 instead of CO_2 to RuBP. This process reduces the efficiency of photosynthesis—it consumes ATP and oxygen, releases CO_2, and produces no sugar. It can waste up to half the carbon fixed by the Calvin cycle. Several mechanisms have evolved in different lineages that raise the carbon dioxide concentration relative to oxygen within the chloroplast, increasing the efficiency of photosynthesis. These mechanisms are called carbon dioxide concentrating mechanisms, or CCMs. These include Crassulacean acid metabolism, C_4 carbon fixation, and pyrenoids. Chloroplasts in C_4 plants are notable as they exhibit a distinct chloroplast dimorphism.

pH

Because of the H^+ gradient across the thylakoid membrane, the interior of the thylakoid is acidic, with a pH around 4, while the stroma is slightly basic, with a pH of around 8. The optimal stroma pH for the Calvin cycle is 8.1, with the reaction nearly stopping when the pH falls below 7.3.

CO_2 in water can form carbonic acid, which can disturb the pH of isolated chloroplasts, interfering with photosynthesis, even though CO_2 is used in photosynthesis. However, chloroplasts in living plant cells are not affected by this as much.

Chloroplasts can pump K^+ and H^+ ions in and out of themselves using a poorly understood light-driven transport system.

In the presence of light, the pH of the thylakoid lumen can drop up to 1.5 pH units, while the pH of the stroma can rise by nearly one pH unit.

Amino Acid Synthesis

Chloroplasts alone make almost all of a plant cell's amino acids in their stroma except the sulfur-containing ones like cysteine and methionine. Cysteine is made in the chloroplast (the proplastid too) but it is also synthesized in the cytosol and mitochondria, probably because it has trouble crossing membranes to get to where it is needed. The chloroplast is known to make the precursors to methionine but it is unclear whether the organelle carries out the last leg of the pathway or if it happens in the cytosol.

Other Nitrogen Compounds

Chloroplasts make all of a cell's purines and pyrimidines—the nitrogenous bases found in DNA and RNA. They also convert nitrite (NO_2^-) into ammonia (NH_3) which supplies the plant with nitrogen to make its amino acids and nucleotides.

Other Chemical Products

Chloroplasts are the site of complex lipid metabolism.

Differentiation, Replication, and Inheritance

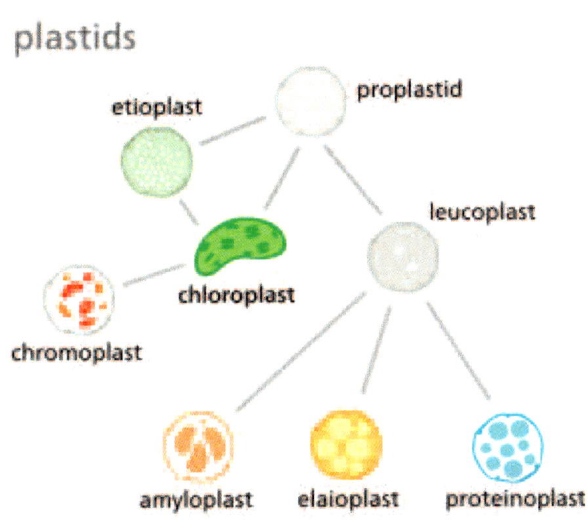

Plastid types *(Interactive diagram)* Plants contain many different kinds of plastids in their cells.

Chloroplasts are a special type of a plant cell organelle called a **plastid**, though the two terms are sometimes used interchangeably. There are many other types of plastids, which carry out various functions. All chloroplasts in a plant are descended from undifferentiated proplastids found in the zygote, or fertilized egg. Proplastids are commonly found in an adult plant's apical meristems. Chloroplasts do not normally develop from proplastids in root tip meristems—instead, the formation of starch-storing amyloplasts is more common.

In shoots, proplastids from shoot apical meristems can gradually develop into chloroplasts in photosynthetic leaf tissues as the leaf matures, if exposed to the required light. This process involves invaginations of the inner plastid membrane, forming sheets of membrane that project into the internal stroma. These membrane sheets then fold to form thylakoids and grana.

If angiosperm shoots are not exposed to the required light for chloroplast formation, proplastids may develop into an etioplast stage before becoming chloroplasts. An etioplast is a plastid that lacks chlorophyll, and has inner membrane invaginations that form a lattice of tubes in their stroma, called a prolamellar body. While etioplasts lack chlorophyll, they have a yellow chlorophyll precursor stocked. Within a few minutes of light exposure, the prolamellar body begins to reorganize into stacks of thylakoids, and chlorophyll starts to be produced. This process, where the etioplast becomes a chloroplast, takes several hours. Gymnosperms do not require light to form chloroplasts.

Light, however, does not guarantee that a proplastid will develop into a chloroplast. Whether a proplastid develops into a chloroplast some other kind of plastid is mostly controlled by the nucleus and is largely influenced by the kind of cell it resides in.

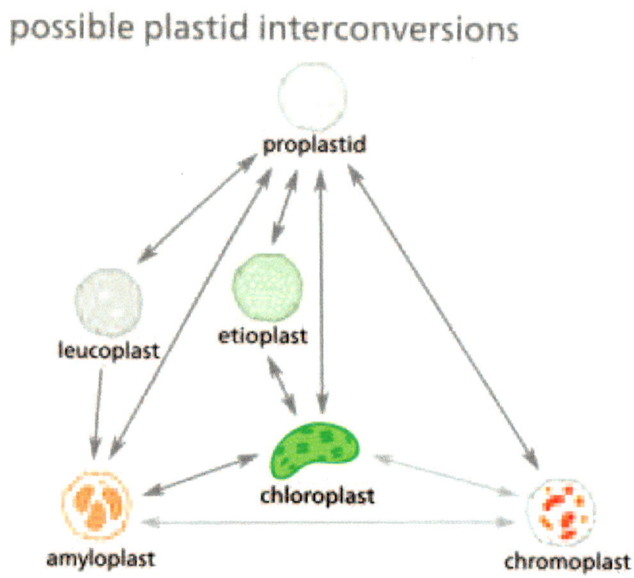

Many plastid interconversions are possible.

Plastid Interconversion

Plastid differentiation is not permanent, in fact many interconversions are possible. Chloroplasts may be converted to chromoplasts, which are pigment-filled plastids responsible for the bright colors seen in flowers and ripe fruit. Starch storing amyloplasts can also be converted to chromoplasts, and it is possible for proplastids to develop straight into chromoplasts. Chromoplasts and amyloplasts can also become chloroplasts, like what happens when a carrot or a potato is illuminated. If a plant is injured, or something else causes a plant cell to revert to a meristematic state, chloroplasts and other plastids can turn back into proplastids. Chloroplast, amyloplast, chromoplast, proplast, etc., are not absolute states—intermediate forms are common.

Chloroplast Division

Most chloroplasts in a photosynthetic cell do not develop directly from proplastids or etioplasts. In fact, a typical shoot meristematic plant cell contains only 7–20 proplastids. These proplastids differentiate into chloroplasts, which divide to create the 30–70 chloroplasts found in a mature photosynthetic plant cell. If the cell divides, chloroplast division provides the additional chloroplasts to partition between the two daughter cells.

In single-celled algae, chloroplast division is the only way new chloroplasts are formed. There is no proplastid differentiation—when an algal cell divides, its chloroplast divides along with it, and each daughter cell receives a mature chloroplast.

Almost all chloroplasts in a cell divide, rather than a small group of rapidly dividing chloroplasts. Chloroplasts have no definite S-phase—their DNA replication is not synchronized or limited to that of their host cells. Much of what we know about chloroplast division comes from studying organisms like *Arabidopsis* and the red alga *Cyanidioschyzon merolæ*.

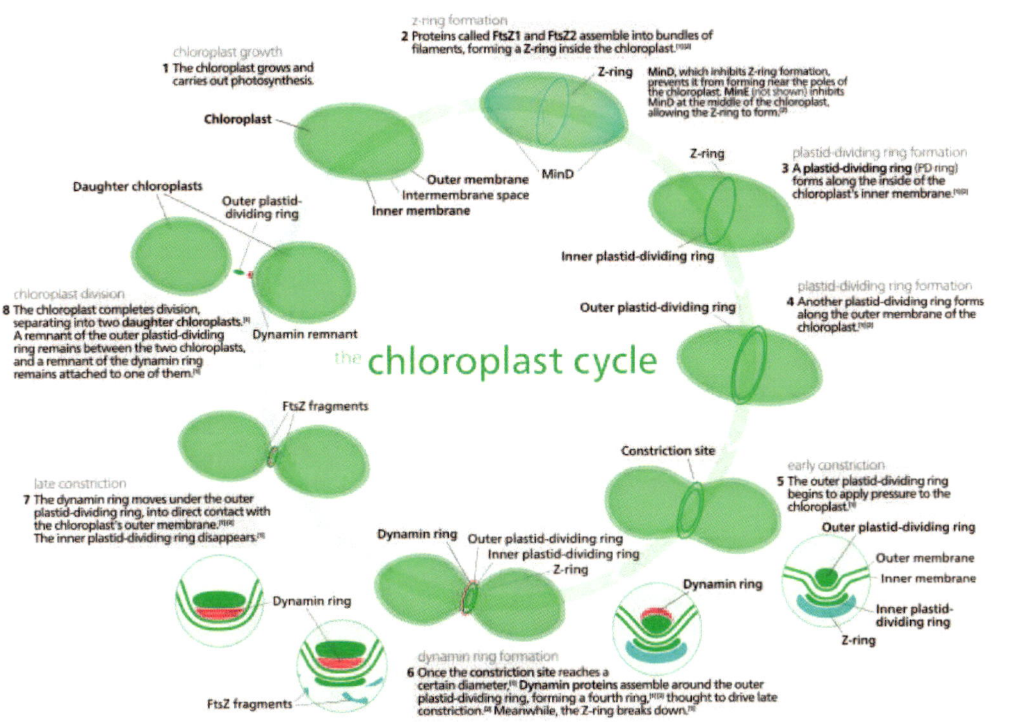

Most chloroplasts in plant cells, and all chloroplasts in algae arise from chloroplast division.
Picture references,

The division process starts when the proteins FtsZ1 and FtsZ2 assemble into filaments, and with the help of a protein ARC6, form a structure called a Z-ring within the chloroplast›s stroma. The Min system manages the placement of the Z-ring, ensuring that the chloroplast is cleaved more or less evenly. The protein MinD prevents FtsZ from linking up and forming filaments. Another protein ARC3 may also be involved, but it is not very well understood. These proteins are active at the poles of the chloroplast, preventing Z-ring formation there, but near the center of the chloroplast, MinE inhibits them, allowing the Z-ring to form.

Next, the two plastid-dividing rings, or PD rings form. The inner plastid-dividing ring is located in the inner side of the chloroplast›s inner membrane, and is formed first. The outer plastid-dividing ring is found wrapped around the outer chloroplast membrane. It consists of filaments about 5 nanometers across, arranged in rows 6.4 nanometers apart, and shrinks to squeeze the chloroplast. This is when chloroplast constriction begins. In a few species like *Cyanidioschyzon merolæ*, chloroplasts have a third plastid-dividing ring located in the chloroplast's intermembrane space.

Late into the constriction phase, dynamin proteins assemble around the outer plastid-dividing ring, helping provide force to squeeze the chloroplast. Meanwhile, the Z-ring and the inner plastid-dividing ring break down. During this stage, the many chloroplast DNA plasmids floating around in the stroma are partitioned and distributed to the two forming daughter chloroplasts.

Later, the dynamins migrate under the outer plastid dividing ring, into direct contact with the chloroplast›s outer membrane,to cleave the chloroplast in two daughter chloroplasts.

A remnant of the outer plastid dividing ring remains floating between the two daughter chloroplasts, and a remnant of the dynamin ring remains attached to one of the daughter chloroplasts.

Of the five or six rings involved in chloroplast division, only the outer plastid-dividing ring is present for the entire constriction and division phase—while the Z-ring forms first, constriction does not begin until the outer plastid-dividing ring forms.

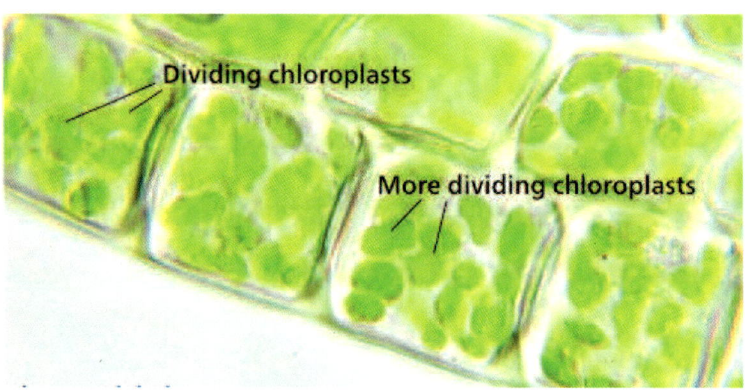

Chloroplast division In this light micrograph of some moss chloroplasts, many dumbbell-shaped chloroplasts can be seen dividing. Grana are also just barely visible as small granules.

In species of algae that contain a single chloroplast, regulation of chloroplast division is extremely important to ensure that each daughter cell receives a chloroplast—chloroplasts can›t be made from scratch. In organisms like plants, whose cells contain multiple chloroplasts, coordination is looser and less important. It is likely that chloroplast and cell division are somewhat synchronized, though the mechanisms for it are mostly unknown.

Light has been shown to be a requirement for chloroplast division. Chloroplasts can grow and progress through some of the constriction stages under poor quality green light, but are slow to complete division—they require exposure to bright white light to complete division. Spinach leaves grown under green light have been observed to contain many large dumbbell-shaped chloroplasts. Exposure to white light can stimulate these chloroplasts to divide and reduce the population of dumbbell-shaped chloroplasts.

Chloroplast Inheritance

Like mitochondria, chloroplasts are usually inherited from a single parent. Biparental chloroplast inheritance—where plastid genes are inherited from both parent plants—occurs in very low levels in some flowering plants.

Many mechanisms prevent biparental chloroplast DNA inheritance, including selective destruction of chloroplasts or their genes within the gamete or zygote, and chloroplasts from one parent being excluded from the embryo. Parental chloroplasts can be sorted so that only one type is present in each offspring.

Gymnosperms, such as pine trees, mostly pass on chloroplasts paternally, while flowering plants often inherit chloroplasts maternally. Flowering plants were once thought to only inherit chloroplasts maternally. However, there are now many documented cases of angiosperms inheriting chloroplasts paternally.

Angiosperms, which pass on chloroplasts maternally, have many ways to prevent paternal inheritance. Most of them produce sperm cells that do not contain any plastids. There are many other documented mechanisms that prevent paternal inheritance in these flowering plants, such as different rates of chloroplast replication within the embryo.

Among angiosperms, paternal chloroplast inheritance is observed more often in hybrids than in offspring from parents of the same species. This suggests that incompatible hybrid genes might interfere with the mechanisms that prevent paternal inheritance.

Transplastomic Plants

Recently, chloroplasts have caught attention by developers of genetically modified crops. Since, in most flowering plants, chloroplasts are not inherited from the male parent, transgenes in these plastids cannot be disseminated by pollen. This makes plastid transformation a valuable tool for the creation and cultivation of genetically modified plants that are biologically contained, thus posing significantly lower environmental risks. This biological containment strategy is therefore suitable for establishing the coexistence of conventional and organic agriculture. While the reliability of this mechanism has not yet been studied for all relevant crop species, recent results in tobacco plants are promising, showing a failed containment rate of transplastomic plants at 3 in 1,000,000.

REFERENCES

1. Jones, Daniel (2003) , Peter Roach, James Hartmann and Jane Setter, eds., English Pronouncing Dictionary, Cambridge: Cambridge University Press, ISBN 3-12-539683-2

2. Chloroplast. Merriam-Webster Dictionary.
3. How large are chloroplasts?. Cell Biology By The Number. Ron Milo and Rob Philips.
4. chloroplast. Online Etymology Dictionary.
5. von Mohl, H. 1835/1837. Ueber die Vermehrung der Pflanzen-Zellen durch Teilung. Dissert. Tubingen 1835. Flora 1837, .
6. Schimper, A.F.W. (1883). Über die Entwicklung der Chlorophyllkörner und Farbkörper [About the development of the chlorophyll grains and stains]. Bot. Zeitung (in German). **41**: 105–14, 121–31, 137–46, 153–62.
7. Strasburger, E. 1884. Das botanische Praktikum, 1. Aufl. Gustav Fischer: Jena
8. Gunning, B., Koenig, F. & Govindjee, P. (2006). A dedication to pioneers of research on chloroplast structure. In: R. R. Wise & J. K. Hoober (ed.). The structure and function of plastids. Springer: Netherlands, pp. xxiii-xxxi, .
9. Hoober, J. K. (1984). Chloroplasts. Plenum: New York, .
10. Biology 8th Edition Campbell & Reece. Benjamin Cummings (Pearson). 2009. p. 516.
11. Mereschkowsky C (1905). Über Natur und Ursprung der Chromatophoren im Pflanzenreiche [About the nature and origin of chromatophores in the vegetable kingdom]. Biol Centralbl (in German). **25**: 593–604.
12. Alberts, Bruce (2002). Molecular biology of the cell (4. ed.). New York [u.a.]: Garland. ISBN 0-8153-4072-9.
13. to:[a] [b] Kumar, K.; Mella-Herrera, R. A.; Golden, J. W. (2010). Cyanobacterial Heterocysts. Cold Spring Harbor Perspectives in Biology. **2** *(4): a000315. doi:10.1101/cshperspect.a000315. PMC 2845205. PMID 20452939.*
14. Biology 8th edition—Campbell&Reece. Benjamin Cummings. 2008. pp. 186–187. ISBN 978-0-321-54325-7.
15. Kim, E.; Archibald, J. M. (2009). Diversity and Evolution of Plastids and Their Genomes. In Sandelius, Anna Stina; Aronsson, Henrik. The Chloroplast. Plant Cell Monographs. **13**. pp. 1–39. doi:10.1007/978-3-540-68696-5_1. ISBN 978-3-540-68692-7.
16. Bryant, Donald A.; Guglielmi, Gérard; Marsac, Nicole Tandeau; Castets, Anne-Marie; Cohen-Bazire, Germaine (1979). The structure of cyanobacterial phycobilisomes: A model. Archives of Microbiology. **123** *(2): 311–34. doi:10.1007/BF00446810.*
17. Keeling, Patrick J. (2004). Diversity and evolutionary history of plastids and their hosts. American Journal of Botany. **91** *(10): 1481–93. doi:10.3732/ajb.91.10.1481. PMID 21652304.*
18. McFadden, Geoffrey I; Van Dooren, Giel G (2004). Evolution: Red Algal Genome Affirms a Common Origin of All Plastids. Current Biology. **14** *(13): R514–6. doi:10.1016/j.cub.2004.06.041. PMID 15242632.*
19. Fuks, Bruno; Fabrice Homblé (8 July 1996). Mechanism of Proton Permeation through Chloroplast Lipid Membranes (PDF). Plant Physiology. **112** *(2): 759–766. doi:10.1104/pp.112.2.759. PMC 158000* . PMID 8883387. Retrieved 25 August 2013.
20. Joyard J Block MA, Douce R; Block; Douce (1991). Molecular aspects of plastid envelope biochemistry. Eur J Biochem. **199** *(3): 489–509. doi:10.1111/j.1432-1033.1991.tb16148.x. PMID 1868841.*

21. Chloroplast. Encyclopedia of Science. Retrieved 27 December 2012.

22. Nakayama, Takuro; Archibald, John M (2012). Evolving a photosynthetic organelle. BMC Biology. **10** *(1): 35. doi:10.1186/1741-7007-10-35. PMC 3337241. PMID 22531210.*

23. McFadden, G. I. (2001). Chloroplast Origin and Integration. Plant Physiology. **125** *(1): 50–3. doi:10.1104/pp.125.1.50. PMC 1539323. PMID 11154294.*

24. Archibald, John M. (2009). The Puzzle of Plastid Evolution. Current Biology. **19** *(2): R81–8. doi:10.1016/j.cub.2008.11.067. PMID 19174147.*

25. Delaye, Luis; Valadez-Cano, Cecilio; Pérez-Zamorano, Bernardo (2016). How Really Ancient Is Paulinella Chromatophora?. PLoS Currents. doi:10.1371/currents.tol.e68a099364bb1a1e129a17b4e06b0c6b.

26. , edited by Robert R.; Hoober, J. Kenneth (2006). The structure and function of plastids. Dordrecht: Springer. pp. 3–21. ISBN 978-1-4020-4061-0.

27. Ball, S.; Colleoni, C.; Cenci, U.; Raj, J. N.; Tirtiaux, C. (2011). The evolution of glycogen and starch metabolism in eukaryotes gives molecular clues to understand the establishment of plastid endosymbiosis. Journal of Experimental Botany. **62** *(6): 1775–801. doi:10.1093/jxb/erq411. PMID 21220783.*

28. Nair, Sethu C.; Striepen, Boris (2011). What Do Human Parasites Do with a Chloroplast Anyway?. PLoS Biology. **9** *(8): e1001137. doi:10.1371/journal.pbio.1001137. PMC 3166169. PMID 21912515.*

29. Keeling, P. J. (2010). The endosymbiotic origin, diversification and fate of plastids. Philosophical Transactions of the Royal Society B: Biological Sciences. **365** *(1541): 729–48. doi:10.1098/rstb.2009.0103. PMC 2817223. PMID 20124341.*

30. Biology 8th Edition Campbell & Reece. Benjamin Cummings (Pearson). 2009. pp. 582–92.

31. rhodo-. The Free Dictionary. Farlex. Retrieved 2013-06-07.

32. Lewis, L. A.; McCourt, R. M. (2004). Green algae and the origin of land plants. American Journal of Botany. **91** *(10): 1535–56. doi:10.3732/ajb.91.10.1535. PMID 21652308.*

33. Machida, M.; Takechi, K; Sato, H; Chung, SJ; Kuroiwa, H; Takio, S; Seki, M; Shinozaki, K; Fujita, T; Hasebe, M; Takano, H (2006). Genes for the peptidoglycan synthesis pathway are essential for chloroplast division in moss. Proceedings of the National Academy of Sciences. **103** *(17): 6753–8. Bibcode:2006PNAS..103.6753M. doi:10.1073/pnas.0510693103. PMC 1458953. PMID 16618924.*

34. Moroney, J. V.; Somanchi, A (1999). How Do Algae Concentrate CO2 to Increase the Efficiency of Photosynthetic Carbon Fixation?. Plant Physiology. **119** *(1): 9–16. doi:10.1104/pp.119.1.9. PMC 1539202. PMID 9880340.*

35. Tartar, AuréLien; Boucias, Drion G. (2004). The non-photosynthetic, pathogenic green alga Helicosporidium sp. Has retained a modified, functional plastid genome. FEMS Microbiology Letters. **233** *(1): 153–7. doi:10.1016/j.femsle.2004.02.006. PMID 15043882.*

36. Chaal, Balbir K.; Green, Beverley R. (February 2005). Protein import pathways in ‹complex› chloroplasts derived from secondary endosymbiosis involving a red algal ancestor. Plant Molecular Biology. **57** *(3): 333–42. doi:10.1007/s11103-004-7848-y.*

PMID 15830125.

37. Moustafa, Ahmed; Beszteri, Bánk; Maier, Uwe G.; Bowler, Chris; Valentin, Klaus; Bhattacharya, Debashish (2009). Genomic Footprints of a Cryptic Plastid Endosymbiosis in Diatoms. Science. **324** *(5935): 1724–6. Bibcode:2009Sci...324.1724M. doi:10.1126/science.1172983. PMID 19556510.*

38. Rogers, M. B.; Gilson, P. R.; Su, V.; McFadden, G. I.; Keeling, P. J. (2006). The Complete Chloroplast Genome of the Chlorarachniophyte Bigelowiella natans: Evidence for Independent Origins of Chlorarachniophyte and Euglenid Secondary Endosymbionts. Molecular Biology and Evolution. **24** *(1): 54–62. doi:10.1093/molbev/msl129. PMID 16990439.*

39. Hackett, J. D.; Anderson, D. M.; Erdner, D. L.; Bhattacharya, D. (2004). Dinoflagellates: A remarkable evolutionary experiment. American Journal of Botany. **91** *(10): 1523–34. doi:10.3732/ajb.91.10.1523. PMID 21652307.*

40. Dorrell, R. G.; Smith, A. G. (2011). Do Red and Green Make Brown?: Perspectives on Plastid Acquisitions within Chromalveolates. Eukaryotic Cell. **10** *(7): 856–68. doi:10.1128/EC.00326-10. PMC 3147421* . PMID 21622904.

41. Meeson, B. W.; Chang, S. S.; Sweeney, B. M. (1982). Characterization of Peridinin-Chlorophyll α-Proteins from the Marine Dinoflagellate Ceratium furca. Botanica Marina. **25** *(8): 347–50. doi:10.1515/botm.1982.25.8.347.*

42. ᶜ Tengs, T; Dahlberg, OJ; Shalchian-Tabrizi, K; Klaveness, D; Rudi, K; Delwiche, CF; Jakobsen, KS (2000). Phylogenetic analyses indicate that the 19›Hexanoyloxy-fucoxanthin-containing dinoflagellates have tertiary plastids of haptophyte origin. Molecular Biology and Evolution. **17** *(5): 718–29. doi:10.1093/oxfordjournals.molbev.a026350. PMID 10779532.*

43. Schnepf, Eberhard; Elbrächter, Malte (1999). Dinophyte chloroplasts and phylogeny – A review. Grana. **38** *(2–3): 81–97. doi:10.1080/00173139908559217.*

44. Nowack, E. C. M.; Vogel, H.; Groth, M.; Grossman, A. R.; Melkonian, M.; Glöckner, G. (2010). Endosymbiotic Gene Transfer and Transcriptional Regulation of Transferred Genes in Paulinella chromatophora. Molecular Biology and Evolution. **28** *(1): 407–22. doi:10.1093/molbev/msq209. PMID 20702568.*

45. Dann, Leighton (2002). Bioscience—Explained (PDF). Green DNA: BIOSCIENCE EXPLAINED.

46. Clegg, M. T.; Gaut, BS; Learn Jr, GH; Morton, BR (1994). Rates and Patterns of Chloroplast DNA Evolution. Proceedings of the National Academy of Sciences. **91** *(15): 6795–801. Bibcode:1994PNAS...91.6795C. doi:10.1073/pnas.91.15.6795. PMC 44285. PMID 8041699.*

47. Shaw, J.; Lickey, E. B.; Schilling, E. E.; Small, R. L. (2007). Comparison of whole chloroplast genome sequences to choose noncoding regions for phylogenetic studies in angiosperms: The tortoise and the hare III. American Journal of Botany. **94** *(3): 275–88. doi:10.3732/ajb.94.3.275. PMID 21636401.*

48. Skovgaard, Alf (1998). Role of chloroplast retention in a marine dinoflagellate. Aquatic Microbial Ecology. **15**: 293–301. doi:10.3354/ame015293.

49. Hogan, C. Michael (2010). Deoxyribonucleic acid. Encyclopedia of Earth. National

Council for Science and the Environment. eds. S.Draggan and C.Cleveland. Washington DC

50. ctDNA — chloroplast DNA. AllAcronyms.com. Retrieved 2 January 2013.

51. The Oxford Dictionary of Abbreviations. ctDNA—Dictionary definition. 1998.

52. Chloroplasts and Other Plastids. University of Hamburg. Retrieved 27 December 2012.

53. [g] Sandelius, Anna Stina (2009). The Chloroplast: Interactions with the Environment. Springer. p. 18. ISBN 978-3-540-68696-5.

54. Burgess, Jeremy (1989). An introduction to plant cell development. Cambridge: Cambridge university press. p. 62. ISBN 0-521-31611-1.

55. Bendich, A. J. (2004). Circular Chloroplast Chromosomes: The Grand Illusion. The Plant Cell Online. **16** *(7): 1661–6. doi:10.1105/tpc.160771. PMC 514151. PMID 15235123.*

56. [c] Kolodner, R.; Tewari, KK (1979). Inverted Repeats in Chloroplast DNA from Higher Plants. Proceedings of the National Academy of Sciences. **76** *(1): 41–5. Bibcode:1979PNAS...76...41K. doi:10.1073/pnas.76.1.41. PMC 382872. PMID 16592612.*

57. Palmer, Jeffrey D.; Thompson, William F. (1982). Chloroplast DNA rearrangements are more frequent when a large inverted repeat sequence is lost. Cell. **29** *(2): 537–50. doi:10.1016/0092-8674(82)90170-2. PMID 6288261.*

58. Plant Biochemistry (3rd ed.). Academic Press. 2005. p. 517. ISBN 978-0-12-088391-2.

59. Kobayashi, T.; Takahara, M; Miyagishima, SY; Kuroiwa, H; Sasaki, N; Ohta, N; Matsuzaki, M; Kuroiwa, T (2002). Detection and Localization of a Chloroplast-Encoded HU-Like Protein That Organizes Chloroplast Nucleoids. The Plant Cell Online. **14** *(7): 1579–89. doi:10.1105/tpc.002717. PMC 150708* . PMID 12119376.

60. Krishnan, Rao BJ, NM (2009). A comparative approach to elucidate chloroplast genome replication. BMC Genomics. **10** *(237). doi:10.1186/1471-2164-10-237.*

61. Heinhorst, Gordon C. Cannon, Sabine (1993). DNA replication in chloroplasts. Journal of Cell Science. **104**: 1–9.

62. Bendich, AJ (2004). Circular Chloroplast Chromosomes: The Grand Illusion. The Plant Cell. **16** *(7): 1661–1666. doi:10.1105/tpc.160771. PMC 514151. PMID 15235123.*

63. Effect of chemical mutagens on nucleotide sequence. Biocyclopedia. Retrieved 24 October 2015.

64. Alberts, Bruce (2002). Molecular biology of the cell (4. ed.). New York [u.a.]: Garland. ISBN 0-8153-4072-9.

65. Huang, Chun Y.; Ayliffe, Michael A.; Timmis, Jeremy N. (2003). Direct measurement of the transfer rate of chloroplast DNA into the nucleus. Nature. **422** *(6927): 72–6. Bibcode:2003Natur.422...72H. doi:10.1038/nature01435. PMID 12594458.*

66. Martin, W.; Rujan, T; Richly, E; Hansen, A; Cornelsen, S; Lins, T; Leister, D; Stoebe, B; Hasegawa, M; Penny, D (2002). Evolutionary analysis of Arabidopsis, cyanobacterial, and chloroplast genomes reveals plastid phylogeny and thousands of cyanobacterial genes in the nucleus. Proceedings of the National Academy of Sciences. **99** *(19): 12246–51. Bibcode:2002PNAS...9912246M. doi:10.1073/pnas.182432999. PMC*

129430. PMID 12218172.

67. Smith, David Roy; Lee, Robert W. (2014-04-01). A Plastid without a Genome: Evidence from the Nonphotosynthetic Green Algal Genus Polytomella. Plant Physiology. **164** *(4): 1812–1819. doi:10.1104/pp.113.233718. PMC 3982744. PMID 24563281.*
68. Archibald, John M. (2006). Algal Genomics: Exploring the Imprint of Endosymbiosis. Current Biology. **16** *(24): R1033–5. doi:10.1016/j.cub.2006.11.008. PMID 17174910.*
69. Koussevitzky, Shai; Nott, A; Mockler, TC; Hong, F; Sachetto-Martins, G; Surpin, M; Lim, J; Mittler, R; Chory, J (2007). Signals from Chloroplasts Converge to Regulate Nuclear Gene Expression. Science. **316** *(5825): 715–9. Bibcode:2007Sci...316..715K. doi:10.1126/science.1140516 (inactive 2015-01-01). PMID 17395793.*
70. Hedtke, B.; Börner, T; Weihe, A (1997). Mitochondrial and Chloroplast Phage-Type RNA Polymerases in Arabidopsis. Science. **277** *(5327): 809–11. doi:10.1126/science.277.5327.809. PMID 9242608.*
71. Harris, EH; Boynton, JE; Gillham, NW (1994). Chloroplast ribosomes and protein synthesis. Microbiological reviews. **58** *(4): 700–54. PMC 372988. PMID 7854253.*
72. Soll, Jürgen; Schleiff, Enrico (2004). Plant cell biology: Protein import into chloroplasts. Nature Reviews Molecular Cell Biology. **5** *(3): 198–208. doi:10.1038/nrm1333. PMID 14991000.*
73. Biology 8th edition—Campbell & Reece. Benjamin Cummings. 2008. p. 340. ISBN 978-0-321-54325-7.
74. Lung, S.-C.; Chuong, S. D. X. (2012). A Transit Peptide-Like Sorting Signal at the C Terminus Directs the Bienertia sinuspersici Preprotein Receptor Toc159 to the Chloroplast Outer Membrane. The Plant Cell. **24** *(4): 1560–78. doi:10.1105/tpc.112.096248. PMC 3398564. PMID 22517318.*
75. Soll, J r.; Soll, J (1996). Phosphorylation of the Transit Sequence of Chloroplast Precursor Proteins. Journal of Biological Chemistry. **271** *(11): 6545–54. doi:10.1074/jbc.271.11.6545. PMID 8626459.*
76. May, T. (2000). 14-3-3 Proteins Form a Guidance Complex with Chloroplast Precursor Proteins in Plants. The Plant Cell Online. **12**: 53–64. doi:10.1105/tpc.12.1.53.
77. Jarvis, Paul; Soll, Jürgen (2001). Toc, Tic, and chloroplast protein import. Biochimica et Biophysica Acta (BBA) - Molecular Cell Research. **1541** *(1–2): 64–79. doi:10.1016/S0167-4889(01)00147-1. PMID 11750663.*
78. Wise, R.R.; Hoober, J.K. (2007). The Structure and Function of Plastids. Springer. pp. 32–33. ISBN 978-1-4020-6570-5.
79. Oedogonium Link ex Hirn, 1900: 17. algaeBASE. Retrieved 19 May 2013.
80. Chlamydomonas Ehrenberg, 1833: 288. algaeBASE. Retrieved 19 May 2013.
81. Spirogyra Link, 1820: 5. algaeBASE. Retrieved 19 May 2013.
82. Sirogonium Kützing, 1843: 278. algaeBASE. Retrieved 19 May 2013.
83. Zygnema C.Agardh, 1817: xxxii, 98. algaeBASE. Retrieved 19 May 2013.
84. Micrasterias C.Agardh ex Ralfs, 1848: 68. algaeBASE. Retrieved 19 May 2013.
85. John, D.M.; Brook, A.J.; Whitton, B.A. (2002). The freshwater algal flora of the British Isles : an identification guide to freshwater and terrestrial algae. Cambridge: Cambridge

University Press. p. 335. ISBN 978-0-521-77051-4.

86. Biology 8th edition—Campbell & Reece. Benjamin Cummings. 2008. pp. 196–197. ISBN 978-0-321-54325-7.

87. Koike, H.; Yoshio, M.; Kashino, Y.; Satoh, K. (1998). Polypeptide Composition of Envelopes of Spinach Chloroplasts: Two Major Proteins Occupy 90% of Outer Envelope Membranes. Plant and Cell Physiology. **39** *(5): 526–32. doi:10.1093/oxfordjournals.pcp.a029400. PMID 9664716.*

88. Köhler RH, Hanson MR (1 January 2000). Plastid tubules of higher plants are tissue-specific and developmentally regulated. J. Cell. Sci. **113** *(Pt 1): 81–9. PMID 10591627. Archived from the original on 2010-11-18.*

89. Gray JC, Sullivan JA, Hibberd JM, Hansen MR (2001). Stromules: mobile protrusions and interconnections between plastids. Plant Biology. **3** *(3): 223–33. doi:10.1055/s-2001-15204.*

90. Schattat, M.; Barton, K.; Baudisch, B.; Klösgen, R. B.; Mathur, J. (2011). Plastid Stromule Branching Coincides with Contiguous Endoplasmic Reticulum Dynamics. Plant Physiology. **155** *(4): 1667–77. doi:10.1104/pp.110.170480. PMC 3091094. PMID 21273446.*

91. Schattat, M. H.; Griffiths, S.; Mathur, N.; Barton, K.; Wozny, M. R.; Dunn, N.; Greenwood, J. S.; Mathur, J. (2012). Differential Coloring Reveals That Plastids Do Not Form Networks for Exchanging Macromolecules. The Plant Cell. **24** *(4): 1465–77. doi:10.1105/tpc.111.095398. PMC 3398557. PMID 22474180.*

92. Brunkard, Jacob O.; Runkel, Anne M.; Zambryski, Patricia C. (2015). Chloroplasts extend stromules independently and in response to internal redox signals. PNAS. **112** *(32): 10044–10049. doi:10.1073/pnas.1511570112.*

93. Burgess,, Jeremy (1989). An introduction to plant cell development (Pbk. ed.). Cambridge: Cambridge university press. p. 46. ISBN 0-521-31611-1.

94. Whatley, Jean M (5 July 1994). The occurrence of a peripheral reticulum in plastids of the gymnosperm Welwitschia mirabilis. New Phytologist. **74** *(2): 215–220. doi:10.1111/j.1469-8137.1975.tb02608.x.*

95. [c] Wise, Robert R (2007). The Structure and Function of Plastids. Springer. pp. 17–18. ISBN 978-1-4020-6570-5.

96. Manuell, Andrea L.; Quispe, Joel; Mayfield, Stephen P. (2007). Structure of the Chloroplast Ribosome: Novel Domains for Translation Regulation. PLoS Biology. **5** *(8): e209. doi:10.1371/journal.pbio.0050209. PMC 1939882. PMID 17683199.*

97. Lim, K.; Kobayashi, I.; Nakai, K. (7 April 2014). Alterations in rRNA-mRNA Interaction during Plastid Evolution. Molecular Biology and Evolution. **31** *(7): 1728–1740. doi:10.1093/molbev/msu120.*

98. Hirose, T; Sugiura, M (January 2004). Functional Shine-Dalgarno-like sequences for translational initiation of chloroplast mRNAs.. Plant & cell physiology. **45** *(1): 114–7. doi:10.1093/pcp/pch002. PMID 14749493.*

99. Ma, J; Campbell, A; Karlin, S (October 2002). Correlations between Shine-Dalgarno sequences and gene features such as predicted expression levels and operon structures.. Journal of Bacteriology. **184** *(20): 5733–45. doi:10.1128/JB.184.20.5733-5745.2002.*

PMID 12270832.

100. Lim, K; Furuta, Y; Kobayashi, I (October 2012). Large variations in bacterial ribosomal RNA genes.. Molecular Biology and Evolution. **29** *(10): 2937–48. doi:10.1093/molbev/mss101. PMID 22446745.*

101. Austin, J. R.; Frost, E; Vidi, PA; Kessler, F; Staehelin, LA (2006). Plastoglobules Are Lipoprotein Subcompartments of the Chloroplast That Are Permanently Coupled to Thylakoid Membranes and Contain Biosynthetic Enzymes. The Plant Cell Online. **18** *(7): 1693–703. doi:10.1105/tpc.105.039859. PMC 1488921. PMID 16731586.*

102. Crumpton-Taylor, M.; Grandison, S.; Png, K. M. Y.; Bushby, A. J.; Smith, A. M. (2011). Control of Starch Granule Numbers in Arabidopsis Chloroplasts. Plant Physiology. **158** *(2): 905–16. doi:10.1104/pp.111.186957. PMC 3271777. PMID 22135430.*

103. Zeeman, Samuel C.; Delatte, Thierry; Messerli, Gaëlle; Umhang, Martin; Stettler, Michaela; Mettler, Tabea; Streb, Sebastian; Reinhold, Heike; Kötting, Oliver (2007). Starch breakdown: Recent discoveries suggest distinct pathways and novel mechanisms. Functional Plant Biology. **34** *(6): 465–73. doi:10.1071/FP06313.*

104. J D, Rochaix (1998). The molecular biology of chloroplasts and mitochondria in Chlamydomonas. Dordrecht [u.a.]: Kluwer Acad. Publ. pp. 550–565. ISBN 978-0-7923-5174-0.

105. Steer, Brian E.S. Gunning, Martin W. (1996). Plant cell biology : structure and function. Boston, Mass.: Jones and Bartlett Publishers. p. 24. ISBN 0-86720-504-0.

106. Hanson, David; Andrews, T. John; Badger, Murray R. (2002). Variability of the pyrenoid-based CO_2concentrating mechanism in hornworts (Anthocerotophyta)". Functional Plant Biology. **29** *(3): 407–16. doi:10.1071/PP01210.*

107. Ma, Y.; Pollock, S. V.; Xiao, Y.; Cunnusamy, K.; Moroney, J. V. (2011). Identification of a Novel Gene, CIA6, Required for Normal Pyrenoid Formation in Chlamydomonas reinhardtii. Plant Physiology. **156** *(2): 884–96. doi:10.1104/pp.111.173922. PMC 3177283. PMID 21527423.*

108. Retallack, B; Butler, RD (1970). The development and structure of pyrenoids in Bulbochaete hiloensis. Journal of Cell Science. **6** *(1): 229–41. PMID 5417694.*

109. Brown, Malcolm R; Howard J Arnott (1970). Structure and Function of the Algal Pyrenoid (PDF). Journal of Phycology. **6**: 14–22. doi:10.1111/j.1529-8817.1970.tb02350.x. Retrieved 31 December 2012.

110. Mustardy, L.; Buttle, K.; Steinbach, G.; Garab, G. (2008). The Three-Dimensional Network of the Thylakoid Membranes in Plants: Quasihelical Model of the Granum-Stroma Assembly. The Plant Cell Online. **20** *(10): 2552–7. doi:10.1105/tpc.108.059147. PMC 2590735. PMID 18952780.*

111. [c] Austin, J. R.; Staehelin, L. A. (2011). Three-Dimensional Architecture of Grana and Stroma Thylakoids of Higher Plants as Determined by Electron Tomography. Plant Physiology. **155** *(4): 1601–11. doi:10.1104/pp.110.170647. PMC 3091084. PMID 21224341.*

112. Infanger, S; Bischof, S; Hiltbrunner, A; Agne, B; Baginsky, S; Kessler, F (2011). The Chloroplast Import Receptor Toc90 Partially Restores the Accumulation of Toc159 Client Proteins in the Arabidopsis thaliana ppi2 Mutant. Molecular Plant. **4** *(2): 252–63. doi:10.1093/mp/ssq071. PMID 21220583.*

113. thylakoid. Merriam-Webster Dictionary. Merriam-Webster. Retrieved 19 May 2013.
114. Biology 8th Edition Campbell & Reece. Benjamin Cummings (Pearson). 2009. pp. 190–193.
115. Australian scientists discover first new chlorophyll in 60 years. University of Sydney. 20 August 2010.
116. Takaichi, Shinichi (15 June 2011). Carotenoids in Algae: Distributions, Biosyntheses and Functions. Marine Drugs. **9** *(12): 1101–1118. doi:10.3390/md9061101. PMC 3131562. PMID 21747749.*
117. Shapley, Dan. Why Do Leaves Change Color in Fall?. News Articles. Retrieved 21 May 2013.
118. Howe, C.J; Barbrook, A.C; Nisbet, R.E.R; Lockhart, P.J; Larkum, A.W.D (2008). The origin of plastids. Philosophical Transactions of the Royal Society B: Biological Sciences. **363** *(1504): 2675–85. doi:10.1098/rstb.2008.0050. PMC 2606771. PMID 18468982.*
119. Introduction to the Rhodophyta. University of California Museum of Paleontology. Retrieved 20 May 2013.
120. Biology 8th Edition Campbell & Reece. Benjamin Cummings (Pearson). 2009. pp. 200–201.
121. Lawton, June R (March 1988). Ultrastructure of Chloroplast Membranes in Leaves of Maize and Ryegrass as Revealed by Selective Staining Methods. New Phytologist. **108** *(3): 277–283. doi:10.1111/j.1469-8137.1988.tb04163.x. JSTOR 2433294.*
122. Roberts, editor, Keith (2007). Handbook of plant science. Chichester, West Sussex, England: Wiley. p. 16. ISBN 978-0-470-05723-0.
123. Biology 8th edition—Campbell&Reece. Benjamin Cummings. 2008. p. 742. ISBN 978-0-321-54325-7.
124. Lawson T. and J. I. L. Morison. Essay 10.1 Guard Cell Photosynthesis. Plant Physiology and Development, Sixth Edition
125. Eckardt, N. A. (2003). Controlling Organelle Positioning: A Novel Chloroplast Movement Protein. The Plant Cell Online. **15** *(12): 2755–7. doi:10.1105/tpc.151210. PMC 540263.*
126. Glynn, Jonathan M.; Miyagishima, Shin-ya; Yoder, David W.; Osteryoung, Katherine W.; Vitha, Stanislav (2007). Chloroplast Division. Traffic. **8** *(5): 451–61. doi:10.1111/j.1600-0854.2007.00545.x. PMID 17451550.*
127. Dong, X.-J.; Nagai, R.; Takagi, S. (1998). Microfilaments Anchor Chloroplasts along the Outer Periclinal Wall in Vallisneria Epidermal Cells through Cooperation of PFR and Photosynthesis. Plant and Cell Physiology. **39** *(12): 1299–306. doi:10.1093/oxfordjournals.pcp.a029334.*
128. [c] Takagi, Shingo (December 2002). Actin-based photo-orientation movement of chloroplasts in plant cells. Journal of Experimental Biology. **206**: 1963–1969. doi:10.1242/jeb.00215. Retrieved 6 March 2013.
129. Zeiger, E.; Talbott, L. D.; Frechilla, S.; Srivastava, A.; Zhu, J. (2002). The guard cell chloroplast: A perspective for the twenty-first century. New Phytologist. **153** *(3): 415–424. doi:10.1046/j.0028-646X.2001.NPH328.doc.x.*

130. Padmanabhan, Meenu S.; Dinesh-Kumar, S. P. (2010). All Hands on Deck—The Role of Chloroplasts, Endoplasmic Reticulum, and the Nucleus in Driving Plant Innate Immunity. Molecular Plant-Microbe Interactions. **23** *(11): 1368–80. doi:10.1094/MPMI-05-10-0113. PMID 20923348.*

131. Katsir, Leron; Chung, Hoo Sun; Koo, Abraham JK; Howe, Gregg A (2008). Jasmonate signaling: A conserved mechanism of hormone sensing. Current Opinion in Plant Biology. **11** *(4): 428–35. doi:10.1016/j.pbi.2008.05.004. PMC 2560989. PMID 18583180.*

132. Schnurr, J. A.; Shockey, JM; De Boer, GJ; Browse, JA (2002). Fatty Acid Export from the Chloroplast. Molecular Characterization of a Major Plastidial Acyl-Coenzyme a Synthetase from Arabidopsis. Plant Physiology. **129** *(4): 1700–9. doi:10.1104/pp.003251. PMC 166758. PMID 12177483.*

133. Biology—Concepts and Connections. Pearson. 2009. pp. 108–118.

134. Campbell, Neil A.; Williamson, Brad; Heyden, Robin J. (2006). Biology: Exploring Life. Boston, Massachusetts: Pearson Prentice Hall. ISBN 978-0-13-250882-7.

135. Jagendorf A. T. and E. Uribe (1966). ATP formation caused by acid-base transition of spinach chloroplasts.. Proc. Natl. Acad. Sci. USA. **55**: 170–177.

136. Clarke, Jeremy M. Berg; John L. Tymoczko; Lubert Stryer. Web content by Neil D. (2002). Biochemistry (5. ed., 4. print. ed.). New York, NY [u.a.]: W. H. Freeman. pp. Section 19.4. ISBN 0-7167-3051-0.

137. Clarke, Jeremy M. Berg; John L. Tymoczko; Lubert Stryer. Web content by Neil D. (2002). Biochemistry (5. ed., 4. print. ed.). New York, NY [u.a.]: W. H. Freeman. pp. Section 20.1. ISBN 0-7167-3051-0.

138. Wample, R. L.; Davis, R. W. (1983). Effect of Flooding on Starch Accumulation in Chloroplasts of Sunflower (Helianthus annuus L.). Plant Physiology. **73** *(1): 195–8. doi:10.1104/pp.73.1.195. PMC 1066435. PMID 16663176.*

139. Carmi, A.; Shomer, I. (1979). Starch Accumulation and Photosynthetic Activity in Primary Leaves of Bean (Phaseolus vulgaris L.). Annals of Botany. **44** *(4): 479–484.*

140. Berg JM; Tymoczko JL; Stryer L. (2002). Biochemistry. (5th ed.). W H Freeman. pp. Section 19.4. Retrieved 30 October 2012.

141. Hauser, M; Eichelmann, H; Oja, V; Heber, U; Laisk, A (1995). Stimulation by Light of Rapid pH Regulation in the Chloroplast Stroma in Vivo as Indicated by CO2 Solubilization in Leaves. Plant Physiology. **108** *(3): 1059–1066. doi:10.1104/pp.108.3.1059. PMC 157457. PMID 12228527.*

142. Werdan, Karl; Heldt, Hans W.; Milovancev, Mirjana (1975). The role of pH in the regulation of carbon fixation in the chloroplast stroma. Studies on CO2 fixation in the light and dark. Biochimica et Biophysica Acta (BBA) - Bioenergetics. **396** *(2): 276–92. doi:10.1016/0005-2728(75)90041-9. PMID 239746.*

143. Burgess,, Jeremy (1989). An introduction to plant cell development. Cambridge: Cambridge university press. p. 56. ISBN 0-521-31611-1.

144. Ferro, Myriam; Daniel Salvi; Hélène Rivière-Rolland; Thierry Vermat; Daphné Seigneurin-Berny; Didier Grunwald; Jérôme Garin; Jacques Joyard; Norbert Rolland (20 August 2002). Integral membrane proteins of the chloroplast envelope: Identification and subcellular localization of new transporters. Proceedings of the National Academy of Sciences. **99** *(17): 11487–11492. Bibcode:2002PNAS...9911487F. doi:10.1073/*

pnas.172390399. PMC 123283. PMID 12177442.

145. Rolland, Norbert; Michel Droux; Roland Douce (March 1992). Subcellular Distribution of O-Acetylserine(thiol)lyase in Cauliflower (Brassica oleracea L.) Inflorescence. Plant Physiology. **98** *(3): 927–935. doi:10.1104/pp.98.3.927.*

146. Ravanel, Stéphane; Bertrand Gakière; Dominique Job; Roland Douce (23 June 1998). The specific features of methionine biosynthesis and metabolism in plants. Proceedings of the National Academy of Sciences. **95** *(13): 7805–7812. Bibcode:1998PNAS...95.7805R. doi:10.1073/pnas.95.13.7805. PMID 9636232.*

147. Joyard, Jacques; Ferro, Myriam; Masselon, Christophe; Seigneurin-Berny, Daphné; Salvi, Daniel; Garin, Jérôme; Rolland, Norbert (2010). Chloroplast proteomics highlights the subcellular compartmentation of lipid metabolism. Progress in Lipid Research. **49** *(2): 128–58. doi:10.1016/j.plipres.2009.10.003. PMID 19879895.*

148. Steer, Brian E.S. Gunning, Martin W. (1996). Plant cell biology : structure and function. Boston, Mass.: Jones and Bartlett Publishers. p. 20. ISBN 0-86720-504-0.

149. Burgess,, Jeremy (1989). An introduction to plant cell development (Pbk. ed.). Cambridge: Cambridge university press. pp. 54–55. ISBN 0-521-31611-1.

150. Burgess,, Jeremy (1989). An introduction to plant cell development (Pbk. ed.). Cambridge: Cambridge university press. p. 57. ISBN 0-521-31611-1.

151. Possingham, J. V.; Rose, R. J. (1976). Chloroplast Replication and Chloroplast DNA Synthesis in Spinach Leaves. Proceedings of the Royal Society B: Biological Sciences. **193** *(1112): 295–305. Bibcode:1976RSPSB.193..295P. doi:10.1098/rspb.1976.0047.*

152. Heinhorst, S; G.C. Cannon (January 1993). DNA replication in chloroplasts. Journal of Cell Science. **104** *(104): 1–9. Retrieved 3 February 2013.*

153. Miyagishima, S.-y.; Nishida, K; Mori, T; Matsuzaki, M; Higashiyama, T; Kuroiwa, H; Kuroiwa, T (2003). A Plant-Specific Dynamin-Related Protein Forms a Ring at the Chloroplast Division Site. The Plant Cell Online. **15** *(3): 655–65. doi:10.1105/tpc.009373. PMC 150020. PMID 12615939.*

154. Hashimoto, H.; Possingham, J. V. (1989). Effect of Light on the Chloroplast Division Cycle and DNA Synthesis in Cultured Leaf Discs of Spinach. Plant Physiology. **89** *(4): 1178–83. doi:10.1104/pp.89.4.1178. PMC 1055993. PMID 16666681.*

155. Hansen, A. K.; Escobar, L. K.; Gilbert, L. E.; Jansen, R. K. (2007). Paternal, maternal, and biparental inheritance of the chloroplast genome in Passiflora (Passifloraceae): Implications for phylogenetic studies. American Journal of Botany. **94** *(1): 42–6. doi:10.3732/ajb.94.1.42. PMID 21642206.*

156. Birky, C. William (1995). Uniparental Inheritance of Mitochondrial and Chloroplast Genes: Mechanisms and Evolution. Proceedings of the National Academy of Sciences. **92** *(25): 11331–8. Bibcode:1995PNAS...9211331B. doi:10.1073/pnas.92.25.11331. PMC 40394. PMID 8524780.*

157. Powell W, Morgante M, McDevitt R, Vendramin GG, Rafalski JA (August 1995). Polymorphic simple sequence repeat regions in chloroplast genomes: applications to the population genetics of pines. Proc. Natl. Acad. Sci. U.S.A. **92** *(17): 7759–63. Bibcode:1995PNAS...92.7759P. doi:10.1073/pnas.92.17.7759. PMC 41225. PMID 7644491. In the pines, the chloroplast genome is transmitted through pollen*

158. Stegemann, S.; Hartmann, S; Ruf, S; Bock, R (2003). High-frequency gene transfer from the chloroplast genome to the nucleus. Proceedings of the National Academy of Sciences. **100** *(15): 8828–33. Bibcode:2003PNAS..100.8828S. doi:10.1073/pnas.1430924100. PMC 166398. PMID 12817081.*

159. Ruf, S.; Karcher, D.; Bock, R. (2007). Determining the transgene containment level provided by chloroplast transformation. Proceedings of the National Academy of Sciences. **104** *(17): 6998–7002. Bibcode:2007PNAS..104.6998R. doi:10.1073/pnas.0700008104. PMC 1849964* . PMID 17420459.

CHAPTER 6

RNA

Chapter Contents

A monthly journal publishing high-quality, peer-reviewed research on all topics related to **RNA** and its metabolism in all organisms.

INTRODUCTION

A hairpin loop from a pre-mRNA. Highlighted are the nucleobases (green) and the ribose-phosphate backbone (blue). Note that this is a single strand of RNA that folds back upon itself.

Ribonucleic acid (RNA) is a polymeric molecule essential in various biological roles in coding, decoding, regulation, and expression of genes. RNA and DNA are nucleic acids, and, along with proteins and carbohydrates, constitute the four major macromolecules essential for all known forms of life. Like DNA, RNA is assembled as a chain of nucleotides, but unlike DNA it is more often found in nature as a single-strand folded onto itself, rather than a paired double-strand. Cellular organisms use messenger RNA (*mRNA*) to convey genetic information (using the letters G, U, A, and C to denote the nitrogenous bases guanine, uracil, adenine, and cytosine) that directs synthesis of specific proteins. Many viruses encode their genetic information using an RNA genome.

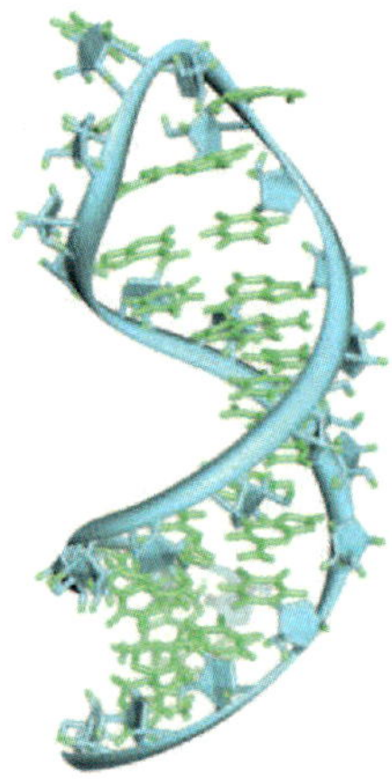

Some RNA molecules play an active role within cells by catalyzing biological reactions, controlling gene expression, or sensing and communicating responses to cellular signals. One of these active processes is protein synthesis, a universal function where RNA molecules direct the assembly of proteins on ribosomes. This process uses transfer RNA (***tRNA***) molecules to deliver amino acids to the ribosome, where ribosomal RNA (***rRNA***) then links amino acids together to form proteins.

COMPARISON WITH DNA

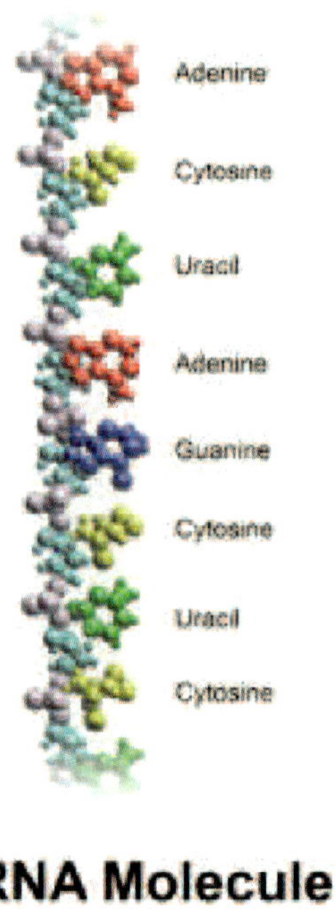

Bases in an RNA molecule.

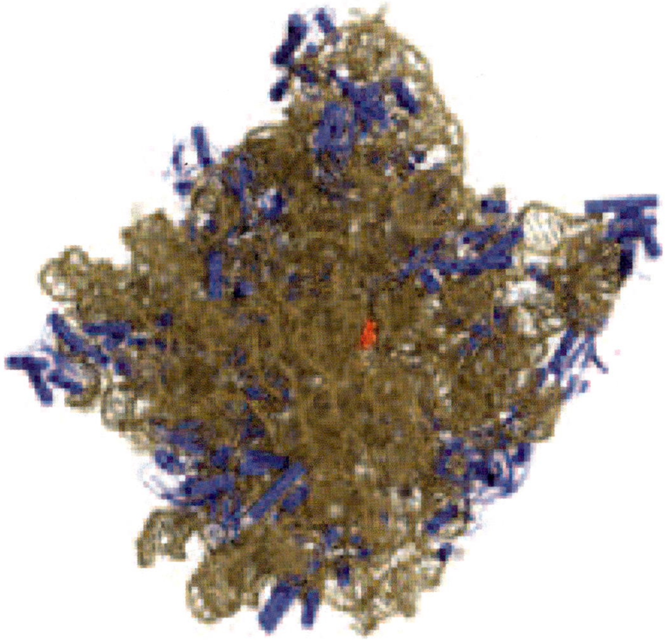

Three-dimensional representation of the 50S ribosomal subunit. Ribosomal RNA is in ochre, proteins in blue. The active site is a small segment of rRNA, indicated in red.

The chemical structure of RNA is very similar to that of DNA, but differs in three main ways:

- Unlike double-stranded DNA, RNA is a single-stranded molecule in many of its biological roles and has a much shorter chain of nucleotides.However, RNA can, by complementary base pairing, form intrastrand (i.e., single-strand) double helixes, as in tRNA.
- While DNA contains *deoxyribose*, RNA contains *ribose* (in deoxyribose there is no hydroxyl group attached to the pentose ring in the 2' position). These hydroxyl groups make RNA less stable than DNA because it is more prone to hydrolysis.
- The complementary base to adenine in DNA is thymine, whereas in RNA, it is uracil, which is an unmethylated form of thymine.

Like DNA, most biologically active RNAs, including mRNA, tRNA, rRNA, snRNAs, and other non-coding RNAs, contain self-complementary sequences that allow parts of the RNA to fold and pair with itself to form double helices. Analysis of these RNAs has revealed that they are highly structured. Unlike DNA, their structures do not consist of long double helices, but rather collections of short helices packed together into structures akin to proteins. In this fashion, RNAs can achieve chemical catalysis (like enzymes). For instance, determination of the structure of the ribosome—an enzyme that catalyzes peptide bond formation—revealed that its active site is composed entirely of RNA.

STRUCTURE

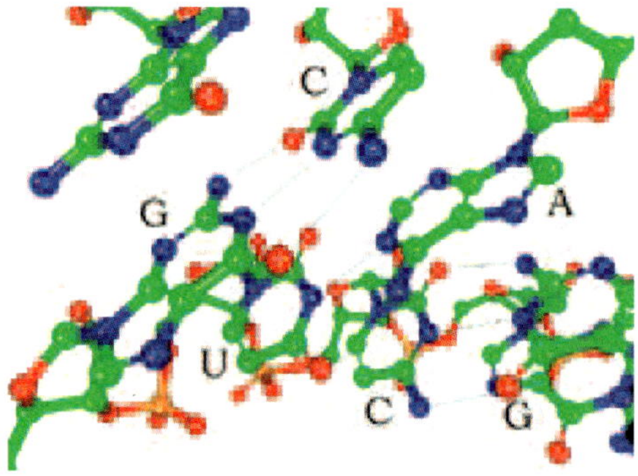

Watson-Crick base pairs in a siRNA (hydrogen atoms are not shown)

Each nucleotide in RNA contains a ribose sugar, with carbons numbered 1' through 5'. A base is attached to the 1' position, in general, adenine (A), cytosine (C), guanine (G), or uracil (U). Adenine and guanine are purines, cytosine and uracil are pyrimidines. A phosphate group is attached to the 3' position of one ribose and the 5' position of the next. The phosphate groups have a negative charge each, making RNA a charged molecule (polyanion). The bases form hydrogen bonds between cytosine and guanine, between adenine and uracil and between guanine and uracil. However, other interactions are possible, such as a group of adenine bases binding to each other in a bulge, or the GNRA tetraloop that has a guanine–adenine base-pair.

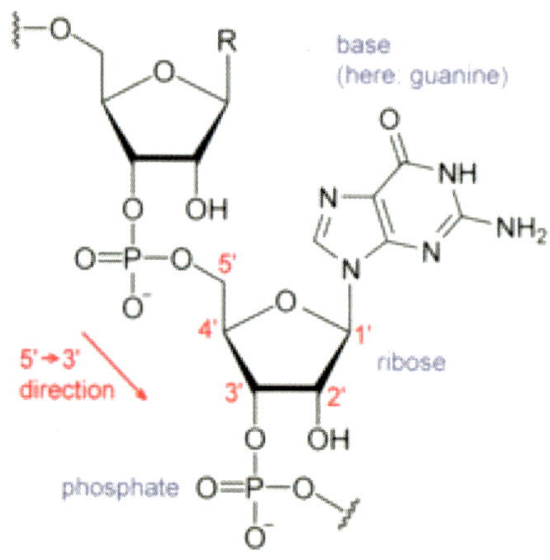

Chemical structure of RNA

An important structural feature of RNA that distinguishes it from DNA is the presence of a hydroxyl group at the 2' position of the ribose sugar. The presence of this functional group causes the helix to mostly adopt the A-form geometry, although in single strand dinucleotide contexts, RNA can rarely also adopt the B-form most commonly observed in DNA. The A-form geometry results in a very deep and narrow major groove and a shallow and wide minor groove. A second consequence of the presence of the 2'-hydroxyl group is that in conformationally flexible regions of an RNA molecule (that is, not involved in formation of a double helix), it can chemically attack the adjacent phosphodiester bond to cleave the backbone.

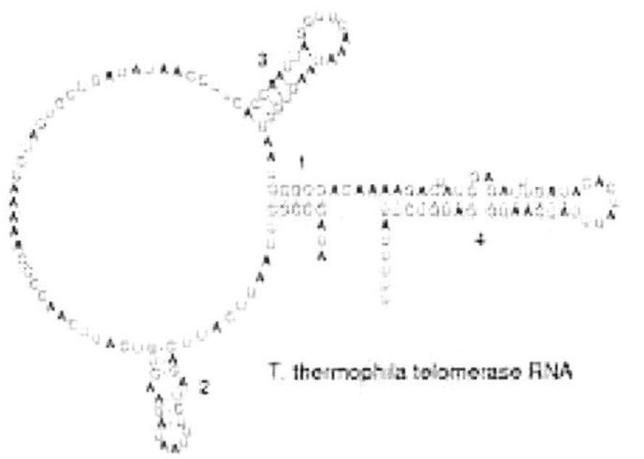

Secondary structure of a telomerase RNA.

RNA is transcribed with only four bases (adenine, cytosine, guanine and uracil), but these bases and attached sugars can be modified in numerous ways as the RNAs mature. Pseudouridine (Ψ), in which the linkage between uracil and ribose is changed from a C–N bond to a C–C bond, and ribothymidine (T) are found in various places (the most notable ones being in the TΨC loop of tRNA). Another notable modified base is hypoxanthine, a deaminated adenine base whose nucleoside is called inosine (I). Inosine plays a key role in the wobble hypothesis of the genetic code.

There are more than 100 other naturally occurring modified nucleosides, The greatest structural diversity of modifications can be found in tRNA, while pseudouridine and nucleosides with 2›-O-methylribose often present in rRNA are the most common. The specific roles of many of these modifications in RNA are not fully understood. However, it is notable that, in ribosomal RNA, many of the post-transcriptional modifications occur in highly functional regions, such as the peptidyl transferase center and the subunit interface, implying that they are important for normal function.

The functional form of single-stranded RNA molecules, just like proteins, frequently requires a specific tertiary structure. The scaffold for this structure is provided by secondary structural elements that are hydrogen bonds within the molecule. This leads to several recognizable domains of secondary structure like hairpin loops, bulges, and internal loops. Since RNA is charged, metal ions such as Mg^{2+} are needed to stabilise many secondary and tertiary structures.

The naturally occurring enantiomer of RNA is D-RNA composed of D-ribonucleotides. All chirality centers are located in the D-ribose. By the use of L-ribose or rather L-ribonucleotides, L-RNA can be synthesized. L-RNA is much more stable against degradation by RNase.

Like other structured biopolymers such as proteins, one can define topology of a folded RNA molecule. This is often done based on arrangement of intra-chain contacts within a folded RNA, termed as circuit topology.

SYNTHESIS

Synthesis of RNA is usually catalyzed by an enzyme—RNA polymerase—using DNA as a template, a process known as transcription. Initiation of transcription begins with the binding of the enzyme to a promoter sequence in the DNA (usually found "upstream" of a gene). The DNA double helix is unwound by the helicase activity of the enzyme. The enzyme then progresses along the template strand in the 3' to 5' direction, synthesizing a complementary RNA molecule with elongation occurring in the 5' to 3' direction. The DNA sequence also dictates where termination of RNA synthesis will occur.

Primary transcript RNAs are often modified by enzymes after transcription. For example, a poly(A) tail and a 5' cap are added to eukaryotic pre-mRNA and introns are removed by the spliceosome.

There are also a number of RNA-dependent RNA polymerases that use RNA as their template for synthesis of a new strand of RNA. For instance, a number of RNA viruses (such as poliovirus) use this type of enzyme to replicate their genetic material. Also, RNA-dependent RNA polymerase is part of the RNA interference pathway in many organisms.

TYPES OF RNA

Overview

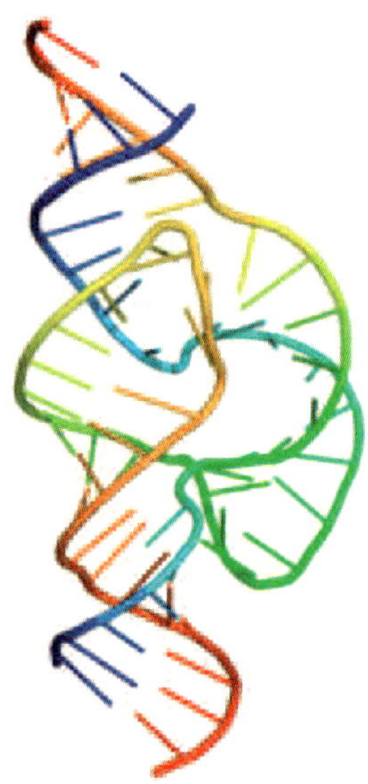

Structure of a hammerhead ribozyme, a ribozyme that cuts RNA

Messenger RNA (mRNA) is the RNA that carries information from DNA to the ribosome, the sites of protein synthesis (translation) in the cell. The coding sequence of the mRNA determines the amino acid sequence in the protein that is produced. However, many RNAs do not code for protein (about 97% of the transcriptional output is non-protein-coding in eukaryotes).

These so-called non-coding RNAs ("ncRNA") can be encoded by their own genes (RNA genes), but can also derive from mRNA introns. The most prominent examples of non-coding RNAs are transfer RNA (tRNA) and ribosomal RNA (rRNA), both of which are involved in the process of translation. There are also non-coding RNAs involved in gene regulation, RNA processing and other roles. Certain RNAs are able to catalyse chemical reactions such as cutting and ligating other RNA molecules, and the catalysis of peptide bond formation in the ribosome; these are known as ribozymes.

In Length

According to the length of RNA chain, RNA includes small RNA and long RNA. Usually, small RNAs are <200 nt in length, and long RNA are >200 nt. Long RNAs, also called large RNAs, mainly include long non-coding RNA (lncRNA) and mRNA. Small RNAs mainly include 5.8S ribosomal RNA (rRNA), 5S rRNA, transfer RNA (tRNA), microRNA (miRNA), small interfering RNA (siRNA), small nucleolar RNA (snoRNAs), Piwi-interacting RNA (piRNA), tRNA-derived small RNA (tsRNA) and small rDNA-derived RNA (srRNA).

In Translation

Messenger RNA (mRNA) carries information about a protein sequence to the ribosomes, the protein synthesis factories in the cell. It is coded so that every three nucleotides (a codon) corresponds to one amino acid. In eukaryotic cells, once precursor mRNA (pre-mRNA) has been transcribed from DNA, it is processed to mature mRNA. This removes its introns—non-coding sections of the pre-mRNA. The mRNA is then exported from the nucleus to the cytoplasm, where it is bound to ribosomes and translated into its corresponding protein form with the help of tRNA. In prokaryotic cells, which do not have nucleus and cytoplasm compartments, mRNA can bind to ribosomes while it is being transcribed from DNA. After a certain amount of time the message degrades into its component nucleotides with the assistance of ribonucleases.

Transfer RNA (tRNA) is a small RNA chain of about 80 nucleotides that transfers a specific amino acid to a growing polypeptide chain at the ribosomal site of protein synthesis during translation. It has sites for amino acid attachment and an anticodon region for codon recognition that binds to a specific sequence on the messenger RNA chain through hydrogen bonding.

Ribosomal RNA (rRNA) is the catalytic component of the ribosomes. Eukaryotic ribosomes contain four different rRNA molecules: 18S, 5.8S, 28S and 5S rRNA. Three of the rRNA molecules are synthesized in the nucleolus, and one is synthesized elsewhere. In the cytoplasm, ribosomal RNA and protein combine to form a nucleoprotein called a ribosome. The ribosome binds mRNA and carries out protein synthesis. Several ribosomes may be attached to a single mRNA at any time. Nearly all the RNA found in a typical eukaryotic cell is rRNA.

Transfer-messenger RNA (tmRNA) is found in many bacteria and plastids. It tags proteins encoded by mRNAs that lack stop codons for degradation and prevents the ribosome from stalling.

Regulatory RNAs

Several types of RNA can downregulate gene expression by being complementary to a part of an mRNA or a gene's DNA. MicroRNAs (miRNA; 21-22 nt) are found in eukaryotes and act through RNA interference (RNAi), where an effector complex of miRNA and enzymes can cleave complementary mRNA, block the mRNA from being translated, or accelerate its degradation.

While small interfering RNAs (siRNA; 20-25 nt) are often produced by breakdown of viral RNA, there are also endogenous sources of siRNAs. siRNAs act through RNA interference in a fashion similar to miRNAs. Some miRNAs and siRNAs can cause genes they target to be methylated, thereby decreasing or increasing transcription of those genes. Animals have Piwi-interacting RNAs (piRNA; 29-30 nt) that are active in germline cells and are thought to be a defense against transposons and play a role in gametogenesis.

Many prokaryotes have CRISPR RNAs, a regulatory system similar to RNA interference. Antisense RNAs are widespread; most downregulate a gene, but a few are activators of transcription. One way antisense RNA can act is by binding to an mRNA, forming double-stranded RNA that is enzymatically degraded. There are many long noncoding RNAs that regulate genes in eukaryotes, one such RNA is Xist, which coats one X chromosome in female mammals and inactivates it.

An mRNA may contain regulatory elements itself, such as riboswitches, in the 5' untranslated region or 3' untranslated region; these cis-regulatory elements regulate the activity of that mRNA. The untranslated regions can also contain elements that regulate other genes.

In RNA Processing

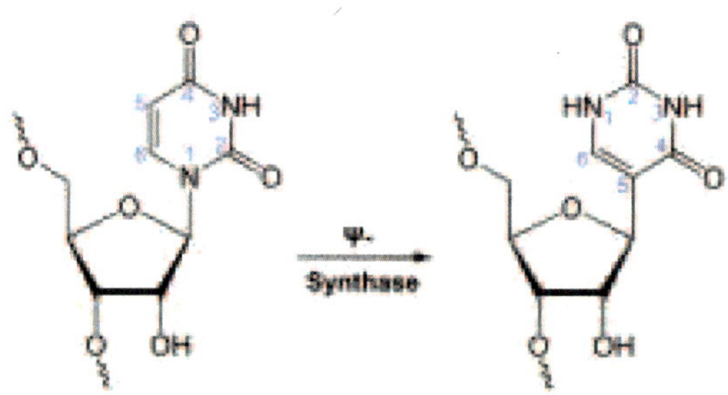

Uridine to pseudouridine is a common RNA modification.

Many RNAs are involved in modifying other RNAs. Introns are spliced out of pre-mRNA by spliceosomes, which contain several small nuclear RNAs (snRNA), or the introns can be ribozymes that are spliced by themselves. RNA can also be altered by having its nucleotides modified to nucleotides other than A, C, G and U. In eukaryotes, modifications of RNA nucleotides are in general directed by small nucleolar RNAs (snoRNA; 60-300 nt), found in the nucleolus and cajal bodies. snoRNAs associate with enzymes and guide them to a spot on an RNA by basepairing to that RNA. These enzymes then perform the nucleotide modification. rRNAs and tRNAs are extensively modified, but snRNAs and mRNAs can also be the target of base modification. RNA can also be methylated.

RNA Genomes

Like DNA, RNA can carry genetic information. RNA viruses have genomes composed of RNA that encodes a number of proteins. The viral genome is replicated by some of those proteins, while other proteins protect the genome as the virus particle moves to a new host cell. Viroids are another group of pathogens, but they consist only of RNA, do not encode any protein and are replicated by a host plant cell's polymerase.

In Reverse Transcription

Reverse transcribing viruses replicate their genomes by reverse transcribing DNA copies from their RNA; these DNA copies are then transcribed to new RNA. Retrotransposons also spread by copying DNA and RNA from one another, and telomerase contains an RNA that is used as template for building the ends of eukaryotic chromosomes.

Double-Stranded RNA

Double-stranded RNA (dsRNA) is RNA with two complementary strands, similar to the DNA found in all cells. dsRNA forms the genetic material of some viruses (double-stranded RNA viruses). Double-stranded RNA such as viral RNA or siRNA can trigger RNA interference in eukaryotes, as well as interferon response in vertebrates.

Circular RNA

Recently, it was shown that there is a single stranded covalently closed, *i.e.* circular form of RNA expressed throughout the animal and plant kingdom (see circRNA). circRNAs are thought to arise via a back-splice reaction where the spliceosome joins a downstream donor to an upstream acceptor splice site. So far the function of circRNAs is largely unknown, although for few examples a microRNA sponging activity has been demonstrated.

KEY DISCOVERIES IN RNA BIOLOGY

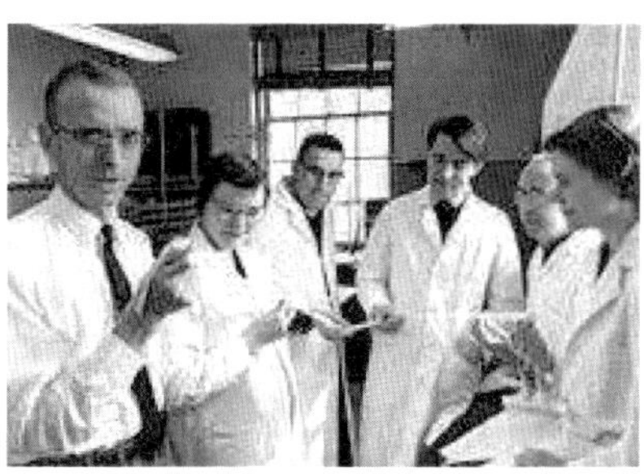

Robert W. Holley, left, poses with his research team.

Research on RNA has led to many important biological discoveries and numerous Nobel Prizes. Nucleic acids were discovered in 1868 by Friedrich Miescher, who called the material 'nuclein' since it was found in the nucleus. It was later discovered that prokaryotic cells, which do not have a nucleus, also contain nucleic acids. The role of RNA in protein synthesis was suspected already in 1939. Severo Ochoa won the 1959 Nobel Prize in Medicine (shared with Arthur Kornberg) after he discovered an enzyme that can synthesize RNA in the laboratory. However, the enzyme discovered by Ochoa (polynucleotide phosphorylase) was later shown to be responsible for RNA degradation, not RNA synthesis. In 1956 Alex Rich and David Davies hybridized two separate strands of RNA to form the first crystal of RNA whose structure could be determined by X-ray crystallography.

The sequence of the 77 nucleotides of a yeast tRNA was found by Robert W. Holley in 1965, winning Holley the 1968 Nobel Prize in Medicine (shared with Har Gobind Khorana and Marshall Nirenberg). In 1967, Carl Woese hypothesized that RNA might be catalytic and suggested that the earliest forms of life (self-replicating molecules) could have relied on RNA both to carry genetic information and to catalyze biochemical reactions—an RNA world.

During the early 1970s, retroviruses and reverse transcriptase were discovered, showing for the first time that enzymes could copy RNA into DNA (the opposite of the usual route for transmission of genetic information). For this work, David Baltimore, Renato Dulbecco and Howard Temin were awarded a Nobel Prize in 1975. In 1976, Walter Fiers and his team determined the first complete nucleotide sequence of an RNA virus genome, that of bacteriophage MS2.

In 1977, introns and RNA splicing were discovered in both mammalian viruses and in cellular genes, resulting in a 1993 Nobel to Philip Sharp and Richard Roberts. Catalytic RNA molecules (ribozymes) were discovered in the early 1980s, leading to a 1989 Nobel award to Thomas Cech and Sidney Altman. In 1990, it was found in *Petunia* that introduced genes can silence similar genes of the plant's own, now known to be a result of RNA interference.

At about the same time, 22 nt long RNAs, now called microRNAs, were found to have a role in the development of *C. elegans*. Studies on RNA interference gleaned a Nobel Prize for Andrew Fire and Craig Mello in 2006, and another Nobel was awarded for studies on the transcription of RNA to Roger Kornberg in the same year. The discovery of gene regulatory RNAs has led to attempts to develop drugs made of RNA, such as siRNA, to silence genes.

EVOLUTION

In March 2015, complex DNA and RNA organic compounds of life, including uracil, cytosine and thymine, were reportedly formed in the laboratory under outer space conditions, using starter chemicals, such as pyrimidine, found in meteorites. Pyrimidine, like polycyclic aromatic hydrocarbons (PAHs), the most carbon-rich chemical found in the Universe, may have been formed in red giants or in interstellar dust and gas clouds, according to the scientists.

REFERENCES

1. RNA: The Versatile Molecule. University of Utah. 2015.
2. Nucleotides and Nucleic Acids (PDF). University of California, Los Angeles.

3. R.N. Shukla. Analysis of Chromosomes. ISBN 9789384568177.

4. Berg JM; Tymoczko JL; Stryer L (2002). Biochemistry (5th ed.). WH Freeman and Company. pp. 118–19, 781–808. ISBN 0-7167-4684-0. OCLC 179705944.

5. I. Tinoco & C. Bustamante (1999). How RNA folds. J. Mol. Biol. 293 *(2): 271–281. doi:10.1006/jmbi.1999.3001. PMID 10550208.*

6. Higgs PG (2000). RNA secondary structure: physical and computational aspects. Quarterly Reviews of Biophysics. 33 *(3): 199–253.* doi:10.1017/S0033583500003620. PMID 11191843.

7. Nissen P; Hansen J; Ban N; Moore PB; Steitz TA (2000). The structural basis of ribosome activity in peptide bond synthesis. Science. 289 *(5481): 920–30. Bibcode*:2000Sci...289..920N. doi:10.1126/science.289.5481.920. PMID 10937990.

8. Lee JC; Gutell RR (2004). Diversity of base-pair conformations and their occurrence in rRNA structure and RNA structural motifs. J. Mol. Biol. 344 *(5): 1225–49. doi:10.1016/j.jmb.2004.09.072. PMID 15561141.*

9. Barciszewski J; Frederic B; Clark C (1999). RNA biochemistry and biotechnology. Springer. pp. 73–87. ISBN 0-7923-5862-7. OCLC 52403776.

10. Salazar M; Fedoroff OY; Miller JM; Ribeiro NS; Reid BR (1992). The DNA strand in DNAoRNA hybrid duplexes is neither B-form nor A-form in solution. Biochemistry. 32 *(16): 4207–15.* doi:10.1021/bi00067a007. PMID 7682844.

11. Sedova A; Banavali NK (2016). RNA approaches the B-form in stacked single strand dinucleotide contexts. Biopolymers. 105 *(2): 65–82. doi:10.1002/bip.22750. PMID 26443416.*

12. Hermann T; Patel DJ (2000). RNA bulges as architectural and recognition motifs. Structure. 8 *(3): R47–R54.* doi:10.1016/S0969-2126(00)00110-6. PMID 10745015.

13. Mikkola S; Stenman E; Nurmi K; Yousefi-Salakdeh E; Strömberg R; Lönnberg H (1999). The mechanism of the metal ion promoted cleavage of RNA phosphodiester bonds involves a general acid catalysis by the metal aquo ion on the departure of the leaving group. Perkin transactions 2 (8): 1619–26. doi:10.1039/a903691a.

14. Jankowski JAZ; Polak JM (1996). Clinical gene analysis and manipulation: Tools, techniques and troubleshooting. Cambridge University Press. p. 14. ISBN 0-521-47896-0. OCLC 33838261.

15. Yu Q; Morrow CD (2001). Identification of critical elements in the tRNA acceptor stem and TΨC loop necessary for human immunodeficiency virus type 1 infectivity. J Virol. 75 *(10): 4902–6. doi:10.1128/JVI.75.10.4902-4906.2001. PMC 114245. PMID 11312362.*

16. Elliott MS; Trewyn RW (1983). Inosine biosynthesis in transfer RNA by an enzymatic insertion of hypoxanthine. J. Biol. Chem. 259 *(4): 2407–10. PMID 6365911.*

17. Cantara, WA; Crain, PF; Rozenski, J; McCloskey, JA; Harris, KA; Zhang, X; Vendeix, FA; Fabris, D; Agris, PF (January 2011). The RNA Modification Database, RNAMDB: 2011 update. Nucleic Acids Research. 39 *(Database issue): D195–201. doi:10.1093/* nar/gkq1028. PMC 3013656. PMID 21071406.

18. Söll D; RajBhandary U (1995). TRNA: Structure, biosynthesis, and function. ASM Press. p. 165. ISBN 1-55581-073-X. OCLC 183036381.

19. Kiss T (2001). Small nucleolar RNA-guided post-transcriptional modification of cellular RNAs. The EMBO Journal. 20 *(14): 3617–22. doi:10.1093/*emboj/20.14.3617. PMC 125535. PMID 11447102.

20. King TH; Liu B; McCully RR; Fournier MJ (2002). Ribosome structure and activity are altered in cells lacking snoRNPs that form pseudouridines in the peptidyl transferase center. Molecular Cell. 11 *(2): 425–35.* doi:10.1016/S1097-2765(03)00040-6. PMID 12620230.

21. Mathews DH; Disney MD; Childs JL; Schroeder SJ; Zuker M; Turner DH (2004). Incorporating chemical modification constraints into a dynamic programming algorithm for prediction of RNA secondary structure. Proc. Natl. Acad. Sci. USA. 101 *(19): 7287–92. Bibcode:*2004PNAS..101.7287M. doi:10.1073/pnas.0401799101. PMC 409911. PMID 15123812.

22. Tan ZJ; Chen SJ (2008). Salt dependence of nucleic acid hairpin stability. Biophys. J. 95 *(2): 738–52. Bibcode:*2008BpJ....95..738T. doi:10.1529/biophysj.108.131524. PMC 2440479. PMID 18424500.

23. Vater A; Klussmann S (January 2015). Turning mirror-image oligonucleotides into drugs: the evolution of Spiegelmer therapeutics. Drug Discovery Today. 20 *(1): 147–155. doi:10.1016/j.drudis.2014.09.004. PMID 25236655.*

24. Nudler E; Gottesman ME (2002). Transcription termination and anti-termination in E. coli. Genes to Cells. 7 *(8): 755–68. doi:10.1046/j.1365-2443.2002.00563.x. PMID 12167155.*

25. Jeffrey L Hansen; Alexander M Long; Steve C Schultz (1997). Structure of the RNA-dependent RNA polymerase of poliovirus. Structure. 5 *(8): 1109–22.* doi:10.1016/S0969-2126(97)00261-X. PMID 9309225.

26. Ahlquist P (2002). RNA-Dependent RNA Polymerases, Viruses, and RNA Silencing. Science. 296 *(5571): 1270–73. Bibcode:*2002Sci...296.1270A. doi:10.1126/science.1069132. PMID 12016304.

27. Cooper GC; Hausman RE (2004). The Cell: A Molecular Approach (3rd ed.). Sinauer. pp. 261–76, 297, 339–44. ISBN 0-87893-214-3. OCLC 174924833.

28. Mattick JS; Gagen MJ (1 September 2001). The evolution of controlled multitasked gene networks: the role of introns and other noncoding RNAs in the development of complex organisms. Mol. Biol. Evol. 18 *(9): 1611–30. doi:10.1093/oxfordjournals.molbev.a003951. PMID 11504843.*

29. Mattick, JS (2001). Noncoding RNAs: the architects of eukaryotic complexity. EMBO Reports. 2 *(11): 986–91. doi:10.1093/*embo-reports/kve230. PMC 1084129. PMID 11713189.

30. Mattick JS (October 2003). Challenging the dogma: the hidden layer of non-protein-coding RNAs in complex organisms (PDF). BioEssays. 25 *(10): 930–9. doi:10.1002/bies.10332. PMID 14505360.*

31. Mattick JS (October 2004). The hidden genetic program of complex organisms. Scientific American. 291 *(4): 60–7.* doi:10.1038/scientificamerican1004-60. PMID 15487671.[dead link]

32. Wirta W (2006). Mining the transcriptome – methods and applications. Stockholm: School of Biotechnology, Royal Institute of Technology. ISBN 91-7178-436-5. OCLC

185406288.

33. Rossi JJ (2004). Ribozyme diagnostics comes of age. Chemistry & Biology. 11 *(7): 894–95. doi:10.1016/j.chembiol.2004.07.002. PMID 15271347.*

34. Storz, G (17 May 2002). An expanding universe of noncoding RNAs.. *Science.* 296 *(5571): 1260–3. doi:10.1126/science.1072249. PMID 12016301.*

35. Fatica, A; Bozzoni, I (January 2014). Long non-coding RNAs: new players in cell differentiation and development.. *Nature Reviews Genetics.* 15 *(1): 7–21.* doi:10.1038/nrg3606. PMID 24296535.

36. Chen, Q; Yan, M; Cao, Z; Li, X; Zhang, Y; Shi, J; Feng, GH; Peng, H; Zhang, X; Zhang, Y; Qian, J; Duan, E; Zhai, Q; Zhou, Q (22 January 2016). Sperm tsRNAs contribute to intergenerational inheritance of an acquired metabolic disorder.. *Science.* 351 *(6271): 397–400. doi:10.1126/science.aad7977. PMID 26721680.*

37. Wei, H; Zhou, B; Zhang, F; Tu, Y; Hu, Y; Zhang, B; Zhai, Q (2013). Profiling and identification of small rDNA-derived RNAs and their potential biological functions.. *PLOS ONE.* 8 *(2): e56842. doi:10.1371/journal.pone.0056842. PMC 3572043. PMID 23418607.*

38. Gueneau de Novoa P, Williams KP; Williams (2004). The tmRNA website: reductive evolution of tmRNA in plastids and other endosymbionts. Nucleic Acids Res. 32 *(Database issue): D104–8. doi:10.1093/*nar/gkh102. PMC 308836. PMID 14681369.

39. Carthew, RW; Sontheimer, EJ. (February 2009). Origins and Mechanisms of miRNAs and siRNAs.. *Cell.* 136: 642–55. doi:10.1016/j.cell.2009.01.035. PMC 2675692. PMID 19239886.

40. Liang, Kung-Hao; Yeh, Chau-Ting (2013). A gene expression restriction network mediated by sense and antisense Alu sequences located on protein-coding messenger RNAs.. *BMC Genomics.* 14: 325. doi:10.1186/1471-2164-14-325. PMC 3655826. PMID 23663499. Retrieved 2013-05-11.

41. Wu L; Belasco JG (January 2008). Let me count the ways: mechanisms of gene regulation by miRNAs and siRNAs. Mol. Cell. 29 *(1): 1–7. doi:10.1016/j.molcel.2007.12.010. PMID 18206964.*

42. Matzke MA; Matzke AJM (2004). Planting the seeds of a new paradigm. PLoS Biology. 2 *(5): e133. doi:10.1371/journal.pbio.0020133. PMC 406394. PMID 15138502.*

43. Vazquez F; Vaucheret H; Rajagopalan R; Lepers C; Gasciolli V; Mallory AC; Hilbert J; Bartel DP; Crété P (2004). Endogenous trans-acting siRNAs regulate the accumulation of Arabidopsis mRNAs. Molecular Cell. 16 *(1): 69–79. doi:10.1016/j.molcel.2004.09.028. PMID 15469823.*

44. Watanabe T, Totoki Y, Toyoda A, et al. (May 2008). Endogenous siRNAs from naturally formed dsRNAs regulate transcripts in mouse oocytes. Nature. 453 *(7194): 539–43. Bibcode:*2008Natur.453..539W. doi:10.1038/nature06908. PMID 18404146.

45. Sontheimer EJ; Carthew RW (July 2005). Silence from within: endogenous siRNAs and miRNAs. Cell. 122 *(1): 9–12. doi:10.1016/j.cell.2005.06.030. PMID 16009127.*

46. Doran G (2007). RNAi – Is one suffix sufficient?. *Journal of RNAi and Gene Silencing.* 3 *(1): 217–19.*

47. Pushparaj PN; Aarthi JJ; Kumar SD; Manikandan J (2008). RNAi and RNAa — The Yin and Yang of RNAome. Bioinformation. 2 *(6): 235–7.* doi:10.6026/97320630002235. PMC 2258431. PMID 18317570.

48. Horwich MD; Li C; Matranga C; Vagin V; Farley G; Wang P; Zamore PD (2007). The Drosophila RNA methyltransferase, DmHen1, modifies germline piRNAs and single-stranded siRNAs in RISC. Current Biology. 17 *(14): 1265–72. doi:10.1016/j.cub.2007.06.030. PMID 17604629.*

49. Girard A; Sachidanandam R; Hannon GJ; Carmell MA (2006). A germline-specific class of small RNAs binds mammalian Piwi proteins. Nature. 442 *(7099): 199–202. Bibcode*:2006Natur.442..199G. doi:10.1038/nature04917. PMID 16751776.

50. Horvath P; Barrangou R (2010). CRISPR/Cas, the Immune System of Bacteria and Archaea. Science. 327 *(5962): 167–70. Bibcode*:2010Sci...327..167H. doi:10.1126/science.1179555. PMID 20056882.

51. Wagner EG; Altuvia S; Romby P (2002). Antisense RNAs in bacteria and their genetic elements. Adv Genet. Advances in Genetics. 46: 361–98. doi:10.1016/S0065-2660(02)46013-0. ISBN 9780120176465. PMID 11931231.

52. Gilbert SF (2003). Developmental Biology (7th ed.). Sinauer. pp. 101–3. ISBN 0-87893-258-5. OCLC 154656422.

53. Amaral PP; Mattick JS (October 2008). Noncoding RNA in development. Mammalian genome : official journal of the International Mammalian Genome Society. 19 *(7–8): 454–92.* doi:10.1007/s00335-008-9136-7. PMID 18839252.

54. Heard E; Mongelard F; Arnaud D; Chureau C; Vourc›h C; Avner P (1999). Human XIST yeast artificial chromosome transgenes show partial X inactivation center function in mouse embryonic stem cells. Proc. Natl. Acad. Sci. USA. 96 *(12): 6841–46. Bibcode*:1999PNAS...96.6841H. doi:10.1073/pnas.96.12.6841. PMC 22003. PMID 10359800.

55. Batey RT (2006). Structures of regulatory elements in mRNAs. Curr. Opin. Struct. Biol. 16 *(3): 299–306. doi:10.1016/j.sbi.2006.05.001. PMID 16707260.*

56. Scotto L; Assoian RK (June 1993). A GC-rich domain with bifunctional effects on mRNA and protein levels: implications for control of transforming growth factor beta 1 expression. Mol. Cell. Biol. 13 *(6): 3588–97. PMC 359828. PMID 8497272.*

57. Steitz TA; Steitz JA (1993). A general two-metal-ion mechanism for catalytic RNA. Proc. Natl. Acad. Sci. U.S.A. 90 *(14): 6498–502. Bibcode*:1993PNAS...90.6498S. doi:10.1073/pnas.90.14.6498. PMC 46959. PMID 8341661.

58. Xie J; Zhang M; Zhou T; Hua X; Tang L; Wu W (2007). Sno/scaRNAbase: a curated database for small nucleolar RNAs and cajal body-specific RNAs. Nucleic Acids Res. 35 *(Database issue): D183–7. doi:10.1093/*nar/gkl873. PMC 1669756. PMID 17099227.

59. Omer AD; Ziesche S; Decatur WA; Fournier MJ; Dennis PP (2003). RNA-modifying machines in archaea. Molecular Microbiology. 48 *(3): 617–29. doi:10.1046/j.1365-2958.2003.03483.x. PMID 12694609.*

60. Cavaillé J; Nicoloso M; Bachellerie JP (1996). Targeted ribose methylation of RNA in vivo directed by tailored antisense RNA guides. Nature. 383 *(6602): 732–5.*

Bibcode:1996Natur.383..732C. doi:10.1038/383732a0. PMID 8878486.

61. Kiss-László Z; Henry Y; Bachellerie JP; Caizergues-Ferrer M; Kiss T (1996). Site-specific ribose methylation of preribosomal RNA: a novel function for small nucleolar RNAs. Cell. 85 *(7): 1077–88.* doi:10.1016/S0092-8674(00)81308-2. PMID 8674114.

62. Daròs JA; Elena SF; Flores R (2006). Viroids: an Ariadne›s thread into the RNA labyrinth. EMBO Rep. 7 *(6): 593–8. doi:10.1038/sj.embor.7400706. PMC 1479586. PMID 16741503.*

63. Kalendar R; Vicient CM; Peleg O; Anamthawat-Jonsson K; Bolshoy A; Schulman AH (2004). Large retrotransposon derivatives: abundant, conserved but nonautonomous retroelements of barley and related genomes. Genetics. 166 *(3): 1437–50. doi:10.1534/genetics.166.3.1437. PMC 1470764. PMID 15082561.*

64. Podlevsky JD; Bley CJ; Omana RV; Qi X; Chen JJ (2008). The telomerase database. Nucleic Acids Res. 36 *(Database issue): D339–43. doi:10.1093/*nar/gkm700. PMC 2238860. PMID 18073191.

65. Blevins, T.; Rajeswaran, R.; Shivaprasad, PV.; Beknazariants, D.; Si-Ammour, A.; Park, HS.; Vazquez, F.; Robertson, D.; Meins, F. (2006). Four plant Dicers mediate viral small RNA biogenesis and DNA virus induced silencing. Nucleic Acids Res. 34 *(21): 6233–46. doi:10.1093/*nar/gkl886. PMC 1669714. PMID 17090584.

66. Jana S; Chakraborty C; Nandi S; Deb JK (2004). RNA interference: potential therapeutic targets. Appl. Microbiol. Biotechnol. 65 *(6): 649–57.* doi:10.1007/s00253-004-1732-1. PMID 15372214.

67. Schultz U; Kaspers B; Staeheli P (2004). The interferon system of non-mammalian vertebrates. Dev. Comp. Immunol. 28 *(5): 499–508. doi:10.1016/j.dci.2003.09.009. PMID 15062646.*

68. Whitehead, K. A.; Dahlman, J. E.; Langer, R. S.; Anderson, D. G. (2011). Silencing or Stimulation? SiRNA Delivery and the Immune System. Annual Review of Chemical and Biomolecular Engineering. 2: 77–96. doi:10.1146/annurev-chembioeng-061010-114133. PMID 22432611.

69. Dahm R (2005). Friedrich Miescher and the discovery of DNA. Developmental Biology. 278 *(2): 274–88. doi:10.1016/j.ydbio.2004.11.028. PMID 15680349.*

70. Caspersson T; Schultz J (1939). Pentose nucleotides in the cytoplasm of growing tissues. Nature. 143 *(3623): 602–3. Bibcode*:1939Natur.143..602C. doi:10.1038/143602c0.

71. Ochoa S (1959). Enzymatic synthesis of ribonucleic acid (PDF). Nobel Lecture.

72. Rich A; Davies, D (1956). A New Two-Stranded Helical Structure: Polyadenylic Acid and Polyuridylic Acid. Journal of the American Chemical Society. 78 *(14): 3548–3549.* doi:10.1021/ja01595a086.

73. Holley RW, et al. (1965). Structure of a ribonucleic acid. Science. 147 *(3664): 1462–65. Bibcode*:1965Sci...147.1462H. doi:10.1126/science.147.3664.1462. PMID 14263761.

74. Siebert S (2006). Common sequence structure properties and stable regions in RNA secondary structures (PDF). Dissertation, Albert-Ludwigs-Universität, Freiburg im Breisgau. p. 1. Archived from the original (PDF) on March 9, 2012.

75. Szathmáry E (1999). The origin of the genetic code: amino acids as cofactors in an

RNA world. Trends Genet. 15 *(6): 223–9.* doi:10.1016/S0168-9525(99)01730-8. PMID 10354582.

76. Fiers W, et al. (1976). Complete nucleotide-sequence of bacteriophage MS2-RNA: primary and secondary structure of replicase gene. Nature. 260 *(5551): 500–7. Bibcode*:1976Natur.260..500F. doi:10.1038/260500a0. PMID 1264203.

77. Napoli C; Lemieux C; Jorgensen R (1990). Introduction of a chimeric chalcone synthase gene into petunia results in reversible co-suppression of homologous genes in trans. Plant Cell. 2 *(4): 279–89. doi:10.1105/tpc.2.4.279. PMC 159885. PMID 12354959.*

78. Dafny-Yelin M; Chung SM; Frankman EL; Tzfira T (December 2007). pSAT RNA interference vectors: a modular series for multiple gene down-regulation in plants. Plant Physiol. 145 *(4): 1272–81. doi:10.1104/pp.107.106062. PMC 2151715. PMID 17766396.*

79. Ruvkun G (2001). Glimpses of a tiny RNA world. Science. 294 *(5543): 797–99. doi:10.1126/science.1066315. PMID 11679654.*

80. Fichou Y; Férec C (2006). The potential of oligonucleotides for therapeutic applications. Trends in Biotechnology. 24 *(12): 563–70. doi:10.1016/j.tibtech.2006.10.003. PMID 17045686.*

81. Marlaire, Ruth (3 March 2015). NASA Ames Reproduces the Building Blocks of Life in Laboratory. NASA. Retrieved 5 March 2015.

CHAPTER 7

Cytoplasm

Chapter Contents

- **Introduction**
- **History**
- **Physical Nature**
- **Constituents**

Cytoplasm is an important part of your cells and serves several functions that are key to cellular health and productivity. Each type of cell serves a specific function, but they all have a very similar structure.

INTRODUCTION

In cell biology, the **cytoplasm** is the material or protoplasm within a living cell, excluding the cell nucleus. It comprises cytosol (the gel-like substance enclosed within the cell membrane) and the organelles – the cell›s internal sub-structures. All of the contents of the cells of prokaryote organisms (such as bacteria, which lack a cell nucleus) are contained within the cytoplasm. Within the cells of eukaryote organisms the contents of the cell nucleus are separated from the cytoplasm, and are then called the nucleoplasm. The cytoplasm is about 80% water and usually colorless.

It is within the cytoplasm that most cellular activities occur, such as many metabolic pathways including glycolysis, and processes such as cell division. The concentrated inner area is called the endoplasm and the outer layer is called the cell cortex or the ectoplasm.

Movement of calcium ions in and out of the cytoplasm is a signaling activity for metabolic processes.

In plants, movement of the cytoplasm around vacuoles is known as cytoplasmic streaming.

HISTORY

The term was introduced by Rudolf von Kölliker in 1863, originally as a synonym for protoplasm, but later it has come to mean the cell substance and organelles outside the nucleus.

There has been certain disagreement on the definition of cytoplasm, as some authors prefer to exclude from it some organelles, especially the vacuoles and sometimes the plastids.

PHYSICAL NATURE

The physical properties of the cytoplasm have been contested in recent years. It remains uncertain how the varied components of the cytoplasm interact to allow movement of particles and organelles while maintaining the cell's structure. The flow of cytoplasmic components plays an important role in many cellular functions which are dependent on the permeability of the cytoplasm. An example of such function is cell signalling, a process which is dependent on the manner in which signaling molecules are allowed to diffuse across the cell. While small signaling molecules like calcium ions are able to diffuse with ease, larger molecules and subcellular structures often require aid in moving through the cytoplasm. The irregular dynamics of such particles have given rise to various theories on the nature of the cytoplasm.

As a sol-gel

There has long been evidence that the cytoplasm behaves like a sol-gel. It is thought that the component molecules and structures of the cytoplasm behave at times like a disordered colloidal solution (sol) and at other times like an integrated network, forming a solid mass (gel). This theory thus proposes that the cytoplasm exists in distinct fluid and solid phases depending on the level of interaction between cytoplasmic components, which may explain the differential dynamics of different particles observed moving through thc cytoplasm.

As a glass

Recently it has been proposed that the cytoplasm behaves like a glass-forming liquid approaching the glass transition. In this theory, the greater the concentration of cytoplasmic components, the less the cytoplasm behaves like a liquid and the more it behaves as a solid glass, freezing larger cytoplasmic components in place (it is thought that the cell's metabolic activity is able to fluidize the cytoplasm to allow the movement of such larger cytoplasmic components). A cell's ability to vitrify in the absence of metabolic activity, as in dormant periods, may be beneficial as a defence strategy. A solid glass cytoplasm would freeze subcellular structures in place, preventing damage, while allowing the transmission of very small proteins and metabolites, helping to kickstart growth upon the cell's revival from dormancy.

Other Perspectives

There has been research examining the motion of cytoplasmic particles independent of the nature of the cytoplasm. In such an alternative approach, the aggregate random forces within the cell caused by motor proteins explain the non-Brownian motion of cytoplasmic constituents.

CONSTITUENTS

The three major elements of the cytoplasm are the cytosol, organelles and inclusions.

Cytosol

The cytosol is the portion of the cytoplasm not contained within membrane-bound organelles. Cytosol makes up about 70% of the cell volume and is a complex mixture of cytoskeleton filaments, dissolved molecules, and water. The cytosol's filaments include the protein filaments such as actin filaments and microtubules that make up the cytoskeleton, as well as soluble proteins and small structures such as ribosomes, proteasomes, and the mysterious vault complexes. The inner, granular and more fluid portion of the cytoplasm is referred to as endoplasm.

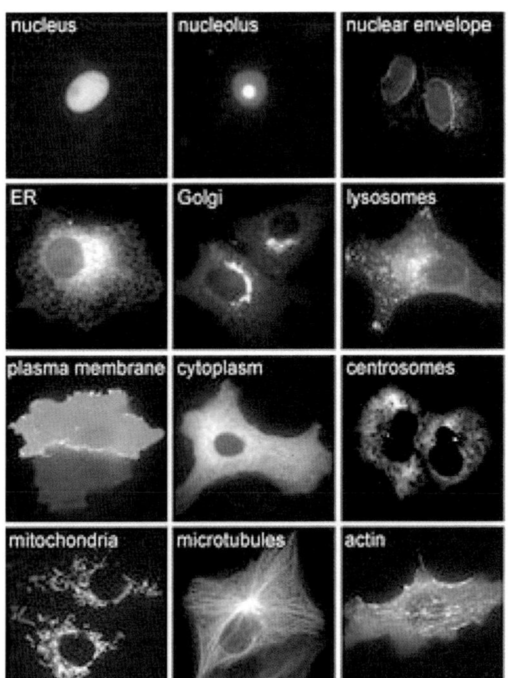

with friendly permission of Jeremy Simpson and Rainer Pepperkok

Proteins in different cellular compartments and structures tagged with green fluorescent protein

Due to this network of fibres and high concentrations of dissolved macromolecules, such as proteins, an effect called macromolecular crowding occurs and the cytosol does not act as an ideal solution. This crowding effect alters how the components of the cytosol interact with each other.

Organelles

Organelles (literally "little organs"), are usually membrane-bound structures inside the cell that have specific functions. Some major organelles that are suspended in the cytosol are the mitochondria, the endoplasmic reticulum, the Golgi apparatus, vacuoles, lysosomes, and in plant cells, chloroplasts.

Cytoplasmic Inclusions

The inclusions are small particles of insoluble substances suspended in the cytosol. A huge range of inclusions exist in different cell types, and range from crystals of calcium oxalate or silicon dioxide in plants, to granules of energy-storage materials such as starch, glycogen, or polyhydroxybutyrate. A particularly widespread example are lipid droplets, which are spherical droplets composed of lipids and proteins that are used in both prokaryotes and eukaryotes as a way of storing lipids such as fatty acids and sterols. Lipid droplets make up much of the volume of adipocytes, which are specialized lipid-storage cells, but they are also found in a range of other cell types.

Controversy and Research

The cytoplasm, mitochondria and most organelles are contributions to the cell from the maternal gamete. Contrary to the older information that disregards any notion of the cytoplasm being active, new research has shown it to be in control of movement and flow of nutrients in and out of the cell by viscoplastic behavior and a measure of the reciprocal rate of bond breakage within the cytoplasmic network.

The material properties of the cytoplasm remain an ongoing investigation. Recent measurements using force spectrum microscopy reveal that the cytoplasm can be likened to an elastic solid, rather than a viscoelastic fluid.

REFERENCES

1. Shepherd, V. A. (2006). «The cytomatrix as a cooperative system of macromolecular and water networks». Current Topics in Developmental Biology. Current Topics in Developmental Biology. 75: 171–223. doi:10.1016/S0070-2153(06)75006-2. ISBN 9780121531751. PMID 16984813.
2. Hogan, C. Michael (2010). "Calcium" in Encyclopedia of Earth. A. Jorgensen, C. Cleveland (eds.). National Council for Science and the Environment.
3. Kölliker, R. A. v. (1863). Handbuch der Gewebelehre des Menschen. 4. Auflage. Leipzig: Wilhelm Engelmann.

4. Bynum, W. F., Browne, E. J. and Porter, Ray (1981). Dictionary of the history of science. Princeton University Press.

5. Parker, J. (1972). "Protoplasmic resistance to water deficits", pp. 125–176 in Kozlowski, T. T. (ed.), Water deficits and plant growth. Vol. III. Plant responses and control of water balance. Academic Press, New York, p. 144, .

6. Strasburger, E. (1882). «Ueber den Theilungsvorgang der Zellkerne und das Verhältnis der Kernteilung zur Zellteilung». Arch Mikr Anat. 21: 476–590.

7. Cowan AE, Moraru II, Schaff JC, Slepchenko BM, Loew LM (2012). «Spatial Modeling of Cell Signaling Networks». Methods in Cell Biology. Methods in Cell Biology. 110: 195–221. doi:10.1016/B978-0-12-388403-9.00008-4. ISBN 9780123884039. PMC 3519356. PMID 22482950.

8. Holcman, David; Korenbrot, Juan I. (2004). «Longitudinal Diffusion in Retinal Rod and Cone Outer Segment Cytoplasm: The Consequence of Cell Structure». Biophysical Journal. 86 *(4): 2566–2582. Bibcode*:2004BpJ....86.2566H. doi:10.1016/S0006-3495(04)74312-X. PMC 1304104. PMID 15041693.

9. Parry, Bradley R.; Surovtsev, Ivan V.; Cabeen, Matthew T.; o›Hern, Corey S.; Dufresne, Eric R.; Jacobs-Wagner, Christine (2014). «The Bacterial Cytoplasm Has Glass-like Properties and is Fluidized by Metabolic Activity». Cell. 156 *(1–2): 183–94. doi:10.1016/j.cell.2013.11.028. PMC 3956598. PMID 24361104.*

10. Taylor, C. V. (1923). «The contractile vacuole in Euplotes: An example of the sol-gel reversibility of cytoplasm». Journal of Experimental Zoology. 37 *(3): 259–289. doi:10.1002/jez.1400370302.*

11. Guo, Ming; Ehrlicher, Allen J.; Jensen, Mikkel H.; Renz, Malte; Moore, Jeffrey R.; Goldman, Robert D.; Lippincott-Schwartz, Jennifer; MacKintosh, Frederick C.; Weitz, David A. (2014). «Probing the Stochastic, Motor-Driven Properties of the Cytoplasm Using Force Spectrum Microscopy». Cell. 158 *(4): 822–32. doi:10.1016/j.cell.2014.06.051. PMC 4183065. PMID 25126787.*

12. van Zon A, Mossink MH, Scheper RJ, Sonneveld P, Wiemer EA (September 2003). «The vault complex». Cell. Mol. Life Sci. 60 *(9): 1828–37.* doi:10.1007/s00018-003-3030-y. PMID 14523546.

13. Prychid, Christina J.; Rudall, Paula J. (1999). «Calcium Oxalate Crystals in Monocotyledons: A Review of their Structure and Systematics». Annals of Botany. 84 *(6): 725–739. doi:10.1006/anbo.1999.0975.*

14. Prychid, C. J.; Rudall, P. J.; Gregory, M. (2004). «Systematics and Biology of Silica Bodies in Monocotyledons». The Botanical Review. 69 *(4): 377–440.* doi:10.1663/0006-8101(2004)069[0377:SABOSB]2.0.CO;2. JSTOR 4354467.

15. Ball SG, Morell MK (2003). «From bacterial glycogen to starch: understanding the biogenesis of the plant starch granule». Annu Rev Plant Biol. 54: 207–33. doi:10.1146/annurev.arplant.54.031902.134927. PMID 14502990.

16. Shearer J, Graham TE (April 2002). «New perspectives on the storage and organization of muscle glycogen». Can J Appl Physiol. 27 *(2): 179–203.* doi:10.1139/h02-012. PMID 12179957.

17. Anderson AJ, Dawes EA (1 December 1990). «Occurrence, metabolism, metabolic

role, and industrial uses of bacterial polyhydroxyalkanoates». Microbiol. Rev. 54 *(4): 450–72. PMC 372789. PMID 2087222.*

18. Murphy DJ (September 2001). «The biogenesis and functions of lipid bodies in animals, growth and microorganisms». Prog. Lipid Res. 40 *(5): 325–438.* doi:10.1016/S0163-7827(01)00013-3. PMID 11470496.
19. Feneberg, Wolfgang; Sackmann, Erich; Westphal, Monika (2001). «Dictyostelium cells› cytoplasm as an active viscoplastic body». European Biophysics Journal. 30 *(4): 284–94.* doi:10.1007/s002490100135. PMID 11548131.

CHAPTER 8

Proteasome

Chapter Contents

- **Introduction**
- **Discovery**
- **Structure and Organization**
- **Assembly**
- **The Protein Degradation Processt**
- **Ubiquitination and Targeting**
- **Evolution**
- **Cell Cycle Control**
- **Proteasome Inhibitors**

The proteasome is a multisubunit enzyme complex that plays a central role in the regulation of proteins that control cell-cycle progression and apoptosis, and has therefore become an important target for anticancer therapy.

INTRODUCTION

Cartoon representation of a proteasome. Its active sites are sheltered inside the tube (blue). The caps (red; in this case, 11S regulatory particles) on the ends regulate entry into the destruction chamber, where the protein is degraded.

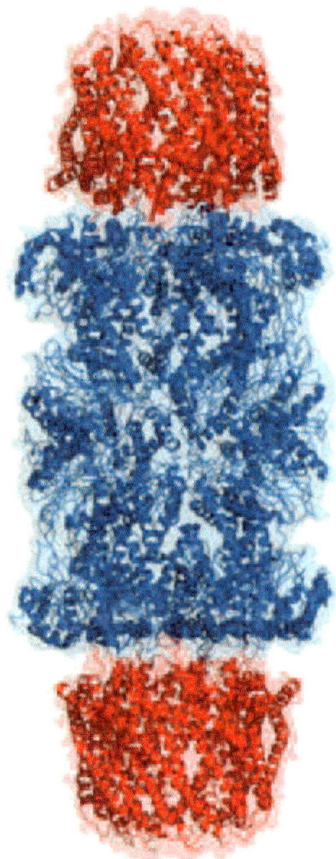

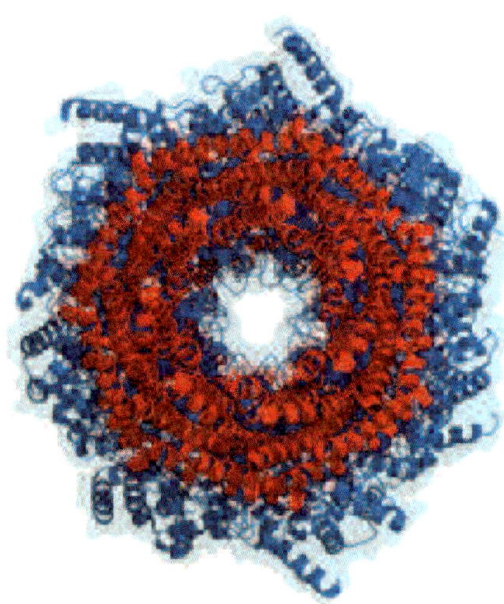

Top view of the proteasome above.

Proteasomes are protein complexes inside all eukaryotes and archaea, and in some bacteria. In eukaryotes, proteasomes are located in the nucleus and the cytoplasm. The main function of the proteasome is to degrade unneeded or damaged proteins by proteolysis, a chemical reaction that breaks peptide bonds. Enzymes that help such reactions are called proteases. Proteasomes are part of a major mechanism by which cells regulate the concentration of particular proteins and degrade misfolded proteins. The degradation process yields peptides of about seven to eight amino acids long, which can then be further degraded into shorter amino acid sequences and used in synthesizing new proteins. Proteins are tagged for degradation with a small protein called ubiquitin. The tagging reaction is catalyzed by enzymes called ubiquitin ligases. Once a protein is tagged with a single ubiquitin molecule, this is a signal to other ligases to attach additional ubiquitin molecules. The result is a polyubiquitin chain that is bound by the proteasome, allowing it to degrade the tagged protein.

In structure, the proteasome is a cylindrical complex containing a «core» of four stacked rings forming a central pore. Each ring is composed of seven individual proteins. The inner two rings are made of seven β subunits that contain three to seven protease active sites. These sites are located on the interior surface of the rings, so that the target protein must enter the central pore before it is degraded. The outer two rings each contain seven α subunits whose function is to maintain a «gate» through which proteins enter the barrel. These α subunits are controlled by binding to «cap» structures or regulatory particles that recognize polyubiquitin tags attached to protein substrates and initiate the degradation process. The overall system of ubiquitination and proteasomal degradation is known as the ubiquitin-proteasome system.

The proteasomal degradation pathway is essential for many cellular processes, including the cell cycle, the regulation of gene expression, and responses to oxidative stress. The importance of proteolytic degradation inside cells and the role of ubiquitin in proteolytic pathways was acknowledged in the award of the 2004 Nobel Prize in Chemistry to Aaron Ciechanover, Avram Hershko and Irwin Rose.

DISCOVERY

Before the discovery of the ubiquitin proteasome system, protein degradation in cells was thought to rely mainly on lysosomes, membrane-bound organelles with acidic and protease-filled interiors that can degrade and then recycle exogenous proteins and aged or damaged organelles. However, work by Alfred Goldberg in 1977 on ATP-dependent protein degradation in reticulocytes, which lack lysosomes, suggested the presence of a second intracellular degradation mechanism. This was shown in 1978 to be composed of several distinct protein chains, a novelty among proteases at the time. Later work on modification of histones led to the identification of an unexpected covalent modification of the histone protein by a bond between a lysine side chain of the histone and the C-terminal glycine residue of ubiquitin, a protein that had no known function. It was then discovered that a previously identified protein associated with proteolytic degradation, known as ATP-dependent proteolysis factor 1 (APF-1), was the same protein as ubiquitin. The proteolytic activities of this system was isolated as a multi-protein complex originally called the multi-catalytic proteinase complex by Sherwin Wilk and Marion Orlowski. Later, the ATP-dependent proteolytic complex that was responsible for ubiquitin-dependent protein degradation was discovered and was called the 26S proteasome.

Much of the early work leading up to the discovery of the ubiquitin proteasome system occurred in the late 1970s and early 1980s at the Technion in the laboratory of Avram Hershko, where Aaron Ciechanover worked as a graduate student. Hershko's year-long sabbatical in the laboratory of Irwin Rose at the Fox Chase Cancer Center provided key conceptual insights, though Rose later downplayed his role in the discovery. The three shared the 2004 Nobel Prize in Chemistry for their work in discovering this system.

Although electron microscopy data revealing the stacked-ring structure of the proteasome became available in the mid-1980s, the first structure of the proteasome core particle was not solved by X-ray crystallography until 1994.

STRUCTURE AND ORGANIZATION

A schematic diagram of the proteasome 20S core particle viewed from one side. The α subunits that make up the outer two rings are shown in green, and the β subunits that make up the inner two rings are shown in blue.

The proteasome subcomponents are often referred to by their Svedberg sedimentation coefficient (denoted S). The proteasome most exclusively used in mammals is the cytosolic 26S proteasome, which is about 2000 kilodaltons (kDa) in molecular mass containing one 20S protein subunit and two 19S regulatory cap subunits. The core is hollow and provides an enclosed cavity in which proteins are degraded; openings at the two ends of the core allow the target protein to enter. Each end of the core particle associates with a 19S regulatory subunit that contains multiple ATPase active sites and ubiquitin binding sites; it is this structure that recognizes polyubiquitinated proteins and transfers them to the catalytic core. An alternative form of regulatory subunit called the 11S particle can associate with the core in essentially the same manner as the 19S particle; the 11S may play a role in degradation of foreign peptides such as those produced after infection by a virus.

20S core Particle

The number and diversity of subunits contained in the 20S core particle depends on the organism; the number of distinct and specialized subunits is larger in multicellular than unicellular organisms and larger in eukaryotes than in prokaryotes. All 20S particles consist of four stacked heptameric ring structures that are themselves composed of two different types of subunits; α subunits are structural in nature, whereas β subunits are predominantly catalytic. The outer two rings in the stack consist of seven α subunits each, which serve as docking domains for the regulatory particles and the alpha subunits N-termini form a gate that blocks unregulated access of substrates to the interior cavity. The inner two rings each consist of seven β subunits and contain the protease active sites that perform the proteolysis reactions. Three distinct catalytic activities were identified in the purified complex: chymotrypsin-like, trypsin-like and peptidylglutamyl-peptide hydrolyzing. The size of the proteasome is relatively conserved and is about 150 angstroms (Å) by 115 Å. The interior chamber is at most 53 Å wide, though the entrance can be as narrow as 13 Å, suggesting that substrate proteins must be at least partially unfolded to enter.

In archaea such as Thermoplasma acidophilum, all the α and all the β subunits are identical, whereas eukaryotic proteasomes such as those in yeast contain seven distinct types of each subunit. In mammals, the β1, β2, and β5 subunits are catalytic; although they share a common mechanism, they have three distinct substrate specificities considered chymotrypsin-like, trypsin-like, and peptidyl-glutamyl peptide-hydrolyzing (PHGH). Alternative β forms denoted β1i, β2i, and β5i can be expressed in hematopoietic cells in response to exposure to pro-inflammatory signals such as cytokines, in particular, interferon gamma. The proteasome assembled with these alternative subunits is known as the immunoproteasome, whose substrate specificity is altered relative to the normal proteasome.Recently an alternative proteasome was identified in human cells that lack the α3 core subunit. These proteasomes (known as the α4-α4 proteasomes) instead form 20S core particles containing an additional α4 subunit in place of the missing α3 subunit. Interestingly, these alternative 'α4-α4' proteasomes have been known previously to exist in yeast. Although the precise function of these proteasome isoforms is still largely unknown, cells expressing these proteasomes show enhanced resistance to toxicity induced by metallic ions such as cadmium.

19S Regulatory Particle

The 19S particle in eukaryotes consists of 19 individual proteins and is divisible into two subassemblies, a 9-subunit base that binds directly to the α ring of the 20S core particle, and a 10-subunit lid. Six of the nine base proteins are ATPase subunits from the AAA Family, and an evolutionary homolog of these ATPases exists in archaea, called PAN (Proteasome-Activating Nucleotidase). The association of the 19S and 20S particles requires the binding of ATP to the 19S ATPase subunits, and ATP hydrolysis is required for the assembled complex to degrade folded and ubiquitinated proteins. Note that only the step of substrate unfolding requires energy from ATP hydrolysis, while ATP-binding alone can support all the other steps required for protein degradation (e.g., complex assembly, gate opening, translocation, and proteolysis). In fact, ATP binding to the ATPases by itself supports the rapid degradation of unfolded proteins. However, while ATP hydrolysis is required for unfolding only, it is not yet clear whether this energy may be used in the coupling of some of these steps.

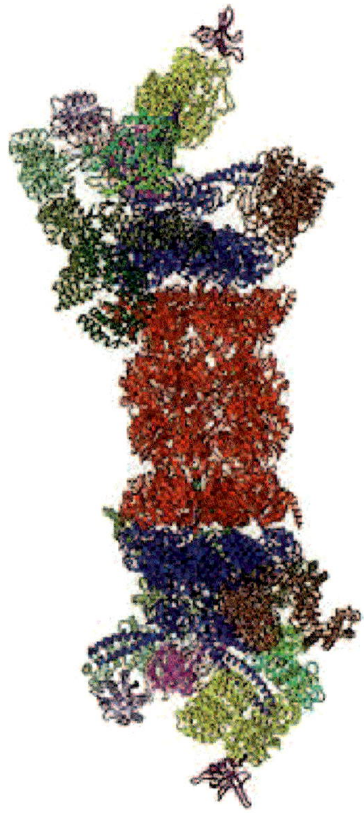

Cartoon representation of the 26S proteasome.

In 2012, two independent efforts have elucidated the molecular architecture of the 26S proteasome by single particle electron microscopy. More recently, a pseudo-atomic atomic model has been built, again using cryo-EM. In the heart of the 19S, directly adjacent to the 20S, are the AAA-ATPases (AAA proteins) that assemble to a heterohexameric ring of the order Rpt1/Rpt2/Rpt6/Rpt3/Rpt4/Rpt5. This ring is a trimer of dimers: Rpt1/Rpt2, Rpt6/Rpt3, and Rpt4/Rpt5 dimerize via their N-terminal coiled-coils. These coiled-coils protrude from the hexameric ring. The largest regulatory particle non-ATPases Rpn1 and Rpn2 bind to the tips of Rpt1/2 and Rpt6/3, respectively. The ubiquitin receptor Rpn13 binds to Rpn2 and completes the base cub-complex. The lid covers one half of the AAA-ATPase hexamer (Rpt6/Rpt3/Rpt4) and, unexpectedly, directly contacts the 20S via Rpn6 and to lesser extent Rpn5. The subunits Rpn9, Rpn5, Rpn6, Rpn7, Rpn3, and Rpn12, which are structurally related among themselves and to subunits of the COP9 complex and the Eukaryotic initiation factor 3 (hence called PCI subunits) assemble to a horseshoe-like structure enclosing the Rpn8/Rpn11 heterodimer. Rpn11, the deubiquinating enzyme, is placed at the mouth of the AAA-ATPase hexamer, ideally positioned to remove ubiquitin moieties immediately before translocation of substrates into the 20S. The second ubiquitin receptor identified to date, Rpn10, is positioned at the periphery of the lid, near subunits Rpn8 and Rpn9.

Conformational changes of 19S

The 19S regulatory particle has been observed in three strongly differing conformational states to date. Realization of all these three conformational states is likely necessary for accomplishing substrate recognition and degradation (see below). A hallmark of the AAA-ATPase configuration in this predominant low-energy state is a staircase- or lockwasher-like arrangement of the AAA-domains.Also in the presence of ATP but absence of substrate an alternative, less abundant conformation of the 19S is adopted primarily differing in the positioning of the lid with respect to the AAA-ATPase module.In the presence of ATP-gammaS or a substrate (stabilized in a 26S mutant with defective Rpn11) a third conformation has been observed displaying a dramatic structural change of the AAA-ATPase module.

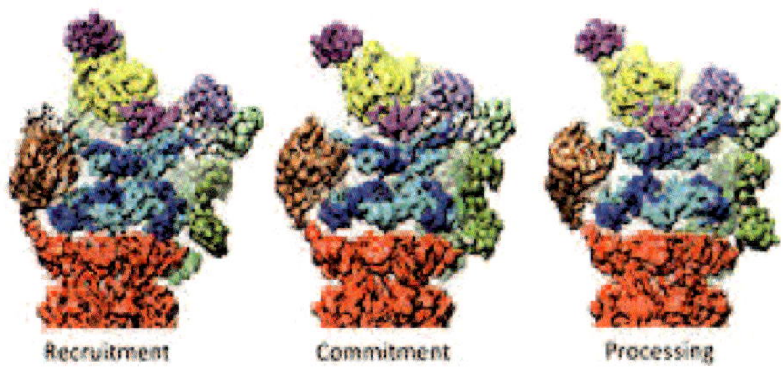

Three distinct conformational states of the 26S proteasome. The conformations are hypothesized to be responsible for recruitment of the substrate, its irreversible commitment, and finally processing and translocation into the core particle, where degradation occurs.

Regulation of the 20S by the 19S

The 19S regulatory particle is responsible for stimulating the 20S to degrade proteins. A primary function of the 19S regulatory ATPases is to open the gate in the 20S that blocks the entry of substrates into the degradation chamber. The mechanism by which the proteasomal ATPase open this gate has been recently elucidated.20S gate opening, and thus substrate degradation, requires the C-termini of the proteasomal ATPases, which contains a specific motif (i.e., HbYX motif). The ATPases C-termini bind into pockets in the top of the 20S, and tether the ATPase complex to the 20S proteolytic complex, thus joining the substrate unfolding equipment with the 20S degradation machinery. Binding of these C-termini into these 20S pockets by themselves stimulates opening of the gate in the 20S in much the same way that a «key-in-a-lock» opens a door. The precise mechanism by which this «key-in-a-lock» mechanism functions has been structurally elucidated.

11S regulatory particle

20S proteasomes can also associate with a second type of regulatory particle, the 11S regulatory particle, a heptameric structure that does not contain any ATPases and can promote the degradation of short peptides but not of complete proteins. It is presumed that this is because the complex cannot unfold larger substrates. This structure is also known as PA28 or REG. The mechanisms by which it binds to the core particle through the C-terminal tails of its subunits and induces α-ring conformational changes to open the 20S gate suggest a similar mechanism for the 19S particle. The expression of the 11S particle is induced by interferon gamma and is responsible, in conjunction with the immunoproteasome β subunits, for the generation of peptides that bind to the major histocompatibility complex.

ASSEMBLY

The assembly of the proteasome is a complex process due to the number of subunits that must associate to form an active complex. The β subunits are synthesized with N-terminal "propeptides» that are post-translationally modified during the assembly of the 20S particle to expose the proteolytic active site. The 20S particle is assembled from two half-proteasomes, each of which consists of a seven-membered pro-β ring attached to a seven-membered α ring. The association of the β rings of the two half-proteasomes triggers threonine-dependent autolysis of the propeptides to expose the active site. These β interactions are mediated mainly by salt bridges and hydrophobic interactions between conserved alpha helices whose disruption by mutation damages the proteasome's ability to assemble. The assembly of the half-proteasomes, in turn, is initiated by the assembly of the α subunits into their heptameric ring, forming a template for the association of the corresponding pro-β ring. The assembly of α subunits has not been characterized.

Only recently, the assembly process of the 19S regulatory particle has been elucidated to considerable extent. The 19S regulatory particle assembles as two distinct subcomponents, the base and the lid. Assembly of the base complex is facilitated by four assembly chaperones, Hsm3/S5b, Nas2/p27, Rpn14/PAAF1, and Nas6/gankyrin (names for yeast/mammals). These assembly chaperones bind to the AAA-ATPase subunits and their main function seems to be to ensure proper assembly of the heterohexameric AAA-ATPase ring. To date it is still under debate whether the base complex assembles separately, whether the assembly is templated by the 20S core particle, or whether alternative assembly pathways exist. In addition to the four assembly chaperones, the deubiquitinating enzyme Ubp6/Usp14 also promotes base assembly, but it is not essential. The lid assembles separately in a specific order and does not require assembly chaperones.

THE PROTEIN DEGRADATION PROCESS

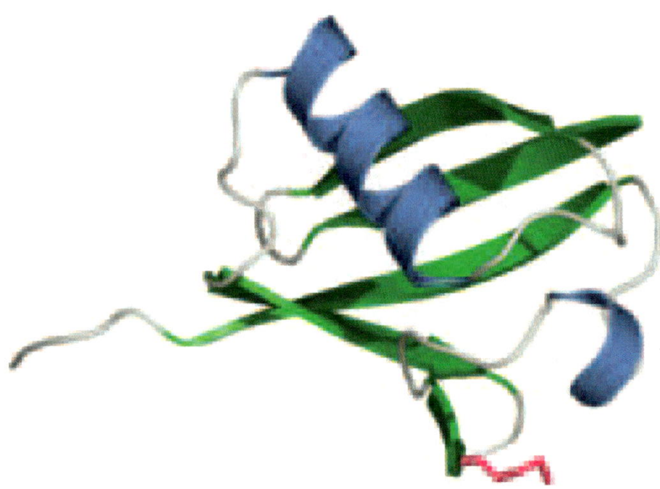

Ribbon diagram of ubiquitin, the highly conserved protein that serves as a molecular tag targeting proteins for degradation by the proteasome

UBIQUITINATION AND TARGETING

Proteins are targeted for degradation by the proteasome with covalent modification of a lysine residue that requires the coordinated reactions of three enzymes. In the first step, a ubiquitin-activating enzyme (known as E1) hydrolyzes ATP and adenylylates a ubiquitin molecule. This is then transferred to E1›s active-site cysteine residue in concert with the adenylylation of a second ubiquitin. This adenylylated ubiquitin is then transferred to a cysteine of a second enzyme, ubiquitin-conjugating enzyme (E2). In the last step, a member of a highly diverse class of enzymes known as ubiquitin ligases (E3) recognizes the specific protein to be ubiquitinated and catalyzes the transfer of ubiquitin from E2 to this target protein. A target protein must be labeled with at least four ubiquitin monomers (in the form of a polyubiquitin chain) before it is recognized by the proteasome lid. It is therefore the E3 that confers substrate specificity to this system. The number of E1, E2, and E3 proteins expressed depends on the organism and cell type, but there are many different E3 enzymes present in humans, indicating that there is a huge number of targets for the ubiquitin proteasome system.

The mechanism by which a polyubiquitinated protein is targeted to the proteasome is not fully understood. Ubiquitin-receptor proteins have an N-terminal ubiquitin-like (UBL) domain and one or more ubiquitin-associated (UBA) domains. The UBL domains are recognized by the 19S proteasome caps and the UBA domains bind ubiquitin via three-helix bundles. These receptor proteins may escort polyubiquitinated proteins to the proteasome, though the specifics of this interaction and its regulation are unclear.

The ubiquitin protein itself is 76 amino acids long and was named due to its ubiquitous nature, as it has a highly conserved sequence and is found in all known eukaryotic organisms. The genes encoding ubiquitin in eukaryotes are arranged in tandem repeats, possibly due to the heavy transcription demands on these genes to produce enough ubiquitin for the cell. It has been proposed that ubiquitin is the slowest-evolving protein identified to date. Ubiquitin contains seven lysine residues to which another ubiquitin can be ligated, resulting in different types of polyubiquitin chains. Chains in which each additional ubiquitin is linked to lysine 48 of the previous ubiquitin have a role in proteasome targeting, while other types of chains may be involved in other processes.

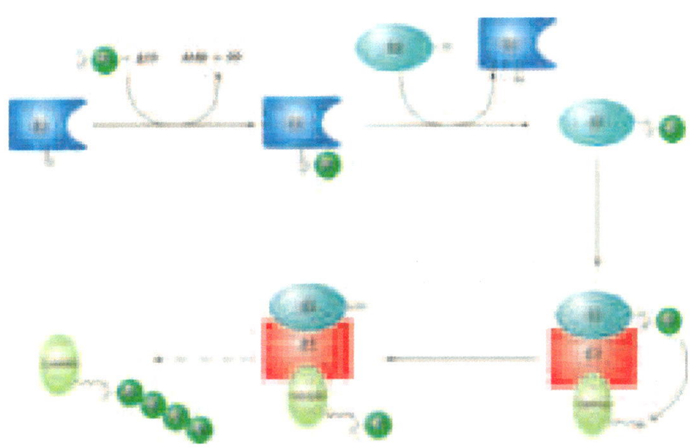

The ubiquitination pathway

Unfolding and Translocation

After a protein has been ubiquitinated, it is recognized by the 19S regulatory particle in an ATP-dependent binding step. The substrate protein must then enter the interior of the 20S particle to come in contact with the proteolytic active sites. Because the 20S particle's central channel is narrow and gated by the N-terminal tails of the α ring subunits, the substrates must be at least partially unfolded before they enter the core. The passage of the unfolded substrate into the core is called translocation and necessarily occurs after deubiquitination. However, the order in which substrates are deubiquitinated and unfolded is not yet clear. Which of these processes is the rate-limiting step in the overall proteolysis reaction depends on the specific substrate; for some proteins, the unfolding process is rate-limiting, while deubiquitination is the slowest step for other proteins. The extent to which substrates must be unfolded before translocation is not known, but substantial tertiary structure, and in particular nonlocal interactions such as disulfide bonds, are sufficient to inhibit degradation. The presence of intrinsically disordered protein segments of sufficient size, either at the protein terminus or internally, has also been proposed to facilitate efficient initiation of degradation.

The gate formed by the α subunits prevents peptides longer than about four residues from entering the interior of the 20S particle. The ATP molecules bound before the initial recognition step are hydrolyzed before translocation. While energy is needed for substrate unfolding, it is not required for translocation. The assembled 26S proteasome can degrade unfolded proteins in the presence of a non-hydrolyzable ATP analog, but cannot degrade folded proteins, indicating that energy from ATP hydrolysis is used for substrate unfolding. Passage of the unfolded substrate through the opened gate occurs via facilitated diffusion if the 19S cap is in the ATP-bound state.

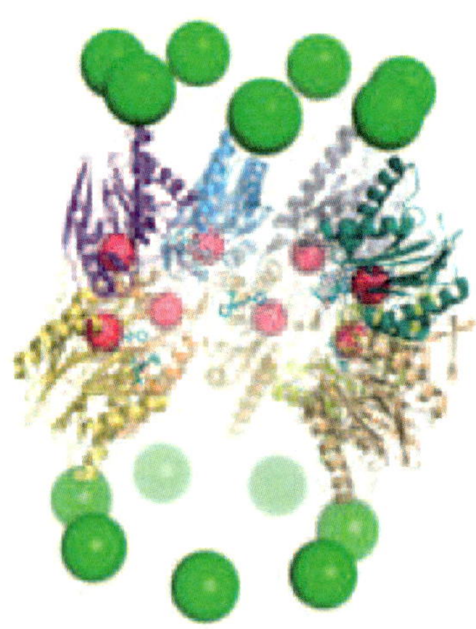

The mechanism for unfolding of globular proteins is necessarily general, but somewhat dependent on the amino acid sequence. Long sequences of alternating glycine and alanine have been shown to inhibit substrate unfolding, decreasing the efficiency of proteasomal degradation; this results in the release of partially degraded byproducts, possibly due to the decoupling of the ATP hydrolysis and unfolding steps. Such glycine-alanine repeats are also found in nature, for example in silk fibroin; in particular, certain Epstein–Barr virus gene products bearing this sequence can stall the proteasome, helping the virus propagate by preventing antigen presentation on the major histocompatibility complex.

A cutaway view of the proteasome 20S core particle illustrating the locations of the active sites. The α subunits are represented as green spheres and the β subunits as protein backbones colored by individual polypeptide chain. The small pink spheres represent the location of the active-site threonine residue in each subunit. Light blue chemical structures are the inhibitor bortezomib bound to the active sites.

Proteolysis

The mechanism of proteolysis by the β subunits of the 20S core particle is through a threonine-dependent nucleophilic attack. This mechanism may depend on an associated water molecule for deprotonation of the reactive threonine hydroxyl. Degradation occurs within the central chamber formed by the association of the two β rings and normally does not release partially degraded products, instead reducing the substrate to short polypeptides typically 7–9 residues long, though they can range from 4 to 25 residues, depending on the organism and substrate. The biochemical mechanism that determines product length is not fully characterized. Although the three catalytic β subunits have a common mechanism, they have slightly different substrate specificities, which are considered chymotrypsin-like, trypsin-like, and peptidyl-glutamyl peptide-hydrolyzing (PHGH)-like. These variations in specificity are the result of interatomic contacts with local residues near the active sites of each subunit. Each catalytic β subunit also possesses a conserved lysine residue required for proteolysis.

Although the proteasome normally produces very short peptide fragments, in some cases these products are themselves biologically active and functional molecules. Certain transcription factors regulating the expression of specific genes, including one component of the mammalian complex NF-κB, are synthesized as inactive precursors whose ubiquitination and subsequent proteasomal degradation converts them to an active form. Such activity requires the proteasome to cleave the substrate protein internally, rather than processively degrading it from one terminus. It has been suggested that long loops on these proteins' surfaces serve as the proteasomal substrates and enter the central cavity, while the majority of the protein remains outside. Similar effects have been observed in yeast proteins; this mechanism of selective degradation is known as regulated ubiquitin/proteasome dependent processing (RUP).

Ubiquitin-independent degradation

Although most proteasomal substrates must be ubiquitinated before being degraded, there are some exceptions to this general rule, especially when the proteasome plays a normal role in the post-translational processing of the protein. The proteasomal activation of NF-κB by processing p105 into p50 via internal proteolysis is one major example. Some proteins that are hypothesized to be unstable due to intrinsically unstructured regions, are degraded in a ubiquitin-independent manner. The most well-known example of a ubiquitin-independent proteasome substrate is the enzyme ornithine decarboxylase. Ubiquitin-independent mechanisms targeting key cell cycle regulators such as p53 have also been reported, although p53 is also subject to ubiquitin-dependent degradation. Finally, structurally abnormal, misfolded, or highly oxidized proteins are also subject to ubiquitin-independent and 19S-independent degradation under conditions of cellular stress.

EVOLUTION

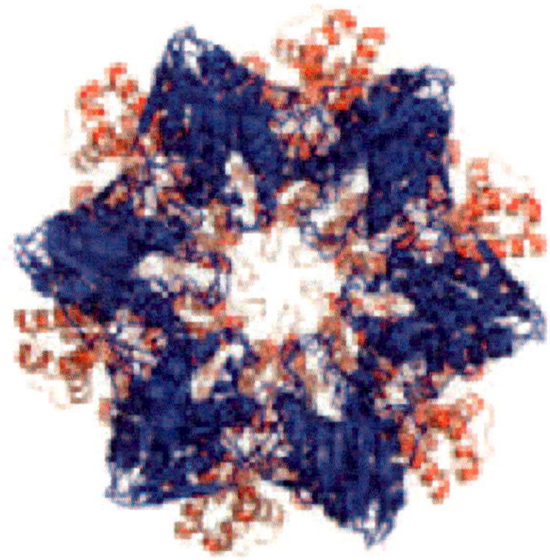

The assembled complex of hslV (blue) and hslU (red) from E. coli. This complex of heat shock proteins is thought to resemble the ancestor of the modern proteasome.

The 20S proteasome is both ubiquitous and essential in eukaryotes. Some prokaryotes, including many archaea and the bacterial order Actinomycetales also share homologs of the 20S proteasome, whereas most bacteria possess heat shock genes hslV and hslU, whose gene products are a multimeric protease arranged in a two-layered ring and an ATPase. The hslV protein has been hypothesized to resemble the likely ancestor of the 20S proteasome. In general, HslV is not essential in bacteria, and not all bacteria possess it, whereas some protists possess both the 20S and the hslV systems. Many bacteria also possess other homologs of the proteasome and an associated ATPase, most notably ClpP and ClpX. This redundancy explains why the HslUV system is not essential.

Sequence analysis suggests that the catalytic β subunits diverged earlier in evolution than the predominantly structural α subunits. In bacteria that express a 20S proteasome, the β subunits have high sequence identity to archaeal and eukaryotic β subunits, whereas the α sequence identity is much lower. The presence of 20S proteasomes in bacteria may result from lateral gene transfer, while the diversification of subunits among eukaryotes is ascribed to multiple gene duplication events.

CELL CYCLE CONTROL

Cell cycle progression is controlled by ordered action of cyclin-dependent kinases (CDKs), activated by specific cyclins that demarcate phases of the cell cycle. Mitotic cyclins, which persist in the cell for only a few minutes, have one of the shortest life spans of all intracellular proteins. After a CDK-cyclin complex has performed its function, the associated cyclin is polyubiquitinated and destroyed by the proteasome, which provides directionality for the cell cycle. In particular, exit from mitosis requires the proteasome-dependent dissociation of the regulatory component cyclin B from the mitosis promoting factor complex. In vertebrate cells, "slippage" through the mitotic checkpoint leading to premature M phase exit can occur despite the delay of this exit by the spindle checkpoint.

Earlier cell cycle checkpoints such as post-restriction point check between G_1 phase and S phase similarly involve proteasomal degradation of cyclin A, whose ubiquitination is promoted by the anaphase promoting complex (APC), an E3 ubiquitin ligase. The APC and the Skp1/Cul1/F-box protein complex (SCF complex) are the two key regulators of cyclin degradation and checkpoint control; the SCF itself is regulated by the APC via ubiquitination of the adaptor protein, Skp2, which prevents SCF activity before the G1-S transition.

Individual components of the 19S particle have their own regulatory roles. Gankyrin, a recently identified oncoprotein, is one of thc 19S subcomponents that also tightly binds the cyclin-dependent kinase CDK4 and plays a key role in recognizing ubiquitinated p53, via its affinity for the ubiquitin ligase MDM2. Gankyrin is anti-apoptotic and has been shown to be overexpressed in some tumor cell types such as hepatocellular carcinoma.

Regulation of Plant Growth

In plants, signaling by auxins, or phytohormones that order the direction and tropism of plant growth, induces the targeting of a class of transcription factor repressors known as Aux/IAA proteins for proteasomal degradation. These proteins are ubiquitinated by SCFTIR1, or SCF in complex with the auxin receptor TIR1. Degradation of Aux/IAA proteins derepresses transcription factors in the auxin-response factor (ARF) family and induces ARF-directed gene expression. The cellular consequences of ARF activation depend on the plant type and developmental stage, but are involved in directing growth in roots and leaf veins. The specific response to ARF derepression is thought to be mediated by specificity in the pairing of individual ARF and Aux/IAA proteins.

Apoptosis

Both internal and external signals can lead to the induction of apoptosis, or programmed cell death. The resulting deconstruction of cellular components is primarily carried out by specialized proteases known as caspases, but the proteasome also plays important and diverse roles in the apoptotic process. The involvement of the proteasome in this process is indicated by both the increase in protein ubiquitination, and of E1, E2, and E3 enzymes that is observed well in advance of apoptosis. During apoptosis, proteasomes localized to the nucleus have also been observed to translocate to outer membrane blebs characteristic of apoptosis.

Proteasome inhibition has different effects on apoptosis induction in different cell types. In general, the proteasome is not required for apoptosis, although inhibiting it is pro-apoptotic in most cell types that have been studied. Apoptosis is mediated through disrupting the regulated degradation of pro-growth cell cycle proteins. However, some cell lines — in particular, primary cultures of quiescent and differentiated cells such as thymocytes and neurons — are prevented from undergoing apoptosis on exposure to proteasome inhibitors. The mechanism for this effect is not clear, but is hypothesized to be specific to cells in quiescent states, or to result from the differential activity of the pro-apoptotic kinaseJNK. The ability of proteasome inhibitors to induce apoptosis in rapidly dividing cells has been exploited in several recently developed chemotherapy agents such as bortezomib and salinosporamide A.

Response to Cellular Stress

In response to cellular stresses – such as infection, heat shock, or oxidative damage – heat shock proteins that identify misfolded or unfolded proteins and target them for proteasomal degradation are expressed. Both Hsp27 and Hsp90—chaperone proteins have been implicated in increasing the activity of the ubiquitin-proteasome system, though they are not direct participants in the process. Hsp70, on the other hand, binds exposed hydrophobic patches on the surface of misfolded proteins and recruits E3 ubiquitin ligases such as CHIP to tag the proteins for proteasomal degradation. The CHIP protein (carboxyl terminus of Hsp70-interacting protein) is itself regulated via inhibition of interactions between the E3 enzyme CHIP and its E2 binding partner.

Similar mechanisms exist to promote the degradation of oxidatively damaged proteins via the proteasome system. In particular, proteasomes localized to the nucleus are regulated by PARP and actively degrade inappropriately oxidized histones. Oxidized proteins, which often form large amorphous aggregates in the cell, can be degraded directly by the 20S core particle without the 19S regulatory cap and do not require ATP hydrolysis or tagging with ubiquitin. However, high levels of oxidative damage increases the degree of cross-linking between protein fragments, rendering the aggregates resistant to proteolysis. Larger numbers and sizes of such highly oxidized aggregates are associated with aging.

Dysregulation of the ubiquitin proteasome system may contribute to several neural diseases. It may lead to brain tumors such as astrocytomas. In some of the late-onset neurodegenerative diseases that share aggregation of misfolded proteins as a common feature, such as Parkinson›s disease and Alzheimer›s disease, large insoluble aggregates of misfolded proteins can form and then result in neurotoxicity, through mechanisms that are not yet well understood. Decreased proteasome activity has been suggested as a cause of aggregation and Lewy body formation in Parkinson's. This hypothesis is supported by the observation that yeast models of Parkinson's are more susceptible to toxicity from α-synuclein, the major protein component of Lewy bodies, under conditions of low proteasome activity.Impaired proteasomal activity may underlie cognitive disorders such as the autism spectrum disorders, and muscle and nerve diseases such as inclusion body myopathy.

Role in the Immune System

The proteasome plays a straightforward but critical role in the function of the adaptive immune system. Peptide antigens are displayed by the major histocompatibility complex class I (MHC) proteins on the surface of antigen-presenting cells. These peptides are products of proteasomal degradation of proteins originated by the invading pathogen. Although constitutively expressed proteasomes can participate in this process, a specialized complex composed of proteins, whose expression is induced by interferon gamma, are the primary producers of peptides which are optimal in size and composition for MHC binding. These proteins whose expression increases during the immune response include the 11S regulatory particle, whose main known biological role is regulating the production of MHC ligands, and specialized β subunits called β1i, β2i, and β5i with altered substrate specificity. The complex formed with the specialized β subunits is known as the immunoproteasome. Another β5i variant subunit, β5t, is expressed in the thymus, leading to a thymus-specific "thymoproteasome" whose function is as yet unclear.

The strength of MHC class I ligand binding is dependent on the composition of the ligand C-terminus, as peptides bind by hydrogen bonding and by close contacts with a region called the "B pocket" on the MHC surface. Many MHC class I alleles prefer hydrophobic C-terminal residues, and the immunoproteasome complex is more likely to generate hydrophobic C-termini.

Due to its role in generating the activated form of NF-κB, an anti-apoptotic and pro-inflammatory regulator of cytokine expression, proteasomal activity has been linked to inflammatory and autoimmune diseases. Increased levels of proteasome activity correlate with disease activity and have been implicated in autoimmune diseases including systemic lupus erythematosus and rheumatoid arthritis.

The proteasome is also involved in Intracellular antibody-mediated proteolysis of antibody-bound virions. In this neutralisation pathway, TRIM21 (a protein of the tripartite motif family) binds with immunoglobulin G to direct the virion to the proteasome where it is degraded.

PROTEASOME INHIBITORS

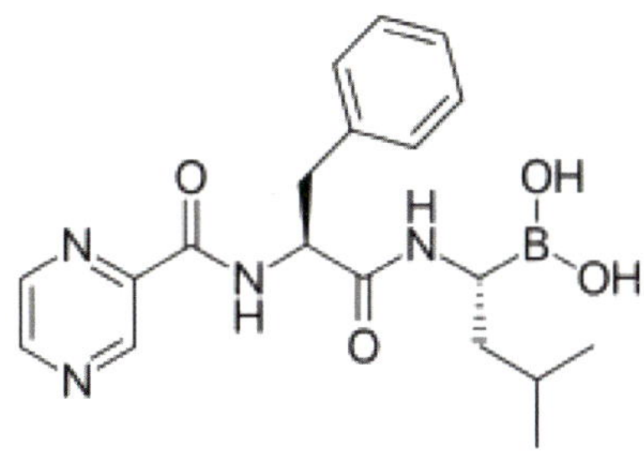

Chemical structure of bortezomib (Boronated form of MG132), a proteasome inhibitor used in chemotherapy that is particularly effective against multiple myeloma

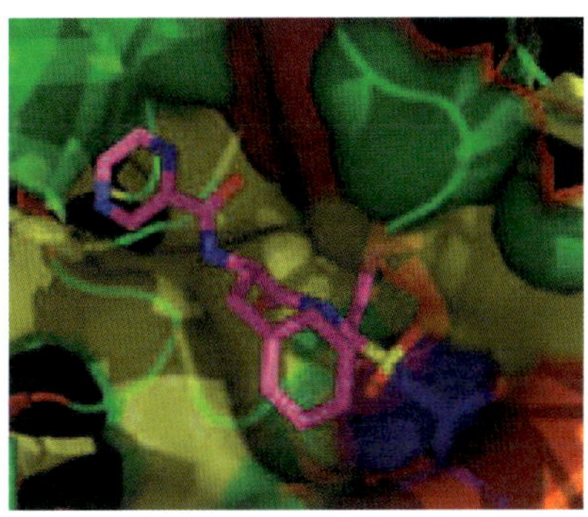

Bortezomib bound to the core particle in a yeast proteasome. The bortezomib molecule is in the center colored by atom type (carbon = pink, nitrogen = blue, oxygen = red, boron = yellow), surrounded by the local protein surface. The blue patch is the catalytic threonine residue whose activity is blocked by the presence of bortezomib.

Proteasome inhibitors have effective anti-tumor activity in cell culture, inducing apoptosis by disrupting the regulated degradation of pro-growth cell cycle proteins. This approach of selectively inducing apoptosis in tumor cells has proven effective in animal models and human trials.

Lactacystin, a natural product synthesized by Streptomycesbacteria, was the first non-peptidic proteasome inhibitor discovered and is widely used as a research tool in biochemistry and cell biology. Lactacystin was licensed to Myogenics/Proscript, which was acquired by Millennium Pharmaceuticals, now part of Takeda Pharmaceuticals. Lactacystin covalently modifies the amino-terminal threonine of catalytic β subunits of the proteasome, particularly the β5 subunit responsible for the proteasome's chymotrypsin-like activity. This discovery helped to establish the proteasome as a mechanistically novel class of protease: an amino-terminal threonine protease.

Bortezomib (Boronated MG132), a molecule developed by Millennium Pharmaceuticals and marketed as Velcade, is the first proteasome inhibitor to reach clinical use as a chemotherapy agent. Bortezomib is used in the treatment of multiple myeloma. Notably, multiple myeloma has been observed to result in increased proteasome-derived peptide levels in blood serum that decrease to normal levels in response to successful chemotherapy. Studies in animals have indicated that bortezomib may also have clinically significant effects in pancreatic cancer. Preclinical and early clinical studies have been started to examine bortezomib›s effectiveness in treating other B-cell-related cancers, particularly some types of non-Hodgkin›s lymphoma. Clinical results also seem to justify use of proteasome inhibitor combined with chemotherapy, for B-cell acute lymphoblastic leukemia Proteasome inhibitor can kill some types of cultured leukemia cells that are resistant to glucocorticoid.

The molecule ritonavir, marketed as Norvir, was developed as a protease inhibitor and used to target HIV infection. However, it has been shown to inhibit proteasomes as well as free proteases; to be specific, the chymotrypsin-like activity of the proteasome is inhibited by ritonavir, while the trypsin-like activity is somewhat enhanced. Studies in animal models suggest that ritonavir may have inhibitory effects on the growth of glioma cells. Proteasome inhibitors have also shown promise in treating autoimmune diseases in animal models. For example, studies in mice bearing human skin grafts found a reduction in the size of lesions from psoriasis after treatment with a proteasome inhibitor. Inhibitors also show positive effects in rodent models of asthma.

Labeling and inhibition of the proteasome is also of interest in laboratory settings for both in vitro and in vivo study of proteasomal activity in cells. The most commonly used laboratory inhibitors are lactacystin and the peptide aldehyde MG132 initially developed by Goldberg lab. Fluorescent inhibitors have also been developed to specifically label the active sites of the assembled proteasome.

Clinical Significance

The Proteasome and its subunits are of clinical significance for at least two reasons: (1) a compromised complex assembly or a dysfunctional proteasome can be associated with the underlying pathophysiology of specific diseases, and (2) they can be exploited as drug targets for therapeutic interventions. More recently, more effort has been made to consider the proteasome for the development of novel diagnostic markers and strategies. An improved and comprehensive understanding of the pathophysiology of the proteasome should lead to clinical applications in the future.

The proteasomes form a pivotal component for the Ubiquitin-Proteasome System (UPS) and corresponding cellular Protein Quality Control (PQC). Protein ubiquitination and subsequent proteolysis and degradation by the proteasome are important mechanisms in the regulation of the cell cycle, cell growth and differentiation, gene transcription, signal transduction and apoptosis.] Subsequently, a compromised proteasome complex assembly and function lead to reduced proteolytic activities and the accumulation of damaged or misfolded protein species. Such protein accumulation may contribute to the pathogenesis and phenotypic characteristics in neurodegenerative diseases, cardiovascular diseases, inflammatory responses and autoimmune diseases, and systemic DNA damage responses leading to malignancies.

Several experimental and clinical studies have indicated that aberrations and deregulations of the UPS contribute to the pathogenesis of several neurodegenerative and myodegenerative disorders, including Alzheimer's disease, Parkinson›s disease and Pick›s disease, Amyotrophic lateral sclerosis (ALS), Huntington›s disease, Creutzfeldt–Jakob disease, and motor neuron diseases, polyglutamine (PolyQ) diseases, Muscular dystrophies and several rare forms of neurodegenerative diseases associated with dementia. As part of the Ubiquitin-Proteasome System (UPS), the proteasome maintains cardiac protein homeostasis and thus plays a significant role in cardiac Ischemic injury, ventricular hypertrophy and Heart failure. Additionally, evidence is accumulating that the UPS plays an essential role in malignant transformation. UPS proteolysis plays a major role in responses of cancer cells to stimulatory signals that are critical for the development of cancer. Accordingly, gene expression by degradation of transcription factors, such as p53, c-Jun, c-Fos, NF-κB, c-Myc, HIF-1α, MATα2, STAT3, sterol-regulated element-binding proteins and androgen receptors are all controlled by the UPS and thus involved in the development of various malignancies. Moreover, the UPS regulates the degradation of tumor suppressor gene products such as adenomatous polyposis coli (APC) in colorectal cancer, retinoblastoma (Rb). and von Hippel-Lindau tumor suppressor (VHL), as well as a number of proto-oncogenes (Raf, Myc, Myb, Rel, Src, Mos, Abl). The UPS is also involved in the regulation of inflammatory responses. This activity is usually attributed to the role of proteasomes in the activation of NF-κB which further regulates the expression of pro inflammatory cytokines such as TNF-α, IL-β, IL-8, adhesion molecules (ICAM-1, VCAM-1, P selectine) and prostaglandins and nitric oxide (NO). Additionally, the UPS also plays a role in inflammatory responses as regulators of leukocyte proliferation, mainly through proteolysis of cyclines and the degradation of CDK inhibitors. Lastly, autoimmune disease patients with SLE, Sjogren›s syndrome and rheumatoid arthritis (RA) predominantly exhibit circulating proteasomes which can be applied as clinical biomarkers.

REFERENCES

1. Peters JM, Franke WW, Kleinschmidt JA (Mar 1994). "Distinct 19 S and 20 S subcomplexes of the 26 S proteasome and their distribution in the nucleus and the cytoplasm". The Journal of Biological Chemistry. 269 (10): 7709–18. PMID 8125997.
2. Lodish H, Berk A, Matsudaira P, Kaiser CA, Krieger M, Scott MP, Zipursky SL, Darnell J (2004). "3". Molecular cell biology (5th ed.). New York: W.H. Freeman and CO. pp. 66–72. ISBN 978-0-7167-4366-8.
3. Nassif, Nicholas D.; Cambray, Samantha E.; Kraut, Daniel A. (May 2014). "Slipping up: Partial substrate degradation by ATP-dependent proteases". IUBMB Life. 66 (5): 309–317. doi:10.1002/iub.1271. Retrieved 15 April 2016.
4. Nobel Prize Committee (2004). "Nobel Prize Awardees in Chemistry, 2004". Retrieved 2006-12-11.
5. Etlinger JD, Goldberg AL (Jan 1977). "A soluble ATP-dependent proteolytic system responsible for the degradation of abnormal proteins in reticulocytes". Proceedings of the National Academy of Sciences of the United States of America. 74 (1): 54–8. doi:10.1073/pnas.74.1.54. PMC 393195Freely accessible. PMID 264694.

6. Ciehanover A, Hod Y, Hershko A (Apr 1978). "A heat-stable polypeptide component of an ATP-dependent proteolytic system from reticulocytes". Biochemical and Biophysical Research Communications. 81 (4): 1100–5. doi:10.1016/0006-291X(78)91249-4. PMID 666810.

7. Goldknopf IL, Busch H (Mar 1977). "Isopeptide linkage between nonhistone and histone 2A polypeptides of chromosomal conjugate-protein A24". Proceedings of the National Academy of Sciences of the United States of America. 74 (3): 864–8. doi:10.1073/pnas.74.3.864. PMC 430507Freely accessible. PMID 265581.

8. Ciechanover A (Sep 2005). "Early work on the ubiquitin proteasome system, an interview with Aaron Ciechanover. Interview by CDD". Cell Death and Differentiation. 12 (9): 1167–77. doi:10.1038/sj.cdd.4401691. PMID 16094393.

9. Wilk S, Orlowski M (Nov 1980). "Cation-sensitive neutral endopeptidase: isolation and specificity of the bovine pituitary enzyme". Journal of Neurochemistry. 35 (5): 1172–82. doi:10.1111/j.1471-4159.1980.tb07873.x. PMID 6778972.

10. Tanaka K, Waxman L, Goldberg AL (Jun 1983). "ATP serves two distinct roles in protein degradation in reticulocytes, one requiring and one independent of ubiquitin". The Journal of Cell Biology. 96 (6): 1580–5. doi:10.1083/jcb.96.6.1580. PMC 2112434Freely accessible. PMID 6304111.

11. Hough R, Pratt G, Rechsteiner M (Jun 1987). "Purification of two high molecular weight proteases from rabbit reticulocyte lysate". The Journal of Biological Chemistry. 262 (17): 8303–13. PMID 3298229.

12. Hershko A (Sep 2005). "Early work on the ubiquitin proteasome system, an interview with Avram Hershko. Interview by CDD". Cell Death and Differentiation. 12 (9): 1158–61. doi:10.1038/sj.cdd.4401709. PMID 16094391.

13. Kopp F, Steiner R, Dahlmann B, Kuehn L, Reinauer H (Aug 1986). "Size and shape of the multicatalytic proteinase from rat skeletal muscle". Biochimica et Biophysica Acta. 872 (3): 253–60. doi:10.1016/0167-4838(86)90278-5. PMID 3524688.

14. Löwe J, Stock D, Jap B, Zwickl P, Baumeister W, Huber R (Apr 1995). "Crystal structure of the 20S proteasome from the archaeon T. acidophilum at 3.4 A resolution". Science. 268 (5210): 533–9. doi:10.1126/science.7725097. PMID 7725097.

15. Wang J, Maldonado MA (Aug 2006). "The ubiquitin-proteasome system and its role in inflammatory and autoimmune diseases". Cellular & Molecular Immunology. 3 (4): 255–61. PMID 16978533.

16. c Smith DM, Chang SC, Park S, Finley D, Cheng Y, Goldberg AL (Sep 2007). "Docking of the proteasomal ATPases' carboxyl termini in the 20S proteasome's alpha ring opens the gate for substrate entry". Molecular Cell. 27 (5): 731–44. doi:10.1016/j.molcel.2007.06.033. PMC 2083707Freely accessible. PMID 17803938.

17. Wilk S, Orlowski M (Mar 1983). "Evidence that pituitary cation-sensitive neutral endopeptidase is a multicatalytic protease complex". Journal of Neurochemistry. 40 (3): 842–9. doi:10.1111/j.1471-4159.1983.tb08056.x. PMID 6338156.

18. Nandi D, Tahiliani P, Kumar A, Chandu D (Mar 2006). "The ubiquitin-proteasome system". Journal of Biosciences. 31 (1): 137–55. doi:10.1007/BF02705243. PMID 16595883.

19. Heinemeyer W, Fischer M, Krimmer T, Stachon U, Wolf DH (Oct 1997). "The active sites of the eukaryotic 20 S proteasome and their involvement in subunit precursor processing". The Journal of Biological Chemistry. 272 (40): 25200–9. doi:10.1074/jbc.272.40.25200. PMID 9312134.

20. Padmanabhan A, Vuong SA, Hochstrasser M (March 2016). "Assembly of an Evolutionarily Conserved Alternative Proteasome Isoform in Human Cells". Cell Reports. 14 (12): 2962–74. doi:10.1016/j.celrep.2016.02.068. PMID 26997268.

21. Velichutina I, Connerly PL, Arendt CS, Li X, Hochstrasser M (February 2004). "Plasticity in eucaryotic 20S proteasome ring assembly revealed by a subunit deletion in yeast". The EMBO Journal. 23 (3): 500–10. doi:10.1038/sj.emboj.7600059. PMID 14739934.

22. Kusmierczyk AR, Kunjappu MJ, Funakoshi M, Hochstrasser M (March 2008). "A multimeric assembly factor controls the formation of alternative 20S proteasomes". Nature Structural & Molecular Biology. 15 (3): 237–44. doi:10.1038/nsmb.1389. PMID 18278055.

23. Zwickl P, Ng D, Woo KM, Klenk HP, Goldberg AL (Sep 1999). "An archaebacterial ATPase, homologous to ATPases in the eukaryotic 26 S proteasome, activates protein breakdown by 20 S proteasomes". The Journal of Biological Chemistry. 274 (37): 26008–14. doi:10.1074/jbc.274.37.26008. PMID 10473546.

24. Smith DM, Kafri G, Cheng Y, Ng D, Walz T, Goldberg AL (Dec 2005). "ATP binding to PAN or the 26S ATPases causes association with the 20S proteasome, gate opening, and translocation of unfolded proteins". Molecular Cell. 20 (5): 687–98. doi:10.1016/j.molcel.2005.10.019. PMID 16337593.

25. e Liu CW, Li X, Thompson D, Wooding K, Chang TL, Tang Z, Yu H, Thomas PJ, DeMartino GN (Oct 2006). "ATP binding and ATP hydrolysis play distinct roles in the function of 26S proteasome". Molecular Cell. 24 (1): 39–50. doi:10.1016/j.molcel.2006.08.025. PMID 17018291.

26. Lam YA, Lawson TG, Velayutham M, Zweier JL, Pickart CM (Apr 2002). "A proteasomal ATPase subunit recognizes the polyubiquitin degradation signal". Nature. 416 (6882): 763–7. doi:10.1038/416763a. PMID 11961560.

27. c Beck F, Unverdorben P, Bohn S, Schweitzer A, Pfeifer G, Sakata E, Nickell S, Plitzko JM, Villa E, Baumeister W, Förster F (Sep 2012). "Near-atomic resolution structural model of the yeast 26S proteasome". Proceedings of the National Academy of Sciences of the United States of America. 109 (37): 14870–5. doi:10.1073/pnas.1213333109. PMID 22927375.

28. Lander GC, Estrin E, Matyskiela ME, Bashore C, Nogales E, Martin A (Feb 2012). "Complete subunit architecture of the proteasome regulatory particle". Nature. 482 (7384). doi:10.1038/nature10774. PMC 3285539Freely accessible. PMID 22237024.

29. Lasker K, Förster F, Bohn S, Walzthoeni T, Villa E, Unverdorben P, Beck F, Aebersold R, Sali A, Baumeister W (Jan 2012). "Molecular architecture of the 26S proteasome holocomplex determined by an integrative approach". Proceedings of the National Academy of Sciences of the United States of America. 109 (5): 1380–7. doi:10.1073/pnas.1120559109. PMC 3277140Freely accessible. PMID 22307589.

30. c Unverdorben P, Beck F, Śledź P, Schweitzer A, Pfeifer G, Plitzko JM, Baumeister W, Förster F (Apr 2014). "Deep classification of a large cryo-EM dataset defines the conformational landscape of the 26S proteasome". Proceedings of the National Academy of Sciences of the United States of America. 111 (15): 5544–9. doi:10.1073/pnas.1403409111. PMC 3992697Freely accessible. PMID 24706844.

31. Śledź P, Unverdorben P, Beck F, Pfeifer G, Schweitzer A, Förster F, Baumeister W (Apr 2013). "Structure of the 26S proteasome with ATP-γS bound provides insights into the mechanism of nucleotide-dependent substrate translocation". Proceedings of the

National Academy of Sciences of the United States of America. 110 (18): 7264–7269. doi:10.1073/pnas.1305782110. PMID 23589842.

32. Matyskiela ME, Lander GC, Martin A (Jul 2013). "Conformational switching of the 26S proteasome enables substrate degradation". Nature Structural & Molecular Biology. 20 (7): 781–788. doi:10.1038/nsmb.2616. PMC 3712289Freely accessible. PMID 23770819.
33. Köhler A, Cascio P, Leggett DS, Woo KM, Goldberg AL, Finley D (Jun 2001). "The axial channel of the proteasome core particle is gated by the Rpt2 ATPase and controls both substrate entry and product release". Molecular Cell. 7 (6): 1143–52. doi:10.1016/S1097-2765(01)00274-X. PMID 11430818.
34. Rabl J, Smith DM, Yu Y, Chang SC, Goldberg AL, Cheng Y (May 2008). "Mechanism of gate opening in the 20S proteasome by the proteasomal ATPases". Molecular Cell. 30 (3): 360–8. doi:10.1016/j.molcel.2008.03.004. PMC 4141531Freely accessible. PMID 18471981.
35. Förster A, Masters EI, Whitby FG, Robinson H, Hill CP (May 2005). "The 1.9 A structure of a proteasome-11S activator complex and implications for proteasome-PAN/PA700 interactions". Molecular Cell. 18 (5): 589–99. doi:10.1016/j.molcel.2005.04.016. PMID 15916965.
36. Witt S, Kwon YD, Sharon M, Felderer K, Beuttler M, Robinson CV, Baumeister W, Jap BK (Jul 2006). "Proteasome assembly triggers a switch required for active-site maturation". Structure. 14 (7): 1179–88. doi:10.1016/j.str.2006.05.019. PMID 16843899.
37. Krüger E, Kloetzel PM, Enenkel C (2001). "20S proteasome biogenesis". Biochimie. 83 (3-4): 289–93. doi:10.1016/S0300-9084(01)01241-X. PMID 11295488.
38. Murata S, Yashiroda H, Tanaka K (Feb 2009). "Molecular mechanisms of proteasome assembly". Nature Reviews Molecular Cell Biology. 10 (2): 104–115. doi:10.1038/nrm2630. PMID 19165213.
39. Sakata E, Stengel F, Fukunaga K, Zhou M, Saeki Y, Förster F, Baumeister W, Tanaka K, Robinson CV (Jun 2011). "The catalytic activity of Ubp6 enhances maturation of the proteasomal regulatory particle". Molecular Cell. 42 (5): 637–649. doi:10.1016/j.molcel.2011.04.021. PMID 21658604.
40. Fukunaga K, Kudo T, Toh-e A, Tanaka K, Saeki Y (Jun 2010). "Dissection of the assembly pathway of the proteasome lid in Saccharomyces cerevisiae". Biochemical and Biophysical Research Communications. 396 (4): 1048–1053. doi:10.1016/j.bbrc.2010.05.061. PMID 20471955.
41. Haas AL, Warms JV, Hershko A, Rose IA (Mar 1982). "Ubiquitin-activating enzyme. Mechanism and role in protein-ubiquitin conjugation". The Journal of Biological Chemistry. 257 (5): 2543–8. PMID 6277905.
42. Thrower JS, Hoffman L, Rechsteiner M, Pickart CM (Jan 2000). "Recognition of the polyubiquitin proteolytic signal". The EMBO Journal. 19 (1): 94–102. doi:10.1093/emboj/19.1.94. PMC 1171781Freely accessible. PMID 10619848.
43. Risseeuw EP, Daskalchuk TE, Banks TW, Liu E, Cotelesage J, Hellmann H, Estelle M, Somers DE, Crosby WL (Jun 2003). "Protein interaction analysis of SCF ubiquitin E3 ligase subunits from Arabidopsis". The Plant Journal. 34 (6): 753–67. doi:10.1046/j.1365-313X.2003.01768.x. PMID 12795696.
44. Elsasser S, Finley D (Aug 2005). "Delivery of ubiquitinated substrates to protein-unfolding machines". Nature Cell Biology. 7 (8): 742–9. doi:10.1038/ncb0805-742. PMID 16056265.

45. Sadanandom A, Bailey M, Ewan R, Lee J, Nelis S (Oct 2012). "The ubiquitin-proteasome system: central modifier of plant signalling". The New Phytologist. 196 (1): 13–28. doi:10.1111/j.1469-8137.2012.04266.x. PMID 22897362.

46. Sharp PM, Li WH (1987). "Ubiquitin genes as a paradigm of concerted evolution of tandem repeats". Journal of Molecular Evolution. 25 (1): 58–64. doi:10.1007/BF02100041. PMID 3041010.

47. Pickart CM, Fushman D (Dec 2004). "Polyubiquitin chains: polymeric protein signals". Current Opinion in Chemical Biology. 8 (6): 610–16. doi:10.1016/j.cbpa.2004.09.009. PMID 15556404.

48. Xu P, Duong DM, Seyfried NT, Cheng D, Xie Y, Robert J, Rush J, Hochstrasser M, Finley D, Peng J (Apr 2009). "Quantitative proteomics reveals the function of unconventional ubiquitin chains in proteasomal degradation". Cell. 137 (1): 133–45. doi:10.1016/j.cell.2009.01.041. PMC 2668214Freely accessible. PMID 19345192.

49. Pickart CM (Nov 2000). "Ubiquitin in chains". Trends in Biochemical Sciences. 25 (11): 544–8. doi:10.1016/S0968-0004(00)01681-9. PMID 11084366.

50. Zhu Q, Wani G, Wang QE, El-mahdy M, Snapka RM, Wani AA (Jul 2005). "Deubiquitination by proteasome is coordinated with substrate translocation for proteolysis in vivo". Experimental Cell Research. 307 (2): 436–51. doi:10.1016/j.yexcr.2005.03.031. PMID 15950624.

51. Wenzel T, Baumeister W (Mar 1995). "Conformational constraints in protein degradation by the 20S proteasome". Nature Structural Biology. 2 (3): 199–204. doi:10.1038/nsb0395-199. PMID 7773788.

52. Inobe T, Fishbain S, Prakash S, Matouschek A (Mar 2011). "Defining the geometry of the two-component proteasome degron". Nature Chemical Biology. 7 (3): 161–7. doi:10.1038/nchembio.521. PMC 3129032Freely accessible. PMID 21278740.

53. van der Lee R, Lang B, Kruse K, Gsponer J, Sánchez de Groot N, Huynen MA, Matouschek A, Fuxreiter M, Babu MM (Sep 2014). "Intrinsically disordered segments affect protein half-life in the cell and during evolution". Cell Reports. 8 (6): 1832–44. doi:10.1016/j.celrep.2014.07.055. PMC 4358326Freely accessible. PMID 25220455.

54. Smith DM, Benaroudj N, Goldberg A (Oct 2006). "Proteasomes and their associated ATPases: a destructive combination". Journal of Structural Biology. 156 (1): 72–83. doi:10.1016/j.jsb.2006.04.012. PMID 16919475.

55. Hoyt MA, Zich J, Takeuchi J, Zhang M, Govaerts C, Coffino P (Apr 2006). "Glycine-alanine repeats impair proper substrate unfolding by the proteasome". The EMBO Journal. 25 (8): 1720–9. doi:10.1038/sj.emboj.7601058. PMC 1440830Freely accessible. PMID 16601692.

56. Zhang M, Coffino P (Mar 2004). "Repeat sequence of Epstein–Barr virus-encoded nuclear antigen 1 protein interrupts proteasome substrate processing". The Journal of Biological Chemistry. 279 (10): 8635–41. doi:10.1074/jbc.M310449200. PMID 14688254.

57. Voges D, Zwickl P, Baumeister W (1999). "The 26S proteasome: a molecular machine designed for controlled proteolysis". Annual Review of Biochemistry. 68 (1): 1015–68. doi:10.1146/annurev.biochem.68.1.1015. PMID 10872471.

58. Rape M, Jentsch S (May 2002). "Taking a bite: proteasomal protein processing". Nature Cell Biology. 4 (5): E113–6. doi:10.1038/ncb0502-e113. PMID 11988749.

59. Rape M, Jentsch S (Nov 2004). "Productive RUPture: activation of transcription factors by proteasomal processing". Biochimica et Biophysica Acta. 1695 (1-3): 209–13.

doi:10.1016/j.bbamcr.2004.09.022. PMID 15571816.

60. Asher G, Reuven N, Shaul Y (Aug 2006). "20S proteasomes and protein degradation "by default"". BioEssays. 28 (8): 844–9. doi:10.1002/bies.20447. PMID 16927316.

61. Zhang M, Pickart CM, Coffino P (Apr 2003). "Determinants of proteasome recognition of ornithine decarboxylase, a ubiquitin-independent substrate". The EMBO Journal. 22 (7): 1488–96. doi:10.1093/emboj/cdg158. PMC 152902Freely accessible. PMID 12660156.

62. Asher G, Shaul Y (Aug 2005). "p53 proteasomal degradation: poly-ubiquitination is not the whole story". Cell Cycle. 4 (8): 1015–8. doi:10.4161/cc.4.8.1900. PMID 16082197.

63. Shringarpure R, Grune T, Mehlhase J, Davies KJ (Jan 2003). "Ubiquitin conjugation is not required for the degradation of oxidized proteins by proteasome". The Journal of Biological Chemistry. 278 (1): 311–8. doi:10.1074/jbc.M206279200. PMID 12401807.

64. c Gille C, Goede A, Schlöetelburg C, Preissner R, Kloetzel PM, Göbel UB, Frömmel C (Mar 2003). "A comprehensive view on proteasomal sequences: implications for the evolution of the proteasome". Journal of Molecular Biology. 326 (5): 1437–48. doi:10.1016/S0022-2836(02)01470-5. PMID 12595256.

65. Bochtler M, Ditzel L, Groll M, Hartmann C, Huber R (1999). "The proteasome". Annual Review of Biophysics and Biomolecular Structure. 28 (1): 295–317. doi:10.1146/annurev.biophys.28.1.295. PMID 10410804.

66. Chesnel F, Bazile F, Pascal A, Kubiak JZ (Aug 2006). "Cyclin B dissociation from CDK1 precedes its degradation upon MPF inactivation in mitotic extracts of Xenopus laevis embryos". Cell Cycle. 5 (15): 1687–98. doi:10.4161/cc.5.15.3123. PMID 16921258.

67. Brito DA, Rieder CL (Jun 2006). "Mitotic checkpoint slippage in humans occurs via cyclin B destruction in the presence of an active checkpoint". Current Biology. 16 (12): 1194–200. doi:10.1016/j.cub.2006.04.043. PMC 2749311Freely accessible. PMID 16782009.

68. Havens CG, Ho A, Yoshioka N, Dowdy SF (Jun 2006). "Regulation of late G1/S phase transition and APC Cdh1 by reactive oxygen species". Molecular and Cellular Biology. 26 (12): 4701–11. doi:10.1128/MCB.00303-06. PMC 1489138Freely accessible. PMID 16738333.

69. Bashir T, Dorrello NV, Amador V, Guardavaccaro D, Pagano M (Mar 2004). "Control of the SCF(Skp2-Cks1) ubiquitin ligase by the APC/C(Cdh1) ubiquitin ligase". Nature. 428 (6979): 190–3. doi:10.1038/nature02330. PMID 15014502.

70. Higashitsuji H, Liu Y, Mayer RJ, Fujita J (Oct 2005). "The oncoprotein gankyrin negatively regulates both p53 and RB by enhancing proteasomal degradation". Cell Cycle. 4 (10): 1335–7. doi:10.4161/cc.4.10.2107. PMID 16177571.

71. Dharmasiri S, Estelle M (2002). "The role of regulated protein degradation in auxin response". Plant Molecular Biology. 49 (3-4): 401–9. doi:10.1023/A:1015203013208. PMID 12036263.

72. Weijers D, Benkova E, Jäger KE, Schlereth A, Hamann T, Kientz M, Wilmoth JC, Reed JW, Jürgens G (May 2005). "Developmental specificity of auxin response by pairs of ARF and Aux/IAA transcriptional regulators". The EMBO Journal. 24 (10): 1874–85. doi:10.1038/sj.emboj.7600659. PMC 1142592Freely accessible. PMID 15889151.

73. Haas AL, Baboshina O, Williams B, Schwartz LM (Apr 1995). "Coordinated induction of the ubiquitin conjugation pathway accompanies the developmentally programmed death of insect skeletal muscle". The Journal of Biological Chemistry. 270 (16): 9407–12. doi:10.1074/jbc.270.16.9407. PMID 7721865.

74. Schwartz LM, Myer A, Kosz L, Engelstein M, Maier C (Oct 1990). "Activation of polyubiquitin gene expression during developmentally programmed cell death". Neuron. 5 (4): 411–9. doi:10.1016/0896-6273(90)90080-Y. PMID 2169771.

75. Löw P, Bussell K, Dawson SP, Billett MA, Mayer RJ, Reynolds SE (Jan 1997). "Expression of a 26S proteasome ATPase subunit, MS73, in muscles that undergo developmentally programmed cell death, and its control by ecdysteroid hormones in the insect Manduca sexta". FEBS Letters. 400 (3): 345–9. doi:10.1016/S0014-5793(96)01413-5. PMID 9009228.

76. Pitzer F, Dantes A, Fuchs T, Baumeister W, Amsterdam A (Sep 1996). "Removal of proteasomes from the nucleus and their accumulation in apoptotic blebs during programmed cell death". FEBS Letters. 394 (1): 47–50. doi:10.1016/0014-5793(96)00920-9. PMID 8925925.

77. Adams J, Palombella VJ, Sausville EA, Johnson J, Destree A, Lazarus DD, Maas J, Pien CS, Prakash S, Elliott PJ (Jun 1999). "Proteasome inhibitors: a novel class of potent and effective antitumor agents". Cancer Research. 59 (11): 2615–22. PMID 10363983.

78. Orlowski RZ (Apr 1999). "The role of the ubiquitin-proteasome pathway in apoptosis". Cell Death and Differentiation. 6 (4): 303–13. doi:10.1038/sj.cdd.4400505. PMID 10381632.

79. Garrido C, Brunet M, Didelot C, Zermati Y, Schmitt E, Kroemer G (Nov 2006). "Heat shock proteins 27 and 70: anti-apoptotic proteins with tumorigenic properties". Cell Cycle. 5 (22): 2592–601. doi:10.4161/cc.5.22.3448. PMID 17106261.

80. Park SH, Bolender N, Eisele F, Kostova Z, Takeuchi J, Coffino P, Wolf DH (Jan 2007). "The cytoplasmic Hsp70 chaperone machinery subjects misfolded and endoplasmic reticulum import-incompetent proteins to degradation via the ubiquitin-proteasome system". Molecular Biology of the Cell. 18 (1): 153–65. doi:10.1091/mbc.E06-04-0338. PMC 1751312Freely accessible. PMID 17065559.

81. Dai Q, Qian SB, Li HH, McDonough H, Borchers C, Huang D, Takayama S, Younger JM, Ren HY, Cyr DM, Patterson C (Nov 2005). "Regulation of the cytoplasmic quality control protein degradation pathway by BAG2". The Journal of Biological Chemistry. 280 (46): 38673–81. doi:10.1074/jbc.M507986200. PMID 16169850.

82. Bader N, Grune T (2006). "Protein oxidation and proteolysis". Biological Chemistry. 387 (10-11): 1351–5. doi:10.1515/BC.2006.169. PMID 17081106.

83. Davies KJ (2003). "Degradation of oxidized proteins by the 20S proteasome". Biochimie. 83 (3-4): 301–10. doi:10.1016/S0300-9084(01)01250-0. PMID 11295490.

84. Lehman NL (Sep 2009). "The ubiquitin proteasome system in neuropathology". Acta Neuropathologica. 118 (3): 329–47. doi:10.1007/s00401-009-0560-x. PMC 2716447Freely accessible. PMID 19597829.

85. McNaught KS, Jackson T, JnoBaptiste R, Kapustin A, Olanow CW (May 2006). "Proteasomal dysfunction in sporadic Parkinson's disease". Neurology. 66 (10 Suppl 4): S37–49. doi:10.1212/01.wnl.0000221745.58886.2e. PMID 16717251.

86. Sharma N, Brandis KA, Herrera SK, Johnson BE, Vaidya T, Shrestha R, Debburman SK (2006). "alpha-Synuclein budding yeast model: toxicity enhanced by impaired proteasome and oxidative stress". Journal of Molecular Neuroscience. 28 (2): 161–78. doi:10.1385/JMN:28:2:161. PMID 16679556.

87. Murata S, Sasaki K, Kishimoto T, Niwa S, Hayashi H, Takahama Y, Tanaka K (Jun 2007). "Regulation of CD8+ T cell development by thymus-specific proteasomes". Science. 316 (5829): 1349–53. doi:10.1126/science.1141915. PMID 17540904.

88. Cascio P, Hilton C, Kisselev AF, Rock KL, Goldberg AL (May 2001). "26S proteasomes and immunoproteasomes produce mainly N-extended versions of an antigenic peptide". The EMBO Journal. 20 (10): 2357–66. doi:10.1093/emboj/20.10.2357. PMC 125470Freely accessible. PMID 11350924.

89. Mallery DL, McEwan WA, Bidgood SR, Towers GJ, Johnson CM, James LC (Nov 2010). "Antibodies mediate intracellular immunity through tripartite motif-containing 21 (TRIM21)". Proceedings of the National Academy of Sciences of the United States of America. 107 (46): 19985–19990. doi:10.1073/pnas.1014074107. PMC 2993423Freely accessible. PMID 21045130.

90. Fenteany G, Standaert RF, Lane WS, Choi S, Corey EJ, Schreiber SL (May 1995). "Inhibition of proteasome activities and subunit-specific amino-terminal threonine modification by lactacystin". Science. 268 (5211): 726–31. doi:10.1126/science.7732382. PMID 7732382.

91. United States Food and Drug Administration press release 13 May 2003. Access date 29 December 2006. See also FDA Velcade information page.

92. Fisher RI, Bernstein SH, Kahl BS, Djulbegovic B, Robertson MJ, de Vos S, Epner E, Krishnan A, Leonard JP, Lonial S, Stadtmauer EA, O'Connor OA, Shi H, Boral AL, Goy A (Oct 2006). "Multicenter phase II study of bortezomib in patients with relapsed or refractory mantle cell lymphoma". Journal of Clinical Oncology. 24 (30): 4867–74. doi:10.1200/JCO.2006.07.9665. PMID 17001068.

93. Jakob C, Egerer K, Liebisch P, Türkmen S, Zavrski I, Kuckelkorn U, Heider U, Kaiser M, Fleissner C, Sterz J, Kleeberg L, Feist E, Burmester GR, Kloetzel PM, Sezer O (Mar 2007). "Circulating proteasome levels are an independent prognostic factor for survival in multiple myeloma". Blood. 109 (5): 2100–5. doi:10.1182/blood-2006-04-016360. PMID 17095627.

94. Shah SA, Potter MW, McDade TP, Ricciardi R, Perugini RA, Elliott PJ, Adams J, Callery MP (2001). "26S proteasome inhibition induces apoptosis and limits growth of human pancreatic cancer". Journal of Cellular Biochemistry. 82 (1): 110–22. doi:10.1002/jcb.1150. PMID 11400168.

95. Nawrocki ST, Sweeney-Gotsch B, Takamori R, McConkey DJ (Jan 2004). "The proteasome inhibitor bortezomib enhances the activity of docetaxel in orthotopic human pancreatic tumor xenografts". Molecular Cancer Therapeutics. 3 (1): 59–70. PMID 14749476.

96. Schenkein D (Jun 2002). "Proteasome inhibitors in the treatment of B-cell malignancies". Clinical Lymphoma. 3 (1): 49–55. doi:10.3816/CLM.2002.n.011. PMID 12141956.

97. O'Connor OA, Wright J, Moskowitz C, Muzzy J, MacGregor-Cortelli B, Stubblefield M, Straus D, Portlock C, Hamlin P, Choi E, Dumetrescu O, Esseltine D, Trehu E, Adams J, Schenkein D, Zelenetz AD (Feb 2005). "Phase II clinical experience with the novel proteasome inhibitor bortezomib in patients with indolent non-Hodgkin's lymphoma and mantle cell lymphoma". Journal of Clinical Oncology. 23 (4): 676–84. doi:10.1200/JCO.2005.02.050. PMID 15613699.

98. Messinger YH, Gaynon PS, Sposto R, van der Giessen J, Eckroth E, Malvar J, Bostrom BC (Jul 2012). "Bortezomib with chemotherapy is highly active in advanced B-precursor acute lymphoblastic leukemia: Therapeutic Advances in Childhood Leukemia & Lymphoma (TACL) Study". Blood. 120 (2): 285–90. doi:10.1182/blood-2012-04-418640. PMID 22653976.

99. Lambrou GI, Papadimitriou L, Chrousos GP, Vlahopoulos SA (Apr 2012). "Glucocorticoid

and proteasome inhibitor impact on the leukemic lymphoblast: multiple, diverse signals converging on a few key downstream regulators". Molecular and Cellular Endocrinology. 351 (2): 142–51. doi:10.1016/j.mce.2012.01.003. PMID 22273806.

100. Schmidtke G, Holzhütter HG, Bogyo M, Kairies N, Groll M, de Giuli R, Emch S, Groettrup M (Dec 1999). "How an inhibitor of the HIV-I protease modulates proteasome activity". The Journal of Biological Chemistry. 274 (50): 35734–40. doi:10.1074/jbc.274.50.35734. PMID 10585454.

101. Laurent N, de Boüard S, Guillamo JS, Christov C, Zini R, Jouault H, Andre P, Lotteau V, Peschanski M (Feb 2004). "Effects of the proteasome inhibitor ritonavir on glioma growth in vitro and in vivo". Molecular Cancer Therapeutics. 3 (2): 129–36. PMID 14985453.

102. Zollner TM, Podda M, Pien C, Elliott PJ, Kaufmann R, Boehncke WH (Mar 2002). "Proteasome inhibition reduces superantigen-mediated T cell activation and the severity of psoriasis in a SCID-hu model". The Journal of Clinical Investigation. 109 (5): 671–9. doi:10.1172/JCI12736. PMC 150886Freely accessible. PMID 11877475.

103. Elliott PJ, Pien CS, McCormack TA, Chapman ID, Adams J (Aug 1999). "Proteasome inhibition: A novel mechanism to combat asthma". The Journal of Allergy and Clinical Immunology. 104 (2 Pt 1): 294–300. doi:10.1016/S0091-6749(99)70369-6. PMID 10452747.

104. Verdoes M, Florea BI, Menendez-Benito V, Maynard CJ, Witte MD, van der Linden WA, van den Nieuwendijk AM, Hofmann T, Berkers CR, van Leeuwen FW, Groothuis TA, Leeuwenburgh MA, Ovaa H, Neefjes JJ, Filippov DV, van der Marel GA, Dantuma NP, Overkleeft HS (Nov 2006). "A fluorescent broad-spectrum proteasome inhibitor for labeling proteasomes in vitro and in vivo". Chemistry & Biology. 13 (11): 1217–26. doi:10.1016/j.chembiol.2006.09.013. PMID 17114003.

105. Kleiger G, Mayor T (Jun 2014). "Perilous journey: a tour of the ubiquitin-proteasome system". Trends in Cell Biology. 24 (6): 352–9. doi:10.1016/j.tcb.2013.12.003. PMC 4037451Freely accessible. PMID 24457024.

106. Goldberg AL, Stein R, Adams J (Aug 1995). "New insights into proteasome function: from archaebacteria to drug development". Chemistry & Biology. 2 (8): 503–8. doi:10.1016/1074-5521(95)90182-5. PMID 9383453.

107. Sulistio YA, Heese K (Jan 2015). "The Ubiquitin-Proteasome System and Molecular Chaperone Deregulation in Alzheimer's Disease". Molecular Neurobiology. doi:10.1007/s12035-014-9063-4. PMID 25561438.

108. Ortega Z, Lucas JJ (2014). "Ubiquitin-proteasome system involvement in Huntington's disease". Frontiers in Molecular Neuroscience. 7: 77. doi:10.3389/fnmol.2014.00077. PMC 4179678Freely accessible. PMID 25324717.

109. Sandri M, Robbins J (Jun 2014). "Proteotoxicity: an underappreciated pathology in cardiac disease". Journal of Molecular and Cellular Cardiology. 71: 3–10. doi:10.1016/j.yjmcc.2013.12.015. PMC 4011959Freely accessible. PMID 24380730.

110. Drews O, Taegtmeyer H (Dec 2014). "Targeting the ubiquitin-proteasome system in heart disease: the basis for new therapeutic strategies". Antioxidants & Redox Signaling. 21 (17): 2322–43. doi:10.1089/ars.2013.5823. PMC 4241867Freely accessible. PMID 25133688.

111. Wang ZV, Hill JA (Feb 2015). "Protein quality control and metabolism: bidirectional control in the heart". Cell Metabolism. 21 (2): 215–26. doi:10.1016/j.cmet.2015.01.016. PMC 4317573Freely accessible. PMID 25651176.

112. Karin M, Delhase M (Feb 2000). "The I kappa B kinase (IKK) and NF-kappa B: key elements of proinflammatory signalling". Seminars in Immunology. 12 (1): 85–98. doi:10.1006/smim.2000.0210. PMID 10723801.

113. Ermolaeva MA, Dakhovnik A, Schumacher B (Jan 2015). "Quality control mechanisms in cellular and systemic DNA damage responses". Ageing Research Reviews. 23: 3–11. doi:10.1016/j.arr.2014.12.009. PMID 25560147.

114. Checler F, da Costa CA, Ancolio K, Chevallier N, Lopez-Perez E, Marambaud P (Jul 2000). "Role of the proteasome in Alzheimer's disease". Biochimica et Biophysica Acta. 1502 (1): 133–8. doi:10.1016/s0925-4439(00)00039-9. PMID 10899438.

115. Chung KK, Dawson VL, Dawson TM (Nov 2001). "The role of the ubiquitin-proteasomal pathway in Parkinson's disease and other neurodegenerative disorders". Trends in Neurosciences. 24 (11 Suppl): S7–14. doi:10.1016/s0166-2236(00)01998-6. PMID 11881748.

116. Ikeda K, Akiyama H, Arai T, Ueno H, Tsuchiya K, Kosaka K (Jul 2002). "Morphometrical reappraisal of motor neuron system of Pick's disease and amyotrophic lateral sclerosis with dementia". Acta Neuropathologica. 104 (1): 21–8. doi:10.1007/s00401-001-0513-5. PMID 12070660.

117. Manaka H, Kato T, Kurita K, Katagiri T, Shikama Y, Kujirai K, Kawanami T, Suzuki Y, Nihei K, Sasaki H (May 1992). "Marked increase in cerebrospinal fluid ubiquitin in Creutzfeldt–Jakob disease". Neuroscience Letters. 139 (1): 47–9. doi:10.1016/0304-3940(92)90854-z. PMID 1328965.

118. Mathews KD, Moore SA (Jan 2003). "Limb-girdle muscular dystrophy". Current Neurology and Neuroscience Reports. 3 (1): 78–85. doi:10.1007/s11910-003-0042-9. PMID 12507416.

119. Mayer RJ (Mar 2003). "From neurodegeneration to neurohomeostasis: the role of ubiquitin". Drug News & Perspectives. 16 (2): 103–8. doi:10.1358/dnp.2003.16.2.829327. PMID 12792671.

120. Calise J, Powell SR (Feb 2013). "The ubiquitin proteasome system and myocardial ischemia". American Journal of Physiology. Heart and Circulatory Physiology. 304 (3): H337–49. doi:10.1152/ajpheart.00604.2012. PMC 3774499Freely accessible. PMID 23220331.

121. Predmore JM, Wang P, Davis F, Bartolone S, Westfall MV, Dyke DB, Pagani F, Powell SR, Day SM (Mar 2010). "Ubiquitin proteasome dysfunction in human hypertrophic and dilated cardiomyopathies". Circulation. 121 (8): 997–1004. doi:10.1161/CIRCULATIONAHA.109.904557. PMC 2857348Freely accessible. PMID 20159828.

122. Powell SR (Jul 2006). "The ubiquitin-proteasome system in cardiac physiology and pathology". American Journal of Physiology. Heart and Circulatory Physiology. 291 (1): H1–H19. doi:10.1152/ajpheart.00062.2006. PMID 16501026.

123. Adams J (Apr 2003). "Potential for proteasome inhibition in the treatment of cancer". Drug Discovery Today. 8 (7): 307–15. doi:10.1016/s1359-6446(03)02647-3. PMID 12654543.

124. Ben-Neriah Y (Jan 2002). "Regulatory functions of ubiquitination in the immune system". Nature Immunology. 3 (1): 20–6. doi:10.1038/ni0102-20. PMID 11753406.

125. Egerer K, Kuckelkorn U, Rudolph PE, Rückert JC, Dörner T, Burmester GR, Kloetzel PM, Feist E (Oct 2002). "Circulating proteasomes are markers of cell damage and immunologic activity in autoimmune diseases". The Journal of Rheumatology. 29 (10): 2045–52. PMID 12375310.

Index